The World of the Siege

History of Warfare

VOLUME 126

The titles published in this series are listed at *brill.com/hw*

The World of the Siege

Representations of Early Modern Positional Warfare

Edited by

Anke Fischer-Kattner
Jamel Ostwald

BRILL

LEIDEN | BOSTON

Cover illustration: Siege and Battle of Isfarah. This scene from Walters manuscript W.596 depicts the siege and battle of Isfarah, during which Babur and his army descend upon the fortress of Ibrahim Saru. Author: Zahir al-Din Muhammad Babur (1483–1530), 10th century AH/AD 16th century (Mughal). Accession Number: W.596.6A. CC: The Walter Art Museum.

Library of Congress Cataloging-in-Publication Data

Names: Fischer-Kattner, Anke, editor. | Ostwald, Jamel, editor.
Title: The world of the siege : representations of early modern positional warfare / edited by Anke Fischer-Kattner, Jamel Ostwald.
Other titles: Representations of early modern positional warfare
Description: Leiden ; Boston : Brill, [2019] | Series: History of warfare ; 126 | Includes bibliographical references and index.
Identifiers: LCCN 2019006192 (print) | LCCN 2019007108 (ebook) | ISBN 9789004395695 (ebook) | ISBN 9789004395688 (hardback : alk. paper)
Subjects: LCSH: Siege warfare--History. | Military history, Modern.
Classification: LCC UG444 (ebook) | LCC UG444 .W67 2019 (print) | DDC 355.4/40903--dc23
LC record available at https://lccn.loc.gov/2019006192

Typeface for the Latin, Greek, and Cyrillic scripts: "Brill". See and download: brill.com/brill-typeface.

ISSN 1385-7827
ISBN 978-90-04-39568-8 (hardback)
ISBN 978-90-04-39569-5 (e-book)

Contents

List of Figures

Notes on Contributors

Tonio Andrade

Ph.D. (2000), Yale, is a professor of Chinese and global history at Emory University. His books include *The Gunpowder Age: China, Military Innovation, and the Rise of the West in World History* (2016), *Lost Colony* (2011), and *How Taiwan became Chinese* (2008). He lives in Decatur, Georgia, USA.

Eric R. Dursteler

Ph.D. (2000), Brown University, is Chair and Professor of History at Brigham Young University. He has published extensively on the early modern Mediterranean, including recently with Monique O'Connell, *The Mediterranean World: From the Fall of Rome to the Rise of Napoleon* (2016).

Anke Fischer-Kattner

Dr. phil. (2011), Ludwig-Maximilians-Universität München, teaches History at Bundeswehr University Munich. Her research focuses on practices and representations of cultural contacts, whether peaceful or violent. Her publications include *Spuren der Begegnung: Europäische Reiseberichte über Afrika 1760–1860* (2015).

Sigrun Haude

Ph.D. (1993), University of Arizona, is Associate Professor of History at the University of Cincinnati. She has published on Anabaptism and Gender, including *In the Shadow of "Savage Wolves": Anabaptist Münster and the German Reformation During the 1530s* (2000) and on the Thirty Years' War, e.g., "The Experience of War" in *The Ashgate Research Companion to the Thirty Years' War* (2014).

Pratyay Nath

Ph.D. (2016), Jawaharlal Nehru University, is Assistant Professor of History at Ashoka University. His first monograph *Climate of Conquest: War, Environment, and Empire in Mughal North India* is forthcoming from Oxford University Press in 2019.

Jamel Ostwald

Ph.D. (2002), Ohio State University, is Professor of History at Eastern Connecticut State University. He has published on early modern European military history, including *Vauban under Siege: Engineering Efficiency and Martial Vigor in the War of the Spanish Succession* (2007).

Kahraman Şakul
Ph.D. (2009), Istanbul Şehir University, is Associate Professor of Ottoman History at that university. He has published many articles on early modern Ottoman military history and political culture and contributed several entries to the *Encyclopaedia of the Ottoman Empire* (2009) and the *Encyclopaedia of Islam* THREE (2007-). He also makes translations.

Brian Sandberg
Ph.D. (2001), University of Illinois at Urbana-Champaign, is Professor of History at Northern Illinois University. He has published a monograph, *Warrior Pursuits: Noble Culture and Civil Conflict in Early Modern France* (2010); an interpretive essay, *War and Conflict in the Early Modern World, 1500-1700* (2016); and numerous articles and chapters.

Philip J. Stern
Ph.D. (2005) Columbia University, is Gilhuly Family Associate Professor of History at Duke University. He has published widely on the legal, political, institutional, and intellectual histories of early modern colonialism, the British Empire, geography and exploration, and corporations, including *The Company-State: Corporate Sovereignty and the Early Modern Foundations of the British Empire in India* (2011).

Peter H. Wilson
Ph.D. (1990), University of Oxford, is Chichele Professor of the History of War and Fellow of All Souls College. His books include *The Thirty Years War* (2009) and *The Holy Roman Empire* (2016).

CHAPTER 1

Introduction

Anke Fischer-Kattner

The importance of siege warfare in early modern Europe has become a historiographical commonplace. One contemporary witness, Roger Boyle, first Earl of Orrery, is universally quoted, stating that in his day commanders made "war more like foxes, than like lyons; and you will have twenty sieges for one battell."[1] Rendered a classic by the venerable precedent of Fernand Braudel and Geoffrey Parker, this memorable phrase has become the inevitable concomitant of any discussion of the history of armed conflict.[2] Boyle's colorful simile has often been used to describe the practice and tactical characteristics of war in Europe between the Reformation and the French Revolution. However, it also points to another crucial issue, which has so far not received comparable attention. By likening siege warfare to the moral qualities and behavior associated with the sly fox of fables and explicitly contrasting it with the figure of the majestic lion, Boyle represents the siege in metaphorical fashion.[3] His words conjure up a lively imagery together with a host of implicit cultural

1 Unfortunately, this quote cannot be found in the current digitized version of Boyle's work, usually based on a copy from the British Library, where the pages in question are missing: Roger Boyle, Earl of Orrery: *A Treatise of the Art of War Dedicated to the Kings Most Excellent Majesty* ([London] In the Savoy: Printed by T.N. for Henry Herringman […], 1677). In this copy, pages 14 and 15 are missing. Both of these are variously given as the origin of the quote in citations.

2 Fernand Braudel, *Civilization and Capitalism, 15th–18th Century*, vol. 3: *The Perspective of the World* (Berkeley, Los Angeles, 1992; French original 1979), p. 61; Geoffrey Parker, *The Military Revolution: Military Innovation and the Rise of the West, 1500–1800* (Cambridge, UK, 1988), p. 16; Geoffrey Parker, "Dynastic War 1494–1660," in *The Cambridge History of Warfare*, ed. Geoffrey Parker (Cambridge, UK, 2005), pp. 148–166, here: p. 148; also quoted e. g. in Martha Pollak, *Cities at War in Early Modern Europe* (Cambridge, UK, 2010), p. 109; Leo Braudy, *From Chivalry to Terrorism: War and the Changing Nature of Masculinity* (New York, 2005; original edition 2003), p. 124; Elaine Murphy, *Ireland and the War at Sea, 1641–1653* (Suffolk, 2012), p. 141; Henrik O. Lunde, *A Warrior Dynasty: The Rise and Fall of Sweden as a Military Superpower 1611–1721* (Philadelphia, Oxford, 2014), p. 246; Ian Morris, *War! What Is It Good for? The Role of Conflict in Civilization from Primates to Robots* (New York, 2014), p. 173 (calling Boyle an "English soldier").

3 The moralizing representation of different military techniques has recently been analyzed by Jamel Ostwald, "Popular English Perceptions of Louis XIV's Way of War," in *Louis XIV Outside In: Images of the Sun King Beyond France, 1661–1715*, ed. Tony Claydon, Charles-Edouard Levillain (London, New York, 2015), pp. 93–110.

 | DOI:10.1163/9789004395695_002

associations and moral interpretations. His short and intriguing comment on siege warfare is a reminder that convincing forms of representation often involve powerful acts of mental image production in order to make sense of complex, sometimes even overwhelming experiences. Whether representation comes in the form of news reporting, memoir, fictionalized account, picture, monument or history writing, it only shows selected elements in a specific light, foregrounding certain aspects and hiding others, telling (or just suggesting) one story and silencing other possible narratives. At the same time, the culturally inflected act of representation is deeply entangled with the physical experience and technological operations of siege warfare.

Military operations pose a fundamental challenge for historiography. They are deadly events and, simultaneously, engender specific types of cultural representations.[4] These characteristics obviously apply to early modern siege warfare, with the added difficulty that a siege brought large numbers of noncombatants literally in the line of fire, thus merging soldiers and civilians[5] in a peculiar way. The course of a siege brought forth a mass of source material – textual, visual, and material artefacts – on which any attempt at historiographical reconstruction and contextualization has to be based. Most of these primary sources already possessed representational character at the moment when they were produced. Fighting and representation did (and still do) go together.[6] Moreover, image-making goes on after the fighting ends. The Earl of Orrery's much-quoted phrase serves as a cautionary example. It contained a statement describing contemporary techniques of war from the personal experience of a veteran military commander, and, at the same time, it offered an interpretation of these techniques, drawing on a rich cultural repertoire. These aspects of Boyle's utterance form two sides of the same coin, so separating them threatens to produce a skewed result. Historiography, from its early modern examples up to those of the present, thus needs to be regarded as a peculiar cultural form of representation, a critically reflected technique of

4 A general assessment of the problematic implications of war, "as an unmaker of truths," (p. 133) for socio-cultural order and knowledge is attempted by Tarak Barkawi and Shane Brighton, "Powers of War: Fighting, Knowledge, and Critique," *International Political Sociology* 5, no. 2 (2011), 126–143. Yet, their overarching conceptualization of war as an anthropological constant threatens to cover up the specifics of time, place, and type of action.

5 Although the terminology was only developing at the very end of the early modern period, the distinction between professional military and non-military persons was becoming increasingly important as the profession of the soldier emerged. (See e.g. Erica Charters, Eve Rosenhaft, Hannah Smith, "Introduction," in *Civilians and War in Europe 1618–1815*, ed. Erica Charters, Eve Rosenhaft, Hannah Smith, pp. 1–16, here: p. 11.)

6 See for this foundational idea also Bernd Hüppauf, *Was ist Krieg? Zur Grundlegung einer Kulturgeschichte des Kriegs* (Bielefeld, 2012).

image-making, different from participants' and contemporaries' testimonies while dependent on them. Analyzing the processes that bring forth these connected representations thus promises to link many important concerns of traditional histories of war and the military to a cultural history approach towards armed conflict.

Without explicit reference to the history of war, Johann Gustav Droysen was laying the theoretical and methodological foundations of professional history in nineteenth-century Germany. He pinned down a fundamental problem in identifying an unavoidable precondition for all historiographical writing: history is mediated to us by sources, and these are themselves human-made conceptions – "Auffassungen" – of what happened.[7] Similar caution has been demanded more recently by proponents of postmodernism and a 'linguistic turn' in the humanities.[8] Written sources describing military actions, such as sieges, offer access to the meanings contemporaries attributed to the raw acts of war. They did this by positioning events in particular contexts, from the strictly military (with its tactical, operational, and strategic levels) to the political, legal and religious, or, more generally, the social and cultural spheres. Contemporaries were already debating among themselves which of these contexts were the most relevant and appropriate. As historians later worked with the remains and memories of this past, taking up and rejecting aspects of it, they added another layer of representation.[9] Classic accounts of the history of war, often recounting the deeds of 'great men' with an application to future armed conflicts in mind,[10] thus contributed to the shaping of cultural images of the siege. These accounts, in turn, exerted their influence on later technological, social, or political histories of war.

7 See Johann Gustav Droysen, "Kunst und Methode," in id., *Grundriss der Historik* (Leipzig, 1868), pp. 75–84, here: p. 80.

8 Some of the foundational texts urging a critical approach to history are united in Keith Jenkins, *The Postmodern History Reader* (London, New York, 1997).

9 Engaging with the challenges of postmodernism and the narrative turn (e. g. Hayden White), historians have rejected the equation of their work with fiction. They have, nonetheless, re-gained a consciousness for the constructionist aspects of historiography (see e. g. Keith Jenkins, *Re-Thinking History* (London, New York, 1991)), which became ever more challenging after the loss of the religious and nationalist certainties of 19th-century Europe.

10 Even in the late 19th century, other questions and approaches were considered important for military history, see e. g. the multifarious works of the Prussian officer and historian Max Jähns, whose main work, *Geschichte der Kriegswissenschaften, vornehmlich in Deutschland*, 3 vols (Munich, Leipzig, 1889–1891), was e. g. accompanied by reflections on the influence of military constitutions on culture, id., *Heeresverfassungen und Völkerleben: Eine Umschau* (Berlin, 1885), and on the relationship of war, peace, and culture, id., *Über Krieg, Frieden und Kultur: Eine Umschau* (2nd ed., Berlin, 1893).

The relation of all these representational acts to early modern siege warfare is at the core of the proposed volume. It seeks to provide a survey of early modern military activity and the ensuing human suffering across different world regions, thereby drawing attention to the usually unseen lenses through which historical siege events are viewed. The individual contributions thus consider how sieges have been given meanings as distinct events by participants and contemporary observers as well as later-day commentators and historians.

This project brings together scholars from different fields of history, namely military, social, and cultural, focusing on several regions of the world. Their joint work builds on a number of historiographical foundations. Initial explorations of siege warfare in the 1970s[11] tended to view sieges mainly in terms of the development of military technology.[12] More recently, 'new military history' has explored social and cultural aspects of early modern military institutions.[13] The experiences of soldiers and civilians, both in peace time and in war, have come under scrutiny.[14] In the last few years, the establishment of a new specialized journal[15] and several book series[16] has indicated that the nexus of war and culture is now recognized as a valuable field of enquiry. While military history has a lot to gain by contextualizing military action within a cultural

11 Obviously, famous siege operations and the actions of 'great' commanders and general-engineers had been written about in classic military history ever since their occurrence.

12 The first comprehensive studies were presented by Christopher Duffy, *Fire and Stone: The Science of Fortress Warfare 1660–1860* (Newton Abbott, 1975); id., *Siege Warfare: The Fortress in the Early Modern World, 1494–1660* (London, 1979), which includes reflections on the socio-cultural meanings of the fortress.

13 See e. g. Robert M. Citino, "Military Histories Old and New: A Reintroduction," *American Historical Review* 112 (October 2007), 1070–1090; Bernhard R. Kroener, *Kriegswesen, Herrschaft und Gesellschaft 1300–1800* (Enzyklopädie deutscher Geschichte), vol. 92 (Munich, 2013); Torbjorn L. Knutsen, "Review Essay: Old, Unhappy, Far-Off Things: The New Military History of Europe," *Journal of Peace Research* 24, no. 1 (1987), 87–98; Peter Paret, "New Military History," *Parameters* 31 (1991), 10–18.

14 The classic work for battle experience has been presented by John Keegan, *The Face of Battle* (London, 1976); see for a connection of military history to gender relations: Karen Hagemann, Ralf Pröve, eds., *Landsknechte, Soldatenfrauen und Nationalkrieger: Militär, Krieg und Geschlechterordnung im historischen Wandel* (Frankfurt am Main, 1998); and Karen Hagemann, Gisela Mettele, Jane Rendall, eds., *Gender, War, and Politics: Transatlantic Perspectives, 1775 – 1830* (Basingstoke, 2010); for social relations in a fortress city of the eighteenth century: Ralf Pröve, *Stehendes Heer und städtische Gesellschaft im 18. Jahrhundert: Göttingen und seine Militärbevölkerung 1713–1756* (Munich, 1995).

15 This is the Britain-based *Journal of War and Culture Studies.*

16 Palgrave Macmillan has "War, Culture and Society, 1750–1850," New York University Press "Warfare and Culture;" Edinburgh University Press offers "Edinburgh Critical Studies in War and Culture" for analyses of war through Anglophone film and literature.

framework,[17] social and cultural historians are rightly reminded of the materially disruptive character of armed conflict.[18]

Sieges and their cultural representations, however, have hardly received appropriate attention within the new research that examines the crossroads of war and culture. Some select studies have started to approach the field by reintegrating military history with histories of (expert) knowledge. They have pointed to the crucial role played by military engineers, often polymaths whose feats comprised military construction, tactics, and strategy, but also to architecture and performative displays of power.[19]

Acknowledging the foundational qualities of these diverse and valuable approaches, this volume concentrates on the actual activities of siege warfare and the representational character of the sources and histories that depict, hint at, or even hide these activities. While some of the volumes' contributors directly engage with the making of representational sources by historical actors, others discuss how historians have represented early modern sieges within historiographical debates about larger trends such as state and empire building or the "military revolution."[20] As they consider the making of representations of the siege across the centuries, the contributions join not only in bringing out a new level of second-order observations, but additional layers of historical and historiographical agency. This paves the way for more exchanges between a new cultural history of warfare and established debates of military history.

---

17 This is stressed by Wayne E. Lee, "Warfare and Culture," in *Warfare and Culture in World History*, ed. Wayne E. Lee (New York, 2011), pp. 1–11. This piece not only introduces the volume, but also NYU's book series.

18 See more generally for the impact of war on ontology in the social sciences: Barkawi and Brighton, "Powers of War," 133–139.

19 E. g. Jamel Ostwald, *Vauban Under Siege: Engineering Efficiency and Martial Vigor in the War of the Spanish Succession* (Leiden, 2006); the works of experts in architecture are currently offering inspiring invitations to consider connections between early modern war, (siege) technology, and culture, see e. g. Martha Pollak, *Cities at War*, and the multifaceted work of Simon Pepper, particularly Simon Pepper, Nicholas Adams, *Firearms and Fortifications: Military Architecture and Siege Warfare in Sixteenth-Century Siena* (Chicago, 1986), but also his articles, e. g. "Artisans, Architects and Aristocrats: Professionalism and Renaissance Military Engineering," in *The Chivalric Ethos and the Development of Military Professionalism*, ed. D. J. B. Trim (Leiden, 2003), pp. 117–147; id., "The Face of the Siege: Fortification, Tactics and Strategy in the Early Italian Wars," in *Italy and the European Powers: The Impact of War, 1500–1530*, ed. Christine Shaw (Leiden, 2006), pp. 33–56.

20 An overview of the debates on a history of war as well as the continued importance of questions of state-building is provided for the European continent by Frank Tallett, D. J. B. Trim, eds., *European Warfare, 1350–1750* (Cambridge, UK, 2010).

As representations of sieges are embedded in, and, at the same time, produce peculiar cultural contexts, they call for comparisons and enquiry into global and further regional histories of positional warfare.[21] Uniting examples from various geographical and cultural environments over the course of three centuries from 1500 to 1800 CE, this volume strives to diversify approaches to the ideas and practices associated with the siege. Following disenchantment with the "new cultural history," criticized for an alleged blindness to issues of power, recent approaches to global history indicate new ways to understand sieges.[22] The following chapters jointly demonstrate that by integrating military, cultural, and global history perspectives, a deeper comparative understanding of the practices of positional warfare around the early modern world is to be gained.

Despite all need for specificity and differentiation, there is also a case to be made for the unity of the early modern world of the siege. Two roughly contemporary paintings depicting the conclusion of sieges in the first half of the seventeenth century serve as a visual reminder for this. The Mughal miniature of the capitulation of the Safavid garrison in Kandahar in 1638 (figure 1.1) and the painting depicting the surrender of the city of Breda by the Dutch to a Spanish army in 1625 (figure 1.2) were produced in completely different contexts and formats. The latter work, an oil painting on canvas by the Spanish court painter Diego Rodríguez de Silva y Velázquez, was created ca. 1635 and covers a wall surface of 307 cm by 367 cm. The former, executed – according to the experts at its current location, the Musée Guimet in Paris – in gouache on paper around 1640, was originally an illustration in the Padshahnama, the official chronicle of the reign of Mughal emperor Shah Jahan. Fitting on a book page, its size is a mere 47.5 cm by 31.7 cm. Art historian Gregory Minissale emphasizes that Mughal art has much too frequently been judged by European standards "as a deficient form of illusionism or naturalism," but should be read within "a system of pictorial order evolved from a close relation with Persian prototypes, and in a tradition of concerns different from those of Western art."[23]

21 See for the concepts of global and entangled histories: Ann Laura Stoler, Frederick Cooper, "Between Metropole and Colony: Rethinking a Research Agenda," in *Tensions of Empire: Colonial Cultures in a Bourgeois World*, ed. Ann Laura Stoler, Frederick Cooper (Berkeley, 1997), pp. 1–58; Shalini Randeria, Sebastian Conrad, "Geteilte Geschichten – Europa in einer postkolonialen Welt," in *Jenseits des Eurozentrismus: Postkoloniale Perspektiven in den Geschichts- und Kulturwissenschaften*, ed. Shalini Randeria, Sebastian Conrad (Frankfurt am Main, 2002), pp. 9–49.

22 Such a reorientation is hinted at by one of the previous main proponents of the new cultural history, Lynn Hunt, *Writing History in the Global Era* (New York, 2014).

23 Gregory Minissale, *Images of Thought: Visuality in Islamic India 1550–1750* (Newcastle upon Tyne, 2009), p. 55.

FIGURE 1.1 La Reddition de Qandahâr [The Surrender of Kandahar] (illustration from the Padshahnama, c. 1640)
MUSÉE GUIMET, © BPK / RMN - GRAND PALAIS / THIERRY OLLIVIER

FIGURE 1.2 Diego Rodríguez de Silva y Velázquez, Las lanzas/La rendición de Breda [The Surrender of Breda] (c. 1635)
MUSEO DEL PRADO

Culture-specific sensitivity was already demanded during the early 20th century for European works of art in the iconological approach developed by Erwin Panofsky.[24] Nonetheless, comparison across cultural backgrounds points to highly interesting parallels.

Both images are centered on a ritual act that seems to be understood across cultural borders, namely the presentation of the keys to the conquered city by the defeated to the victorious commanders. Moreover, both paintings depict the object of the contest, the besieged fortress, in the background. In both cases, the time of the armed struggle and surrender are united in the visual composition, as indications of ongoing fighting – e. g. troops storming the walls in the Mughal painting, fire and smoke in the Spanish one – are included in the picture. Dress and demeanor of the portrayed protagonists clearly demarcate variations in cultural context. Thus, the Spanish-Italian general Ambrogio Spínola (coming from the right side) approaches his defeated opponent on foot, answering the latter's bow with a conciliatory gesture that displays magnanimity. In contrast, the victorious Mughal commander Kilij Khan (likewise right) appears dignified and somewhat aloof, mounted on his stately white horse. Yet, both images use a striking combination of massed troops and individualized characters, particularly the famous commanders at the center of the respective compositions, to indicate agency at different social levels in siege warfare. While contrasting features of the paintings are certainly not to be discounted, especially in analyses concerned with the details of representation, there obviously existed cross-culturally recognizable elements of sieges and their orderly termination.

An analysis that encompasses comparison and contrast, shifting between general features and specifics, enables us to better understand the interplay of what might be called the formulaic and particularistic elements of siege representations. On the one hand, recognizable acts, frequently interpreted in historiography as 'rituals' of siege warfare,[25] allowed for relatively easy categorization of a military action as a siege. This identification offered participants and observers a rough script for actions and understanding. On the other hand, local, regional, national, or continental collective specificities as well as individual quirks or military genius might fundamentally alter the situation and its interpretational options. This paradox, mirroring the essential difficulties of any attempt at writing world history, is attenuated by a focus on representation. Investigation into the reasons for which one or the other aspect has been

24 Erwin Panofsky, *Meaning in the Visual Arts* (New York, 1955).

25 Simon Pepper, "Siege Law, Siege Ritual, and the Symbolism of City Walls in Renaissance Europe," in *City Walls: The Urban Enceinte in Global Perspective*, ed. James D. Tracy (Cambridge, UK, 2000), pp. 573–604. See also below, in Jamel Ostwald's chapter, particularly footnote 3.

stressed in sources and secondary accounts promises to enrich our understandings of early modern military history.

Much valuable work has recently been done regarding global histories of warfare, whether across time or specifically in the early modern period.[26] These works necessarily paint a picture of the world in somewhat coarse brush strokes. In contrast, the chapters of this volume take fine-grained representations, engendered by what were often traumatic military, political and social ordeals, as their starting points. During any conflict, representational strategies were already playing significant roles. They were not only intended for the gaze of the distant observer. The opposing sides had to take care to define the limits of their identities, separating "us" and "them" in their descriptions and plans. For this, they relied on widely shared, sometimes even transcultural conceptions of what a siege was and how it should be conducted. Moreover, communication between the contending parties was ensured in ritualistic formats that could prevent extremities with their brutal consequences by opening up possibilities for negotiation. Within this framework, competing representations of the situation were exchanged when besiegers and besieged were sounding out chances of surrender. Shared scripts of behavior helped participants to minimize the insecurities inherent in capitulation.[27]

Over and above the representational acts of those directly engaged in the operations, verbal and visual accounts of the action appeared. These involved a wider audience with the events and promoted certain views of what was going on. The examples of Kandahar and Breda remind us how pictorial depiction of military action was entangled with written, and sometimes published, sources. Generally, representations in and of a siege – from the most functionalistic and strictly military to the overtly propagandistic – gained meaning from an interplay of the formulaic and the individual. The format of an edited collection that unites a variety of case studies thus seems very fitting for a first exploration into the subject of sieges and their representations throughout the early modern world.

26 See for example some of the numerous publications by Jeremy Black, *Introduction to Global Military History: 1775 to the Present Day* (London, New York, 2005); id., *War in the World: A Comparative History, 1450–1600* (Basingstoke, 2011); id., *Beyond the Military Revolution: War in the Seventeenth Century World* (Basingstoke, 2011); overviews are also offered by Gérard Chaliand, *A Global History of War: From Assyria to the Twenty-First Century* (Oakland, 2014); Wayne E. Lee, *Waging War: Conflict, Culture, and Innovation in World History* (Oxford, New York, 2016).

27 See e. g. for the role of time as a shared material problem (limited resources being used up) and a means of both provocative and conciliatory communication: Anke Fischer-Kattner, "Zeit-Not/Not-Zeit: Temporale Perspektiven auf den Belagerungskrieg im 17. Jahrhundert," in *Militär und Gesellschaft in der Frühen Neuzeit*, ed. Achim Landwehr (Potsdam, 2017), pp. 57–95.

The project of this book started with the organization of a workshop entitled "The World of the Siege," held in September 2014 at Duke University.[28] For this event, experts from different disciplines and areas of specialization were invited to re-/present the representational activities in and around early modern sieges. Within the context of the workshop, imbalances within the field of world military history became obvious. It proved extremely difficult, even impossible, to identify specialists for the history of warfare in large world-regions, especially sub-Saharan Africa and the Pacific. Inherent in analyses of global positional warfare in the early modern period is, therefore, a great danger of generalizing from the peculiar examples of Europe and "gunpowder empires"[29] around the world. Yet, the productive exchanges during the conference, which brought together practitioners of military and cultural, general and art history, of area studies and transcultural approaches, demonstrated the value of initiating such discussions. Drawing on the circle of workshop participants and other invited specialists, this volume is meant as an inspiration for further exchanges rather than a definitive collection. At the same time, it is meant to demonstrate the great potential inherent in a focus on issues of representation. Technological, social, and political historiographies of sieges thus also reveal their constructed character, which links them to cultural-history analyses of contemporary images and descriptions of sieges. It is one goal of this collaborative work to encourage more discussions between and across academic fields.

Out of endless possibilities for organizing this collection, we have chosen an approach that clusters the contributions around thematic axes in three sections. The first of these takes a closer look at the complex relations between "Participants and Audiences" of sieges. Working on different European examples of positional warfare, Sigrun Haude, Anke Fischer-Kattner, and Jamel Ostwald focus on representations in different kinds of source material. Ego-documents produced by clerics and nuns in Bavaria during the Thirty Years' War provide the basis for Haude's investigation into non-military experiences of military operations. Her chapter highlights how German civilian audiences

28 This productive event could only take place thanks to the gracious support of the Volkswagen and Josiah Charles Trent Memorial Foundations.

29 See for an extension of this concept, developed originally for the major Islamic powers of the Safavids, Mughals, and Ottomans (this definition put forward by Marshall Hodgson is e. g. retained by Douglas E. Streusand, *Islamic Gunpowder Empires: Ottomans, Safavids, and Mughals* (Boulder, 2011)), to a comparative world history approach: William H. McNeill, "The Age of Gunpowder Empires, 1450–1800," in *Islamic & European Expansion: The Forging of a Global Order*, ed. Michael Adas (Philadelphia, 1993), pp. 103–139 (reprint of a pamphlet originally distributed by the American Historical Association published in Washington, DC, 1989).

represented sieges primarily in terms of their lived experience before, during and after the event. As unwilling participants, the clerical writers might see their identities shaken as well as reinforced, when actual and potential siege actions threatened their localities. Fischer-Kattner investigates how a broadsheet print of what came to be known as the "Siege of Colchester" of 1648 was connected to printed siege representations from continental Europe. The comparative approach demonstrates how English soldiers, printers and readers in the British Civil Wars were drawing on common European conceptions of the conduct and depiction of sieges for their representations of this military event. With recourse to the concept of 'media systems,' this chapter more broadly analyzes the construction of images and narratives of the siege through printed publications for local, (proto-)national, and European audiences. In his contribution, Ostwald draws on written capitulations from the War of the Spanish Succession at the beginning of the 18th century in order to critically engage with the historiographical notion that surrenders were 'mere rituals,' centering on symbolic acts. He shows that participants (besiegers and besieged) were only one of many audiences for early modern siege representations. Thus, the rites of capitulation emphasized by historians were only a small part of a much larger discourse over how garrisons could represent their honor in a siege. All of the contributions in this section engage with source materials which gave sieges meanings that went beyond the actual events. These sources addressed and sometimes constituted audiences transcending (proto-)national confines in Europe.

The second section expands the geographical scope by taking a closer look at "Imperial Boundaries" in the sixteenth and seventeenth centuries. Eric Dursteler and Philip Stern both build on notions of the porousness of imperial boundaries. Inspired by postcolonial critique, their works question conventional historiographical representations of inter-imperial conflict as they undermine clear-cut dichotomies between attackers and defenders. Dursteler disentangles multiple interests from the node of the Adriatic fortress of Clissa, situated in a region where stakes of the Ottoman and Habsburg Empires overlapped with Venetian possessions and papal political influence. Although a succession of violent actions at Clissa in the late sixteenth century has often been stylized as a conflict between Muslims and Christians, closer investigation reveals surprising alliances between the supposed adversaries. More than two parties were also concerned in the late-seventeenth-century siege of the British East India Company's stronghold of Bombay. Stern demonstrates that a differentiating look at acts of boundary-crossing allows for a repositioning of the siege in the historical narrative about the development of the British Empire.

The other two contributions in this section focus more on intra-imperial consequences of siege warfare, namely its propagandistic representation in poetic praise for the victorious ruler. Pratyay Nath concentrates on literary descriptions of the exploits of the Mughal emperor Jalaluddin Muhammad Akbar in the second half of the sixteenth century. By a close reading of the laudatory texts, he shows how these contributed to the construction of a claim to universal sovereignty. Taking into account the chroniclers' position in Akbar's army and/or court, Nath is able to draw out the double importance of warfare for Mughal empire building: besides its physical expansionist drive, it engendered a universalistic imperial ideology. Kahraman Şakul on the one hand shares traditional military history's concerns with the technological and logistical conduct of the siege of the Polish-Lithuanian fortress at Kamianets-Podilsky by an Ottoman army in the later seventeenth century. His analysis of the descriptions of the operations in verse and prose on the other hand uncovers the rich literary imagery employed by writers of the courtly elite to enhance Ottoman military success. In spite of the brevity of Ottoman rule in Kamianets (1672–1699), Şakul argues, the image of this impressive rock fortress and cultural memories of the siege still resonate with modern Turks. Taken together, these four chapters highlight the somewhat paradoxical double consequence of sieges in erecting or strengthening clear lines of imperial demarcation while at the same undermining them – metaphorically, literally, and physically.

The contributions of the third section engage with problems of "Definitions" for siege warfare. They are concerned with contested understandings of the siege – whether in contemporary sources or historians' representations. Tonio Andrade critically examines the Euro-centric concept of the Artillery Revolution by testing its premises within the context of early modern Chinese warfare. He offers a technological explanation for the fact that the Chinese, despite being the inventers of firearms, did not develop heavy siege artillery on a scale comparable to that of fourteenth- to sixteenth-century Europe: Chinese fortification walls were traditionally thick and sloped, thus all but impregnable for early gunpowder weapons. Only in the sixteenth and seventeenth centuries were Chinese armies adopting and adapting to European siege artillery. Yet, an examination of sample siege operations by the Chinese against Dutch and Russian artillery fortresses shows that the possession of similar technological instruments did not determine a similar approach to the craft of besieging. Obviously, mental representations of the military activity in which the contenders were engaged were not necessarily congruent. This contribution thus questions the notion of linear progress inherent in a technologically determined European historiography of the siege by means of cross-cultural comparison. Yet, even within one cultural and linguistic context, there was not necessarily a strict definitional order to the semantic fields of siege or, more generally, positional warfare. Brian Sandberg's investigation into the language of *prises* and *suprises* in the French Wars of Religion of the sixteenth and early seventeenth centuries shows the fluidity and changing political connotations of the military terminology. As towns and noble *seigneurs* were struggling to survive in this era of insecurity, fears and hopes of conquest brought forth contradictory discourses of attacking places. Yet, whether the taking of a place was seen as an honorable act of liberation or a guileful attack on an unsuspecting victim, positional warfare in this context was seen as extremely dynamic.

Faced with indeterminacy in both contemporary and historiographical accounts, Peter Wilson's concluding chapter attempts to give a definition of the siege that relies on a combination of five factors. Using examples beyond the traditional heyday of the siege in early modern Europe, particularly from the nineteenth and twentieth century, this contribution strives to enhance the value of the term siege as an analytical category. In the context of the other chapters of this volume, Wilson's definition makes clear that conceptions shaped by early modern (and even earlier) representations of positional warfare remained highly influential through the modern era. His thoughts raise the question how the early modern period might connect to a comprehensive world history of the siege.

The fundamental problem raised in the last chapter, i.e., the difficulty of a definition of siege across cultures and possibly epochs, is an overarching concern, which links individual chapters and approaches across the thematic sections of this volume. In order to establish which discourses, rhetorical as well as visual and symbolic, demarcate a discernible field of siege warfare for historic actors, it might be helpful to go back to etymological roots. In major European languages, the term for what is called a siege in English is derived from the idea of an attacking force taking a seat (Latin: *obsidio*, French: *siège*, Spanish: *asedio/sitio*, Italian: *assedio*) or setting up camp (German: *Belagerung*, Dutch: *beleg*) around a place. An idea of positional stability seems inherent in this linguistic approach to the problem. Yet, Sandberg and Dursteler demonstrate that speedy operations and surprise attacks were not to be separated from conceptions of the siege in 16th-century Europe. Moreover, Fischer-Kattner's examination of the actions around Colchester in 1648 raises the question of how a blockade relates to the concept of siege. In this case, contemporary usage appears to differentiate a popular, broad category of siege (warfare) from a narrower, technical understanding that opposes tactics of siege from those of blockade. The latter conception would probably have been considered more correct by European military experts in the wake of Vauban, as consulted by

Jamel Ostwald for an earlier attempt at definition.[30] Yet, as his chapter shows, even military specialists after 1700 were still debating evaluations of siege practices, in this particular case, the correct and necessary features of an "honorable defense." Haude's contribution further complicates the problem by bringing in the voices and experiences of clerical civilians. These case studies from early modern Europe indicate that conceptions of the siege, while they might have overlapped around a core of 'attacking a defended place,' varied considerably across time and social group.

The issue of definition becomes even more complex when one draws on extra-European examples, such as in the chapters by Andrade, Nath, Şakul, and Stern. Further comparative analyses of tactical practices and historically specific concepts (*Begriffsgeschichte*) detailing contemporary connotations of the military terminology are in order. Moreover, a truly global history of the siege will have to take into account interdependencies and mutual influence engendered by military confrontations. How did understandings of siege warfare change, if they did change at all, when the Chinese faced Dutch and Russians (Andrade), or when Muslim Mughals attacked Rajput fortresses (Nath)? Did shared meanings emerge or were they a prerequisite, when an Ottoman army conquered a Polish-Lithuanian stronghold defended by a mixed central-European garrison (Şakul), or when the British East India Company's soldiers desperately tried to ward off the forces of an African-Indian[31] naval commander in Mughal imperial service (Stern)? Wilson reminds us that it is important to distinguish the analytical category of the siege, ideally employed in historians' representations of positional warfare, from complex, sometimes even confusing historical usage. Moreover, the former has to be disentangled as far as possible from the dense web of current popular and metaphorical employment of the term. Wilson's contribution thereby sheds light on the wider, trans-chronological implications of the definitional problem – connecting the siege's early modern heyday to the *longue durée* of military history.

Another axis cutting across a number of the contributions in this volume is the relationship between the siege and power. In sieges, authority might be reinforced – thus, military governors of besieged places could exert measures that would have been inconceivable otherwise. Yet, as the positions of besiegers and besieged became increasingly uncomfortable because local resources

30 Ostwald, *Vauban Under Siege*, pp. 348–358.

31 The African connection of the Indian Siddis is explored in the dissertation by Ababu Minda Yimene, *An African Indian Community in Hyderabad: Siddi Identity, Its Maintenance and Change* (Göttingen, 2004).

were running out,[32] military and political authority might also be endangered. Mutiny and riot of badly supplied troops and battered civilian populations always threatened. Representations of siege warfare created by dissimilar sorts of historical actors underline these opposing effects in diverse ways. Nath's analysis of Mughal court chronicles demonstrates how the rhetoric of siege description was used in a project of imperial legitimization. By contrast, Sandberg's French examples indicate that the prospect of fundamental religious and thus political transformation inherent in conquest during the Wars of Religion created an atmosphere of insecurity among town administrators and citizens. The clerical diarists of Haude's chapter were similarly struggling with feeling uprooted and experiencing loss of control. As religious orders inculcated the ideals of *stabilitas loci* ("remaining in place") and obedience, flight and disconnection from superiors were extremely hard on their members. In these cases, military conflict in general and the particular threat of local attacks in a siege were represented as shattering experiences, in which it was hard to uphold authority.

Three contributions connect such empirical findings with a more abstract concept of necessity (as invoked by contemporary Europeans) or even the modern theory of a "state of exception."[33] Dursteler's study of Clissa exemplifies how the practical necessities created by the imperial borderland situation in Dalmatia are nowadays receiving more attention from historians, who are no longer taking clear-cut political dichotomies between Ottomans and Christians for granted. Fischer-Kattner finds that a rhetoric of necessity links legitimizations of concrete acts in the siege to wider claims for exceptional political authority in the British Civil Wars. Stern questions whether "state of exception" is a fitting historiographical label in trying to account for changes allegedly wrought by the siege of Bombay in the East India Company's local and general policies. While these reflections appear too diverse for a unified research agenda, they do indicate that early modern sieges are intimately connected to trans-epochal questions of state-building and of institutionalization or erosion of power.

32 For the logistical Achilles' heel of early modern warfare, see John Lynn, "Food, Funds, and Fortresses: Resource Mobilisation and Positional Warfare in the Campaigns of Louis XIV," in *Feeding Mars: Logistics in Western Warfare from the Middle Ages to the Present*, ed. John A. Lynn (Boulder, 1993), pp. 137–159; its dire consequences are painted in vivid colors by Lauro Martines, *Furies: War in Europe, 1450–1700* (New York, 2013), pp. 103–176.

33 Written from a politically controversial standpoint, but still theoretically engaging through his seminal conception of the state of exception: Carl Schmitt, *Die Diktatur: Von den Anfängen des modernen Souveränitätsgedankens bis zum proletarischen Klassenkampf* (6th ed., Berlin, 1994; original 1921).

All in all, this collaborative project has uncovered large gaps in terms of comparative studies of extra-European and European siege technologies and understandings. More research in the military histories of world regions before contact with or beyond the early modern reach of gunpowder powers is needed in order to better understand the general and specific features of positional warfare and its representation around the world. Yet, as is, this joint effort of military and cultural historians will hopefully engender more discussion and further attention to the early modern world of the siege.

Bibliography

Printed Sources

Boyle, Roger, Earl of Orrery, *A Treatise of the Art of War Dedicated to the Kings Most Excellent Majesty* ([London] In the Savoy [...], 1677).

Secondary Sources

Barkawi, Tarak; Brighton, Shane, "Powers of War: Fighting, Knowledge, and Critique," *International Political Sociology* 5, no. 2 (2011), 126–143.

Black, Jeremy, *Introduction to Global Military History: 1775 to the Present Day* (London, New York, 2005).

Black, Jeremy, *War in the World: A Comparative History, 1450–1600* (Basingstoke, 2011).

Black, Jeremy, *Beyond the Military Revolution: War in the Seventeenth Century World* (Basingstoke, 2011).

Braudel, Fernand, *Civilization and Capitalism, 15th–18th Century*, vol. 3: *The Perspective of the World* (Berkeley, Los Angeles, 1992; French original 1979).

Braudy, Leo, *From Chivalry to Terrorism: War and the Changing Nature of Masculinity* (New York, 2005; original edition 2003).

Chaliand, Gérard, *A Global History of War: From Assyria to the Twenty-First Century* (Oakland, 2014).

Charters, Erica; Rosenhaft, Eve; Smith, Hannah, "Introduction," in *Civilians and War in Europe 1618–1815*, ed. Erica Charters, Eve Rosenhaft, Hannah Smith, pp. 1–16.

Citino, Robert M., "Military Histories Old and New: A Reintroduction," *American Historical Review* 112 (October 2007), 1070–1090.

Droysen, Johann Gustav, "Kunst und Methode," in id., *Grundriss der Historik* (Leipzig, 1868), pp. 75–84.

Duffy, Christopher, *Siege Warfare: The Fortress in the Early Modern World, 1494–1660* (London, 1979).

Duffy, Christopher, *Fire and Stone: The Science of Fortress Warfare 1660–1860* (Newton Abbott, 1975).

Fischer-Kattner, Anke, "Zeit-Not/Not-Zeit: Temporale Perspektiven auf den Belagerungskrieg im 17. Jahrhundert," in *Militär und Gesellschaft in der Frühen Neuzeit*, ed. Achim Landwehr (Potsdam, 2017), pp. 57–95.

Hagemann, Karen; Mettele, Gisela; Rendall, Jane, eds., *Gender, War, and Politics: Transatlantic Perspectives, 1775–1830* (Basingstoke, 2010).

Hagemann, Karen; Pröve, Ralf, eds., *Landsknechte, Soldatenfrauen und Nationalkrieger: Militär, Krieg und Geschlechterordnung im historischen Wandel* (Frankfurt am Main, 1998).

Hüppauf, Bernd, *Was ist Krieg? Zur Grundlegung einer Kulturgeschichte des Kriegs* (Bielefeld, 2012).

Hunt, Lynn, *Writing History in the Global Era* (New York, 2014).

Jähns, Max, *Geschichte der Kriegswissenschaften, vornehmlich in Deutschland*, 3 vols (Munich, Leipzig, 1889–1891).

Jähns, Max, *Heeresverfassungen und Völkerleben: Eine Umschau* (Berlin, 1885).

Jähns, Max, *Über Krieg, Frieden und Kultur: Eine Umschau* (2nd ed., Berlin, 1893).

Jenkins, Keith, *Re-Thinking History* (London, New York, 1991).

Jenkins, Keith, *The Postmodern History Reader* (London, New York, 1997).

Keegan, John, *The Face of Battle* (London, 1976).

Knutsen, Torbjorn L., "Review Essay: Old, Unhappy, Far-Off Things: The New Military History of Europe," *Journal of Peace Research* 24, no. 1 (1987), 87–98.

Kroener, Bernhard R., *Kriegswesen, Herrschaft und Gesellschaft 1300–1800* (Enzyklopädie deutscher Geschichte), vol. 92 (Munich, 2013).

Lee, Wayne E., "Warfare and Culture," in *Warfare and Culture in World History*, ed. Wayne E. Lee (New York, 2011), pp. 1–11.

Lee, Wayne E., *Waging War: Conflict, Culture, and Innovation in World History* (Oxford, New York, 2016).

Lunde, Henrik O., *A Warrior Dynasty: The Rise and Fall of Sweden as a Military Superpower 1611–1721* (Philadelphia, Oxford, 2014).

Lynn, John, "Food, Funds, and Fortresses: Resource Mobilisation and Positional Warfare in the Campaigns of Louis XIV," in *Feeding Mars: Logistics in Western Warfare from the Middle Ages to the Present*, ed. John A. Lynn (Boulder, 1993), pp. 137–159.

McNeill, William H., "The Age of Gunpowder Empires, 1450–1800," in *Islamic & European Expansion: The Forging of a Global Order*, ed. Michael Adas (Philadelphia, 1993), pp. 103–139.

Martines, Lauro, *Furies: War in Europe, 1450–1700* (New York, 2013).

Minissale, Gregory, *Images of Thought: Visuality in Islamic India 1550–1750* (Newcastle upon Tyne, 2009).

Morris, Ian, *War! What Is It Good for? The Role of Conflict in Civilization from Primates to Robots* (New York, 2014).

Murphy, Elaine, *Ireland and the War at Sea, 1641–1653* (Suffolk, 2012).

Ostwald, Jamel, "Popular English Perceptions of Louis XIV's Way of War," in *Louis XIV Outside In: Images of the Sun King Beyond France, 1661–1715*, ed. Tony Claydon, Charles-Edouard Levillain (London, New York, 2015), pp. 93–110.

Ostwald, Jamel, *Vauban Under Siege: Engineering Efficiency and Martial Vigor in the War of the Spanish Succession* (Leiden, 2006).

Panofsky, Erwin, *Meaning in the Visual Arts* (New York, 1955).

Paret, Peter, "New Military History," *Parameters* 31 (1991), 10–18.

Parker, Geoffrey, *The Military Revolution: Military Innovation and the Rise of the West, 1500–1800* (Cambridge, UK, 1988).

Parker, Geoffrey, "Dynastic War 1494–1660," in *The Cambridge History of Warfare*, ed. Geoffrey Parker (Cambridge, UK, 2005), pp. 148–166.

Pepper, Simon; Adams, Nicholas, *Firearms and Fortifications: Military Architecture and Siege Warfare in Sixteenth-Century Siena* (Chicago, 1986).

Pepper, Simon, "Artisans, Architects and Aristocrats: Professionalism and Renaissance Military Engineering," in *The Chivalric Ethos and the Development of Military Professionalism*, ed. D. J. B . Trim (Leiden, 2003), pp. 117–147.

Pepper, Simon, "Siege Law, Siege Ritual, and the Symbolism of City Walls in Renaissance Europe," in *City Walls: The Urban Enceinte in Global Perspective*, ed. James D. Tracy (Cambridge, UK, 2000), pp. 573–604.

Pepper, Simon, "The Face of the Siege: Fortification, Tactics and Strategy in the Early Italian Wars," in *Italy and the European Powers: The Impact of War, 1500–1530*, ed. Christine Shaw (Leiden, 2006), pp. 33–56.

Pollak, Martha, *Cities at War in Early Modern Europe* (Cambridge, UK, 2010).

Pröve, Ralf, *Stehendes Heer und städtische Gesellschaft im 18. Jahrhundert: Göttingen und seine Militärbevölkerung 1713–1756* (Munich, 1995).

Randeria, Shalini; Conrad, Sebastian, "Geteilte Geschichten – Europa in einer postkolonialen Welt," in *Jenseits des Eurozentrismus: Postkoloniale Perspektiven in den Geschichts- und Kulturwissenschaften*, ed. Shalini Randeria, Sebastian Conrad (Frankfurt am Main, 2002), pp. 9–49.

Schmitt, Carl, *Die Diktatur: Von den Anfängen des modernen Souveränitätsgedankens bis zum proletarischen Klassenkampf* (6th ed., Berlin, 1994; original 1921).

Stoler, Ann Laura; Cooper, Frederick, "Between Metropole and Colony: Rethinking a Research Agenda," in *Tensions of Empire: Colonial Cultures in a Bourgeois World*, ed. Ann Laura Stoler, Frederick Cooper (Berkeley, 1997), pp. 1–58.

Streusand, Douglas E., *Islamic Gunpowder Empires: Ottomans, Safavids, and Mughals* (Boulder, 2011).

Tallett, Frank; Trim, D. J. B., eds., *European Warfare, 1350–1750* (Cambridge, UK, 2010).

Yimene, Ababu Minda, *An African Indian Community in Hyderabad: Siddi Identity, Its Maintenance and Change* (Göttingen, 2004).

PART 1

Participants and Audiences

CHAPTER 2

The World of the Siege in New Perspective: the Populace during the Thirty Years' War (1618–1648)

Sigrun Haude

Early modern military history has seen stimulating changes in the last few decades. After the traditional focus on battles and military leaders, and a lively debate about the so-called 'Military Revolution' – if and when it occurred, and what its principle features were – military scholars are increasingly turning their attention to sieges. Frank Tallett provides a useful summary of this historiography, which contends that early modern warfare "was not primarily characterized by pitched battles at all, but rather by sieges and what would now be called 'low intensity operations': skirmishes, ambushes, raids, forays against civilians and prisoner-taking expeditions."[1] Our love affair with big battles, such as at Breitenfeld (1631) and Lützen (1632), Tallett suggests, has blinded us to the fact that "[b]attles remained relatively infrequent events, and it was the siege that predominated."[2] Tallett points to new-style fortifications – especially the *trace italienne* – that changed warfare forever, demanding ever larger armies and tying up a commander's energy and resources for months. His examples from throughout Europe describe sieges that lasted from ten months to over three years, but even blockades as 'brief' as forty to sixty days were thoroughly taxing.[3]

In our heightened attention to sieges, the focus is generally on the logistics of the operation, its strategists, and the complex conditions (political, geographical, etc.) under which it unfolds. The following analysis reflects on these military maneuvers from a point of view that is rarely considered in siege warfare studies: the perspective of the populace. This investigation explores how people experienced these sieges, and whether their testimonies can expand our understanding of siege warfare. Our exploration will start by listening to several contemporary narratives about advancing troops and what this meant for the authors and the people around them. In a second step, we will examine

1 Frank Tallett, *War and Society in Early Modern Europe, 1495–1715* (London, 1992), p. 3.

2 Ibid., p. 66.

3 Ibid., p. 51.

 | DOI:10.1163/9789004395695_003

how these experiences reflect on our perspective of sieges. Insights on how society suffered through the various military campaigns are less readily available than testimonies about the war's predominant tacticians and commanders, but in recent years ever more scholars, including myself, have turned toward the war up close to comprehend people's experiences of and agency during the conflict.[4] This initiative has centered on 'autobiographical' accounts.[5]

Notably, in the first half of the seventeenth century the lines between the literary forms of autobiographies, diaries, and other chronological accounts were fluid.[6] Scholars have underlined that such narratives, unlike their post-Enlightenment counterparts, focused little on self-reflection or the construction of an individual personality. Rather than confidently arranging one's personal history from hindsight, these texts frequently listed events or experiences chrono-

4 See especially the work of Hans Medick, particularly the volume co-edited with Benigna von Krusenstjern, *Zwischen Alltag und Katastrophe: Der Dreißigjährige Krieg aus der Nähe* (Veröffentlichungen des Max-Planck-Instituts für Geschichte), vol. 148 (Göttingen, 1999). See also Sigrun Haude, "The Experience of War," in *Research Companion to the Thirty Years War*, ed. Olaf Asbach, Peter Schroeder (Farnham, 2014), pp. 257–268.

5 For literature foregrounding these autobiographical texts, see, for example, Benigna von Krusenstjern, "Buchhalter ihres Lebens: Über Selbstzeugnisse aus dem 17. Jahrhundert," in *Das dargestellte Ich: Studien zu Selbstzeugnissen des späteren Mittelalters und der frühen Neuzeit*, ed. Klaus Arnold et al. (Selbstzeugnisse des Mittelalters und der beginnenden Neuzeit), vol. 1 (Bochum, 1999), pp. 139–146; eadem, *Selbstzeugnisse der Zeit des Dreißigjährigen Krieges: Beschreibendes Verzeichnis* (Selbstzeugnisse der Neuzeit), vol. 6 (Berlin, 1997); eadem, "Was sind Selbstzeugnisse? Begriffskritische und quellenkundliche Überlegungen anhand von Beispielen aus dem 17. Jahrhundert," *Historische Anthropologie* 2 (1994), 462–471; and Geoff Mortimer, *Eyewitness Accounts of the Thirty Years War, 1618–1648* (New York, 2002). There has also been ample discussion over the terminology of these texts. James S. Amelang describes autobiographical texts as "any literary form that expresses lived experience from a first person point of view." Idem, "*Vox Populi*: Popular Autobiographies as Sources for Early Modern Urban History," *Urban History* 20 (1993), 30–42, here p. 33. See also Kaspar von Greyerz's assessment of autobiographical sources (in this case autobiographies and diaries) in "Religion in the Life of German and Swiss Autobiographers (Sixteenth and Early Seventeenth Centuries)," in *Religion and Society in Early-Modern Europe 1500–1800*, ed. Kaspar von Greyerz (Winchester, 1984), pp. 223–241, and Benigna von Krusenstjern, "Was sind Selbstzeugnisse?," pp. 462–471. Winfried Schulze, following the Dutch scholars Jacob Presser and Rudolf Decker, prefers the term 'Ego-Dokumente' to 'Selbstzeugnisse.' The argument is that 'Ego-Dokumente' include both free and forced testimonies. Thus, tax records, interrogations, visitations, and court records can also be utilized as documents pointing toward the self. See his "Vorbemerkung" and his *Ego-Dokumente: Annäherung an den Menschen in der Geschichte* (Berlin, 1996), pp. 9–30. For more recent discussions, see the special issue of *German History* 28, no. 3 (2010) that is dedicated to ego-documents.

6 Kaspar von Greyerz, *Vorsehungsglaube und Kosmologie: Studien zu englischen Selbstzeugnissen des 17. Jahrhunderts* (Göttingen, 1990), p. 16.

logically, without explanatory connections or transitions. With the adoption of such a serial structure (*Reihenstruktur*), many seventeenth-century 'autobiographies' were still indebted to the medieval recording tradition.[7] Often the author stepped back, while he or she recounted events of the day and other facets of life around him or her.[8] Until the end of the seventeenth century, then, these accounts narrated personal facts in their external rather than internal contexts.[9]

That being said, early modern autobiographical narratives show a wider variety of traits than they have traditionally been credited with. Some are in character very close to the medieval chronicle, where descriptions of chores, religious rituals, hardships, finances, the weather, etc. follow one another with no interpretive frame (for example the diary of the Augustinian prioress, Clara Staiger), while others provide much more than the typical series of personal facts in their external contexts (the account of the Benedictine abbot Veit Höser, for example).[10] Their extensive, sometimes funny, often rhetorically gripping comments, explanations, musings, and dramatizations point in the direction of modern autobiographies. Seventeenth-century autobiographical accounts then spanned a broad range of characteristics and reflect the markings of a transitional period in the genre.

The personal accounts around which the following analysis is built come from members of religious orders in and around modern-day Bavaria. These narratives vividly describe experiences connected with sieges, and the

7 Ingrid Schiewek, "Zur Manifestation des Individuellen in der frühen deutschen Selbstdarstellung: Eine Studie zum Autobiographen Bartholomäus Sastrow (1520–1603)," *Weimarer Beiträge* 13 (1967), 885–915, here p. 893.

8 See Stephan Pastenaci's pithy comment: "Die Lektüre frühneuhochdeutscher Autobiographien ist aufgrund der Heterogenität der Texte, oberflächlich betrachtet, gewiß kein Lesevergnügen. Intimes, Persönliches wechselt mit unpersönlichen Berichten aus der Ereignisgeschichte, es werden Anekdoten aus der Stadt, Biographien von anderen Persönlichkeiten breit ausgemalt. Private Briefe wechseln mit der ausführlichen Wiedergabe zeitgenössischer Dokumente. Es dominiert das Faktische vor der Reflektion." ["The reading of Early-Modern High German autobiographies is – superficially regarded – certainly no pleasure due to the heterogeneity of the texts. Intimate, personal matters alternate with impersonal reports of events; civic anecdotes and biographies of other personalities are described in great detail. Private letters alternate with full accounts of contemporary documents. Facts dominate over reflection."] "Probleme der Edition und Kommentierung deutsch-sprachiger Autobiographien und Tagebücher der Frühen Neuzeit, dargestellt anhand dreier Beispiele," in *Edition von autobiographischen Schriften und Zeugnissen zur Biographie*, ed. Jochen Golz (Beihefte zu Editio), vol. 7 (Tübingen, 1995), pp. 10–26, here p. 11.

9 Amelang, "*Vox populi*," p. 33.

10 See below for a discussion of these documents.

consequences and realities of war more generally. The situation of the *Religiosi* differed from that of the rest of the population in several ways. Monks and nuns lived behind walls and formed a community that could sustain them and in which they could find companionship. Typically these communities were linked to a wider network of convents and monastic houses, to which they could flee or to whom they could appeal for help. Many of them had resources and were the recipients of alms. Moreover, women religious often were under the 'protection' of a male guardian, although in reality he could offer very little physical safety. These conceivable advantages were counterbalanced by more problematic consequences of their status: *Religiosi* – in fact, the clergy in general – were frequently the target of violence. In some cases, they were suspected of being more affluent; in others, they were taken hostage to extract money from cities and communities. Confessional animosities added to the antagonism toward them. Still, although members of religious orders were set apart from the rest of the population in a number of ways, they also suffered some of the same experiences. Moreover, the *Religiosi*'s narratives reflect not only their own experiences but also those of the world around them. Like other early modern narratives, they offer more than one singular viewpoint; by depicting the society and culture in which these persons lived, they tell us something about the larger community. As Kaspar von Greyerz emphasizes: "personal narratives, both in reproducing and in creating discourse, are deeply embedded in a collective context."[11]

Besides the close observations and fascinating vignettes left behind by these men and women, there is another compelling reason why their testimonies have attracted such scholarly attention: at a time when literacy was still largely confined to the upper and middle classes, these religious were some of the most prolific recorders of the days' events. Their position as leader or adminis-

11 "Ego-Documents," p. 276. See also Mary Fulbrook's and Ulinka Rublack's comment: "[o]ne does not have to follow down a post-modernist route to realize the significance of the fact that no account of the self can be produced which is not constructed in terms of social discourses: that the very concepts people use to describe themselves, the ways in which they choose to structure and to account for their past lives, the values, norms, and common-sense explanations to which they appeal in providing meaning to their narratives, are intrinsically products of the times through which they have lived." "In Relation: The 'Social Self' and Ego-Documents," *German History* 28 (2010), 263–272, here p. 267; and Gabriele Jancke, Claudia Ulbrich, eds., *Vom Individuum zur Person: Neue Konzepte im Spannungsfeld von Autobiographietheorie und Selbstzeugnisforschung* (Querelles: Jahrbuch für Frauen- und Geschlechterforschung 2005), vol. 10 (Göttingen, 2005), who also emphasize the author's embeddedness in his or her social context.

trator of their order encouraged and sometimes demanded a daily record, and a good part of their writings were preserved by their orders.[12]

The geographic parameters of this inquiry are the south-east German lands. Although Bavaria had been spared military action and the worst of the repercussions of war until the end of the 1620s, the region suffered severely from the early 1630s on. The chronological focus is on the war of the early 1630s from the landing of the Swedish king, Gustavus Adolphus (d. 1632), in Peenemünde, Pomerania, in mid-1630 to the Peace of Prague in mid-1635. While the Catholics had been the clear winners of the first twelve years of the conflict, the intervention of the Swedish king reversed this course until Gustavus Adolphus was killed at the Battle of Lützen in November 1632, after which gains and losses were more evenly distributed among the warring parties. In January 1633 Chancellor Axel Oxenstierna, the commander of the Swedish/Protestant troops after King Gustavus Adolphus' death, gave orders to his chief generals, Bernhard of Saxe-Weimar and Gustav Horn, to mount a concerted attack on Bavaria, one of the emperor's chief allies.[13] Regensburg stood as entry point to Bohemia, lower Bavaria, and Austria and, because of its strategic importance, became the target of Bernhard's military campaign, followed shortly thereafter by his siege and attack of Straubing further down the Danube.

A gripping description of this advance comes from the pen of Abbot Veit (Vitus) Höser (1577–1634) of the Benedictine abbey Oberaltaich in Bogen near Straubing.[14] Abbot since 1614, Höser became a renowned reformer who took great strides in strengthening monastic discipline and theological training among his brethren. A virtuoso of the word, he chronicles the Swedish

12 Often the leader of the order (abbot/abbess, prior/prioress) or their record keeper or administrator kept these records of the *Religiosi's* activities, but sometimes the recorder remained anonymous so that we do not know what position, if any, he or she held. On female religious women and their writing, see Eva Kormann, *Ich, Welt und Gott: Autobiographik im 17. Jahrhundert* (Selbstzeugnisse der Neuzeit), vol. 13 (Cologne, 2004). See also Mortimer, *Eyewitness Accounts*, on the motivations for writing.

13 The epitaph 'Swedish' should not be taken literally. Especially after 1632 it means little more than 'Protestant.' Different from Gustavus Adolphus, Axel Oxenstierna, a consummate diplomat, never stood on a battle line. Moreover, soldiers regularly chose armies not for confessional reasons, and frequently, upon the victory of one party, the defeated soldiers switched sides – either voluntarily or not.

14 Rupert Sigl, *"Wallensteins Rache an Bayern. Der Schwedenschreck." Veit Hösers Kriegstagebuch* (Grafenau, 1984). There exists an earlier, less accurate German rendering by Joseph von Mußinan, *Über das Schicksal Straubings und des baierischen Waldes während des dreyßigjährigen Krieges vom Oktober 1633 bis April 1634* (Straubing, 1813). The Latin original can be found in the Staatsbibliothek Munich (Clm 1326, "Viti Hoeseri Abbatis Hist. Miscella PEREGRINATIONIS Durante per inferiorem Bauariam [...]"). This is the third volume of Höser's three-volume work entitled "Monomonastikon."

encroachment on the region in 1633 and 1634 that eventually forced the Benedictines to leave their monastery. For himself and the people he observed, Höser identifies uncertainty and instability as central experiences of the war.

After he left the *stabilitas* of his Benedictine abbey,[15] Höser recounts:

> Thus I wandered constantly, waveringly without aim, and nowhere did I find a minute of peace because the daily and nightly, irregular, sudden, and unpredictable attacks and persecutions of the Weimar riders and soldiers did not leave me a single corner where I could have rested with any feeling of security or even taken a short respite, let alone used it as a hiding place. O good God, how often during the course of a day or night was I forced to get up from bed, prayer, or work, after I just started something, to flee helter-skelter? How often was I discouraged by the terror of the Swedes? And still I see no end.[16]

Aimlessness and restlessness, however, did not only characterize the abbot's life but that of the people he encountered on his flights as well. In shelters there was a continuous coming and going. Wandering refugees warmed themselves for a while at a campfire to dispel the wintery cold but then left again. They changed places constantly and moved from one hiding place in the wilderness to the next since they did not know where to find safety.[17] Even in the most inhospitable places Höser regularly came across wandering people.

For lay and religious individuals and communities, their flights often led to temporary exile, not permanent emigration, although they rarely knew when or if they could return.[18] Contemporaries described this experience as "going

15 *Stabilitas loci* (stability of place), i.e., remaining within the walls of the monastery and not wandering about, was one of the foundational principles of the Benedictine order.

16 Sigl, *Veit Hösers Kriegstagebuch*, p. 145. "Riders and soldiers" referred to units of cavalry and infantry. All translations are mine.

17 Ibid., p. 136.

18 Among the burgeoning literature on exile and migration, see especially Matthias Asche et al., eds., *Krieg, Militär und Migration in der Frühen Neuzeit* (Herrschaft und Soziale Systeme in der Frühen Neuzeit), vol. 9 (Berlin, 2008); Joachim Bahlcke, ed., *Glaubensflüchtlinge: Ursachen, Formen und Auswirkungen frühneuzeitlicher Konfessionsmigration in Europa* (Religions- und Kulturgeschichte in Ostmittel- und Südosteuropa), vol. 4 (Berlin, 2008); Manfred Briegel, Wolfgang Frühwald, eds., *Die Erfahrung der Fremde: Kolloquium des Schwerpunktprogramms "Exilforschung" der Deutschen Forschungsgemeinschaft* (Weinheim, 1988); Johannes F. Evelein, ed., *Exiles Traveling: Exploring Displacement, Crossing Boundaries in German Exile Arts and Writings 1933–1945* (Amsterdamer Beiträge zur Neueren Germanistik), vol. 68 (Amsterdam, 2009); Helmut Koopmann, "Exil als geistige Lebensform," in *Exil: Transhistorische und transnationale Perspektiven*, ed. Helmut Koopmann, Klaus D. Post (Paderborn, 2001), pp. 1–19; Marita Krauss, "Heimat – Begriff

into misery," *ins Elend gehen*. This standard phrase, found throughout narratives of the laity and the clergy, implied having to leave one's place for a region that was not home. This strange land could be a distant county or a foreign country, but it could also be as close as the nearby forest, the next city, or, in the case of many *Religiosi*, another monastery or convent. In short, it was an area beyond the immediate boundaries of one's home. The expression has its roots in the middle-high German word "ellende" (old-high German: alilanti/elilenti) and connotes an alien country, *Fremde*, *Verbannung*, exile, as well as the notion of suffering and desolation.[19] Early modern narratives continue the conceptual association between exile, misery, and hardship.

Besides volatility and uncertainty, Veit Höser stresses the fear engendered by the war. After fleeing the troops' advance, he and his companions[20] eventually came back to the provostry

> trembling of fear. Our heart raced when we entered. In fearful expectation we ate our evening soup and stayed overnight without finding peace. With fear and anxiety we anticipated dawn. The lords and barons of Neuhaus and Au fled to us who were refugees ourselves. The former had crawled out of his hiding place in a cavern which he could no longer endure. The latter wandered aimlessly for weeks since the Swedes had robbed him of his *Heimat* and everything else. Countless people from the countryside of all ages were there who could not find safety anywhere.[21]

Höser provides close-ups of what leaving one's home meant for the populace around him. After the Swedish troops had overthrown Neuburg on the Danube, Abensberg, and Kelheim and were advancing on Regensburg while committing many acts of violence, the people were so horrified that they left their "Heimat" and "sought their salvation in flight."[22] The Benedictine abbot

---

und Erfahrung," in *Heimat, liebe Heimat: Exil und Innere Emigration*, ed. Herrmann Haarmann (*1933–1945*) (Berlin, 2004), pp. 11–27.

19 Otto Eberhardt, "Exil im Mittelalter: Einige Streiflichter," in *Weltanschauliche Orientierungsversuche im Exil*, ed. Reinhard Andress (Amsterdamer Beiträge zur neueren Germanistik), vol. 76 (Amsterdam, 2010), pp. 13–36, here p. 15; and *Deutsches Wörterbuch von Jacob und Wilhelm Grimm*, vol. 3 (Munich, 1999 [1862]), p. 406: "Elend" – "*exilium, captivitas, miseria* ... 1) urbedeutung dieses schönen, vom heimweh eingegebenen wortes ist das wohnen im ausland, in der fremde."

20 On his flight Höser frequently crossed paths with people he knew and with whom he might travel for a while. In this case, he met the provost (of the following provostry) and a master builder named Ulrich. Sigl, *Veit Hösers Kriegstagebuch*, p. 136.

21 Ibid., p. 137.

22 Ibid., p. 84.

describes with great sensitivity the suffering of the refugees he sees and meets, and is cognizant of their burden: "No one can see the daily and night-long treks of refugees passing before his eyes without the deepest pity."[23] He highlights the suddenness of the assault that made any planning impossible. People had to vacate their homes at once and had to abandon family, friends, and possessions. "Without hesitating for a second and without a glance back they had to leave for the unknown and start a miserable life (i.e., a life away from home)."[24] Helplessly Höser had to watch as parents and children were torn apart and everyone, whether elderly, half-naked, injured or crippled, wandered around aimlessly among strangers. War became the great equalizer: "They who previously had plenty were now forced to go begging."[25] The paths of clergy and nobles in disguise, of lay people and commoners crossed as they were filling up the houses and every last hiding place of the Bavarian forest.[26]

For religious communities, an approaching army usually led to an involved discussion about what they should do, whether they should stay or flee, and, if the latter, whether some should stay behind to ensure the continued viability of the cloister or all together go into exile. Like many religious houses, the Capuchins of Landshut, situated about 42 miles south-west of Straubing, were in a precarious situation since their monastery was located outside the city walls.[27] When the Swedes under Bernhard of Saxe-Weimar advanced on Landshut in July 1634, the friars decided it was best that all friars move to the city and reside at the vicarage of St. Jobst, from which the priest had already fled.[28] Chronicler

---

23 Ibid., p. 93.

24 Ibid., p. 94.

25 Ibid.

26 Ibid., pp. 94–96.

27 During Landshut's first occupation in 1632, Gustavus Adolphus seized the city by accord and demanded 100,000 Taler. Since the city gave him only 50,000, he took several citizens (ecclesiastics, nobles, councilors, and other eminent lords) as hostages to Augsburg. Landshut's first occupation lasted eight days. After the Catholics regained Augsburg and Gustavus Adolphus died, the hostages returned to Landshut. Bayerische Staatsbibliothek München, Cgm 2943, fol. 2r–3r.

28 "[...] und Wir unß in unseren armen Klösterl ausser der Stadt befunden haben, hat alsobald P: quardians, P: Stanislaus sein liebe ganze familia zusammen berufen, und einen jeden befragt, was hierinnen zu thun ware, haben zwar etliche vermaint, es wäre gut, das ein theil in dem Klösterl verbleibe, und der andere theil sich in die Stadt begeben soll: andere haben vermaint, es wäre gut, das Wir alle beisamen verbleiben, es wäre gleich in dem Klösterl ausser der Stadt, oder aber das Wir bey ainander bleiben in der Stadt in dem Pfarrhof bey S: Jobst, dann selbiger Pfarrer in ein andere Stadt sich reteriert hat, wenn Wir uns in die Stadt begeben wolten, wir solten doch in seinem Pfarrhof unsere Wohnung nemmen: welches Wir ihme auch Versprochen haben. Darauf hat sich P: quardian resolviert, und für gut gehalten, es werde das beste seyn, das Wir alle samt bey einander bleiben, und unß zugleich in die Stadt verfügten." Ibid., fol. 3r–v.

Melchior von Straubing explained that there were many reasons for this decision, but one in particular: if the city was taken by storm, they – being located outside the city walls – would be the first victims, although he seemed to be worried more about the loss of their liturgical equipment than of their lives. If they moved into the city, however, they could not only save their paraphernalia, but also be of use to the distressed citizenry. Melchior added in a critical tone that many spirituals had beaten an early retreat.[29] Landshut eventually was besieged for three days until an accord was reached.

For many women religious, the decision to flee or stay was made by their male superior. Such was the case with the Augustinian nuns of Mariastein near Eichstätt but some female communities had no guardian (like the Brigidine cloister in Altomünster in the district of Dachau close to Munich) or a guardian of no consequence (such as the Dominican nuns of Heiliggrab near Bamberg, who had no help in Bamberg's bishop).[30] In contrast, the Dominican nuns of St. Katharina in Augsburg represent a convent with options and the power to make its own decisions. This is certainly not unrelated to the fact that many of its members were connected to influential families in the city.[31] As the *Kurze Beschreibung* (Short Description) of St. Katharina's chronicler makes clear, however, the convent's decision making process proved to be an exasperating affair, and assessments of the situation changed constantly. The back and forth between its prioress, Maria Magdalena Kurz (not the recorder), her convent, and a wide range of secular and religious authorities affords a close-up of

29 They should move into the city "um vieler Ursachen willen, unter welchen ich nur eine erzellen will, das ist gesetzt, man wolte die Stadt mit Gewalt, und stürmender Hand einnemmen, so wären wir ausser der Stadt in unserem armen Klösterl die erste am spiez; ihre Kirchensachen wuerden den Soldaten anheimfallen. Wenn allerdings die Stadt erhalten bliebe, wuerden auch ihre Kirchensachen erhalten bleiben, und sonderlich auch, damit Wir dem betrengten Volk im fall der Noth mit allen Christlichen Catholischen bystand kunten zu hilf kommen sowohl mit administrierung der heiligen Sacraments als auch Predigen, und heiligen Messlesen, dann gar viel der Geistlichen sich bey Zeiten aus dem Staub gemacht haben. Alsdan haben Wir im Namen Gottes unsere arme klösterliche sachen in die Stadt im Pfarrhof bey S: Jobst getragen." Ibid., fol. 3v.

30 For Mariastein, see the discussion below. For Altomünster, see Wilhelm Liebhart, *Altbayerisches Klosterleben: Das Birgittenkloster Altomünster 1496–1841* (Münchener Theologische Studien), vol. I/30 (St. Ottilien, 1987). For Heiliggrab, see Friedrich K. Hümmer, ed., "Bamberg im Schweden-Kriege. Nach einem Manuscripte (Mittheilungen über die Jahre 1622–1634)," in *Bericht über Bestand und Wirken des historischen Vereins zu Bamberg* 52 (1890), 1–168 (part 1); and 53 (1891), 169–224 (part 2). On all of these cases, see my forthcoming monograph on the Thirty Years' War."

31 Bernd Roeck calls St. Katharina a "Versorgungsinstitut für den Adel." In *Eine Stadt in Krieg und Frieden: Studien zur Geschichte der Reichsstadt Augsburg zwischen Kalenderstreit und Parität*, vol. 1 (Göttingen, 1989), p. 91.

the confusion, disbelief, shock, and uncertainty among the Catholic leaders when confronted with the Swedish advance in the early 1630s, and of their reluctant realization that their fortune was about to change. Exchanges range from convictions that all was still well to the slow recognition of potentially very bad outcomes, and the prioress is caught up in the leaders' attempts to find their way in the thicket of war.

When Gustavus Adolphus took Würzburg in 1631, Prioress Kurz was anxious about what would happen next and how best to protect her convent. The nuns had heard that elsewhere many spirituals had fled, and the Swedes had taken hostage any nun they could find. "Therefore (Kurz) suffered the greatest fears, concerns, and doubts whether she should send her convent away or keep (everyone) together in the cloister."[32] As Gustavus Adolphus slowly closed in on Augsburg, the prioress repeatedly turned to the city's secular and religious leaders for guidance, but their advice for the convent kept changing. First, the Father Provincial and several of Augsburg's secular notables counseled Kurz to take all her nuns to the Tyrol. The leaders feared more for the nuns' honor (i.e., that they would be raped) than for their lives.[33]

Accordingly, Prioress Kurz made every effort to find a refuge in the Tyrol that was large enough to house the entire convent.[34] In the meantime, however, she was advised that it would be better if the convent fled to Munich, where they could also take their sacred treasure (*Heiltumber vnd kürchen schäz*).[35] This suggestion was founded on the conviction that Bavaria was safe since Elector Maximilian would not let anything happen to it.[36] The prioress

32 Diözesanarchiv Augsburg, HS 97, fol. 2r.

33 The prioress "ist also in gresten ängsten, vnd Sorgen gestanden, vnd in grossem Zweiffel, ob Sy den Conuent solte verschükhen, oder beysamen im Closster behalten, vnd wie wol Sy ein hoch verstendtige, weiße, vnd fürsichtige fraw gewest, so hat Sy doch in so schweren, vnd hoch geferlichen Sachen, Rath gesuecht, wie Sye sich mit dem Conuent solte verhalten. Alß nemlich, bey vnsserm hochwürdigen pater prouinciall: welcher der Zeit eben alhie gewest, vnd etlichen in der gleichen Sachen erfahrnen weltlichen Herren, so ist ihr gerathen worden, Sy sole mit sambt dem ganzen Conuent, an ein sicher orth fliehen, vnd khein ainige alhie Lassen verbleiben, von wegen vieller geferligkheiten, die zu befürchten gewest seindt, nicht allein des Lebens, sunder, vnd vil mehr, der ehren verletzung, sambt anderen viellen grossen gefahren ..." Ibid., fol. 2r–v.

34 The convent counted 42 women – 31 nuns and 11 lay sisters, plus 6 nuns from the convent S. Marx in Würzburg who had fled to them.

35 The narrator does not identify whose advice this was, but subsequently Kurz asked her spiritual counselors what they thought of this recommendation, which suggests that the persons in question were secular.

36 "Auff solchen Rath, hat Sy unsser vil Ew: fraw Priorin, sehr vil bemieht, ein gelegenheit zu bekhumben in Tyrol: das Sy mit sambt dem ganzen Conuent an ainem Orth beysamen khundten sein: Vnder dessen aber ist jhr auch gerathen worden, es wehre bösser, wan Sy

then asked her spiritual counselors whether they agreed, and, if so, when this should occur. The religious advisors did not think Bavaria in danger either since one would not let the Swedish king advance this far. But, if it happened after all, "which God may prevent mercifully," Munich and Bavaria were no safer than Augsburg. In this case, the counselors hoped that an accord (*Contract*) with the occupiers would ensure their protection. In the end, the Father Provincial recommended to stay in Augsburg since it would also be a heavy burden on the prioress to sustain her large convent in exile for a potentially very long time. In addition, several of her sisters had relatives in Augsburg and could stay with them.[37] Their sacred treasure, however, should be sent to Munich.

The convent followed this advice, but when rumors (*Geschrei*) arrived in late January 1632 that the Swedes were about to take Augsburg, the question was again whether they should flee or remain, and the consultations started once more. After all the hearsay, reliable news (*unfehlbare Zeitung*) arrived in late June 1632 that Gustavus Adolphus had taken Nuremberg and was headed toward Augsburg. This frightening piece of information prompted many spiritual and lay Catholics to flee the city. The distressed prioress appealed once again to the administrators of the city and the cloister, and was told that they should all stay. Kurz was shrewd enough to ask for this advice in writing in case things went badly and no one remembered that they had counseled the prioress to remain in Augsburg.[38] The document, however, did not come as quickly as she had hoped, and the Swedish king was already at Donauwörth. Thus Kurz convened her sisters to prepare them for the struggles ahead and offer them

---

in dem Bayrlandt Zu München, ein guette gelegenheit khündte bekhumben, mit dem Conuent aldort sicherlich zu wohnen, wie auch die Heiltumber, vnd kürchen schäz, dahin zu flehnen, vnd zu bewahren, dieweil diß mahl, von menigklich ist vermaint worden, es werde in Bayrn am allersichersten sein, vnd ihr ChurFürstl: Durchl: an seinem Landt nichts geschehen." Diözesanarchiv Augsburg, HS 97, fol. 2v.

37 "Darauff jhr von beden Herren die Antwortt erfolgt, zu München wuß man ganz vnd gar nichts von denen Sachen, die Sy geschriben, vermainen auch nit, das soliche gefahr zu besorgen sey, dieweil man darfür halt, den Schwedischen Khönig, werd man so weit nit herauff khumben Lassen, sunder werde wol khünden vorgschlagen werden, zue dem wan es Je solte geschehen (das doch Gott gnedigklich verhietten wölle:) so würde München vnd das Bayrlandt, eben so wenig sücher sein, alß Augspurg, Jedoch wehr den Contract: mit dem Bayrlandt vnd München gemacht, werde Augspurg nit vergessen haben: Weitter hat derweil Ew. Fraw priorin, gemelte geistliche person den Rath geben, das Sy ihr mit nichten ein solche schwere burden, solte Jber ihren Halß bünden, den Conuent, auß der Stadt Augspurg, in fremde Landt Zu schickhen, der vrsachen dieweil man nit wüste, wie lang das Kruegs weßen mechte wehren, vnd Sy einen so grossen Conuent nit würde khunden in die Lenge, in der fremde erhalten." Ibid., fol. 3r–v.

38 Ibid., fol. 5v–6v.

spiritual comfort and strength. She counseled them to pray to God for guidance whether "we should remain in the cloister or flee." In the meantime she would try to find out about the plans for the city's takeover.[39]

The next communication from Munich described Bavaria as in great danger and urged the convent to move their belongings from Munich to the Tyrol.[40] The city's and the cloister's administrators (Bernhardt Rehlinger and David Welser), however, sent word to the convent that they should stay because, once the city was taken by accord, they would not have anything to fear. Kurz only told the older nuns of this 'mandate' and remarked that some still hoped the enemy would be beaten at Donauwörth. Mostly she did not want to worry the rest of the convent during the Easter celebration, but if further things would occur, she had to tell them.[41]

On Easter Day, finally, there was no more escaping the reality: the Swedes had surrounded the city on three sides. Although highly alarmed, the prioress and the oldest convent mothers did not want to frighten the younger sisters "since it was before dinner time and because they were rather weak and tired from the long fasting and strenuous worship, also on account of the many sorrows."[42] Still, it could not be helped, and with great care she gathered her charges around her and let them know that the enemy was about to take the city and that "she desired to know from us what we determined, whether we wanted to remain in the cloister or not."[43] The prioress presented her nuns with several reasons why it was beneficial to remain in the cloister: first, it was advisable to heed the counsel of their administrators, since, if those who left the cloister should fall on hard times, they may not expect any help since they acted against the guidance of "judicious people."[44] In other words, if nuns who had chosen to go into exile needed funds and provisions, they could not necessarily count on the support of those leaders who had counseled against flight.

More than on the authority and wisdom of their political and spiritual advisors, Prioress Kurz counted on the persuasiveness of the convent's own history to convince the sisters to stay.[45] Similar to their own situation, a hundred years ago most *Religiosi* in Augsburg had fled, while St. Katharina's nuns had remained in their cloister. The prioress, however, did not downplay the situation

39 Ibid., fol. 7r–v.

40 Which they did, only to retrieve them eight days later.

41 Diözesanarchiv Augsburg, HS 97, fol. 7v–8v.

42 Ibid., fol. 9r–v.

43 "[…] nach dem sich der Conuent gesamlet, hat Sy vns erünert was Sy vor .4. Tagen, mit vns geredt hab, vnd darneben vermeldt, das es nun mehr an dem sey, das die Stadt werde von dem feindt eingenumben warden, Sy beger von vns zue wüssen, was wür bedacht seindt, ob mir wöllen in dem Closster verbleiben oder nit." Ibid., fol. 9v.

44 Ibid. The requested written advice had come on Maundy Thursday.

45 Ibid., fol. 10r.

in 1632, and instead declared that those who dared to stay could expect not only their forbears' deprivations and troubles, but also "much hunger, sorrow, fright, and other kinds of misery."[46] Still, she believed that such and more challenges could be borne with God's help, and urged her sisters to remain in the cloister.[47]

This is the end of the entangled consultations on whether the convent ought to flee or stay. It shows that, in this militarily volatile situation, confusion was widespread not only among the populace and the *Religiosi*, but also among the political and religious leaders. After more than a dozen years of being victorious, the Catholic leadership had difficulties coming to terms with the altered political and religious landscape. Tellingly, the prioress found her greatest inspiration and hope in the past courage of their predecessors and in God's protection. Augsburg's leadership surrendered with an accord, and next we see Prioress Kurz negotiating with the Swedes to protect her convent – thus the sisters (or at least most of them) must have decided to remain.

The testimonies above do not only give insight into contemporaries' experiences of instability and fear during the war; they also reflect on the shape of warfare, particularly with regard to sieges. While 'siege' as in the literature discussed earlier conjures up scenarios of stalemate, digging in one's heels, settling in for the long haul, and testing the endurance of the other that could last for months or years, many of the cases in this southern region are rather short-term affairs. Veit Höser's graphic portrait of the sieges of Regensburg and Straubing is a case in point:

> Unexpectedly and lightning fast the commander of the Swedish party, Duke Bernhard of Weimar, invades northern Bavaria with his war machinery. In late October 1633 he swiftly conquers the feebly fortified towns Neuburg at the Danube, Abensberg, and Kelheim. On November 1st he advances on *Regensburg*, commences the siege on the 2nd, and greets us with his cannons on November 3rd. Then he continues the siege for nine days with incessant bombardment from his twelve cannons [...] With this he shook the massive city walls, breached them, and finally caused them to collapse. Since the forces that had defended the city heroically had been let down and betrayed by the promised rescue troops, the city was compelled to surrender.[48]

During their siege of Regensburg, the Weimar troops made forays into lower Bavaria, devastating and depopulating the countryside. Daily Höser observes

46 Ibid.
47 Ibid., fol. 10r–v.
48 Sigl, *Veit Hösers Kriegstagebuch*, p. 83.

streams of refugees arriving in his neighborhood in search for quarters and deliverance from the troops' violence. The abbot sees this as a foretaste of what their own territory will have to suffer next.[49] And, indeed, shortly after the Swedes took Regensburg, they advanced on Straubing just six miles from the abbey.

After an unsuccessful attack, the troops returned two days later with reinforcements. Violent raids on neighboring districts were designed to pressure the inhabitants to hand over their city voluntarily. But with the support of 600 men under Colonel Georg Rudolf von Haslang and several other contingents,[50] the Straubingers held out, and thus the Swedes began their siege by laying mines and raising defense works (*Schutzwälle*) ever closer to the city. They seized the cloister of the Capuchins near the city walls and from there bombarded Straubing's fortifications until they were able to breach the walls and force the citizens' surrender.[51]

These descriptions evoke images of short sieges, usually combined with forceful attacks, not two enemies facing each other across entrenchments and blockades for long periods of time. In the latter scenario, the goals were to force the besieged opponent to his knees by cutting off supplies, thwart any attempts of rescue troops to aid the beleaguered, and find a way to breach the fortifications. Except for foraging, then, there is little movement. The warfare perceived by southern Germans was very different. Moreover, as noted earlier, with respect to the effects of these operations on the populace, accounts underline the ever changing situations in which contemporaries found themselves. After a short siege, the enemy might be able to occupy the city for a week or so until he is called elsewhere or driven out by his opponent's troops. Thus the populace had to live with a great deal of uncertainty and continually had to adjust to new realities.

Höser's lines draw attention to some of the realities of warfare during the Thirty Years' War. Few cities were fortified heavily enough to withstand the force of a besieging army for long. While the siege of Regensburg lasted ten days, Straubing's was over in just three. These were not exceptional cases. Many cities resisted for only short periods of time or avoided a siege altogether by negotiating an accord from the start. There were exceptions to this general rule of inadequately secured cities, such as Nuremberg and Ingolstadt, as well as heavily fortified castles, such as Forchheim and Kronach. Because of Nuremberg's political and commercial importance, the imperial city had

49 Ibid., p. 84.

50 Ibid., p. 57 (the editor, Rupert Sigl)

51 Ibid., pp. 96–97.

begun upgrading its walls according to the latest technological insights already during the sixteenth century – a process that was largely completed by the beginning of the Thirty Years' War, but defense efforts reached a new level when Gustavus Adolphus arrived in Franconia (today part of northern Bavaria) and made Nuremberg the base of his operations. In Adolphus' plan, a belt of entrenchments was to encircle not only the city's suburbs but also the closest small towns. For sixteen days in mid-1632, six thousand citizens and peasants worked day and night to finish the fortifications.[52] Because of its state-of-the-art defense system, Nuremberg was neither besieged nor taken during the war, although its countryside suffered tremendously when the imperial and Swedish armies faced each other at the Alte Veste three miles south-west of Nuremberg in September 1632.

Ingolstadt, Bavaria's stronghold and university town, also had strong fortifications.[53] Bavaria's elector Maximilian (r. 1598–1651) had been able to strengthen Ingolstadt's defenses just in time for the Swedish attacks, and, much like Nuremberg, Ingolstadt was never taken, although its fortifications were thoroughly tested. Gustavus Adolphus tried to take the city in spring 1632, but after a week-long siege and several unsuccessful assaults, the frustrated king gave up and instead set his eyes on Munich, Maximilian's court city. Munich fell into Gustavus Adolphus' hands in May 1632 without a siege because of its weak fortifications. The king demanded the enormous ransom of 300,000 Reichstaler to ensure that the city would not be plundered by his troops. Since Munich could not even raise a third of this sum, Adolphus took 42 secular and ecclesiastical hostages to guarantee remuneration. In the end the amount was never fully paid, and several hostages died in captivity.[54]

The case of Munich and its lack of adequate fortifications well into the 1630s illustrate the problems many governments faced. Maximilian fully understood the importance of military readiness, but his efforts throughout the 1610s to improve the medieval walls around Munich were largely in vain since he could not convince the city's magistrate or the country estates to finance

52 Emil Reicke, *Geschichte der Reichsstadt Nürnberg von dem ersten urkundlichen Nachweis ihres Bestehens bis zu ihrem Uebergang an das Königreich Bayern (1806)* (Nuremberg, 1896), p. 977; Hanns H. Hofmann, *Die Nürnberger Stadtmauer* (Nuremberg, 1967), pp. 67–87. For a detailed discussion of the fortification efforts of 1632, see Franz Willax, "Die Befestigungsanlagen Gustav Adolfs von Schweden um Nürnberg 1632," *Mitteilungen des Vereins für Geschichte der Stadt Nürnberg* 82 (1995), pp. 185–235.

53 On Ingolstadt, see Tobias Schönauer, *Ingolstadt in der Zeit des Dreißigjährigen Krieges: Soziale und wirtschaftliche Aspekte der Stadtgeschichte* (Ingolstadt, 2007).

54 For a succinct overview of Munich during the Thirty Years' War, see Manfred P. Heimers, *Krieg, Hunger, Pest und Glaubenszwist: München im Dreißigjährigen Krieg* (Veröffentlichungen des Stadtarchivs München) (Munich, 1998).

such an extensive project. Although Munich's magistrate finally agreed in 1615 to strengthen its fortifications, the planning progressed only slowly. The beginning of the war in 1618 accelerated preparations, but by the middle of the 1620s, after the Battle of the White Mountain (November 1620) and the positive results for Bavaria (for his troubles Maximilian received the Upper-Palatinate and was made a member of the imperial electoral college), work on entrenchments was slacking off again. Since Bavaria was spared immediate military action during the first decade of the war, people did not see a pressing need for such a considerable effort and expense.[55] Thus, when in the spring of 1632 Gustavus Adolphus marched through Bavaria, taking city after city (except for Ingolstadt!) and aiming for Munich, the court city was ready for the taking.

State of the art fortifications were expensive and were carried out on the backs of the people. One could argue that this is precisely why governments had to spend ever more resources on these projects, but to what extent did this actually occur throughout the Thirty Years' War with its heavy financial toll especially in the areas of troop movement?[56] Strong physical defenses became indeed an ever greater concern from the 1630s on. The lack of physical preparedness and its consequences in 1632 were certainly crucial lessons for Maximilian. After the Swedes had left, Maximilian ordered his court city to complete its fortifications at its own expense, but Munich's city fathers begged him to spare them with his demand because of their miserable financial situation.[57] Unperturbed, the elector directed the council in April 1633 to have sixty people ready to begin building fortifications at the Anger gate, and the city responded

---

55 Ibid., pp. 11–15.

56 On the differential population loss in the Holy Roman Empire and the particularly high death rate along the path of troop movement roughly from north-east to south-west, see Günther Franz, *Der Dreißigjährige Krieg und das deutsche Volk: Untersuchungen zur Bevölkerungs- und Agrargeschichte* (Jena, 1940). Critical of Franz: Wolfgang Behringer, "Von Krieg zu Krieg: Neue Perspektiven auf das Buch von Günther Franz 'Der Dreißigjährige Krieg und das deutsche Volk' (1940)," in *Zwischen Alltag und Katastrophe: Der Dreißigjährige Krieg aus der Nähe*, ed. Benigna von Krusenstjern, Hans Medick (Göttingen, 1999), pp. 543–591, and John Theibault, "The Demography of the Thirty Years War Re-Revisited: Günther Franz and his Critics," *German History* 15 (1997), 1–21. Some correctives to Franz's study and numbers are offered by Christian Pfister, *Bevölkerungsgeschichte und Historische Demographie 1500–1800* (Munich, 1994), Quentin Outram, "The Socio-Economic Relations of Warfare and the Military Mortality Crises of the Thirty Years' War," *Medical History* 45 (2001), 151–184, and a summary of the results of regional studies in Peter H. Wilson, *The Thirty Years War: Europe's Tragedy* (Cambridge, MA, 2009), pp. 786–795.

57 Helmuth Stahleder, *Chronik der Stadt München: Belastungen und Bedrückungen*, vol. 2: *Die Jahre 1506–1705* (Hamburg, 2005), p. 462 (22 February 1633).

with several calls to the populace to contribute to this effort.[58] Once Maximilian returned to Munich in 1635 after a three-year absence,[59] he accelerated his efforts to fortify the city and ordered a special tax to pay for the measures. He was so upset about the slow progress and the councilors' lackluster supervision of the work that in 1638 he fined every councilmember ten Reichstaler. Two months later, however, he revoked the fine because its threat alone had apparently had the desired effect.[60] Work on fortifications now gathered speed, and when the Swedish-French armies stood before Munich in 1646, they were confronted with a defensive system constructed in accordance with the latest technology. This time, its fortifications contributed to preventing a hostile siege and takeover of the city. Much like Nuremberg in the early 1630s, however, Munich could not prevent the devastation of its countryside.[61]

Besides the caliber of the fortification, other factors played a role in influencing the occurrence and length of a siege. The prospect of relief troops from the outside made cities hang on during a siege or, conversely – as in the case of Regensburg's dashed hopes – give in to the besiegers. The degree of resistance also depended on the city's reading of the situation: whether they had any chance of success or whether their resistance would only make the takeover more hurtful for them. In other words, would the occupiers make them pay for their resistance and for the losses they had inflicted on the besiegers?

One way to avoid or shorten a siege was to negotiate an accord, which was a negotiated settlement between the occupiers and the occupied that regulated, for example, how much money and/or goods the city had to pay if it wanted to circumvent being ransacked. Such an agreement also regulated conduct on both sides with regard to matters of religion, i.e., whether the inhabitants could continue their religious practices or had to follow the confession of the victor. Many, as in the cases of Augsburg and Landshut above, expected the siege to end in an accord; therefore, both parties expended much energy on bargaining for the most advantageous arrangement.

An accord, however, was not necessarily comforting, as the testimony of Clara Staiger (1588–1656), prioress of the Augustinian convent Mariastein near Eichstätt makes clear. Her diary-style *Verzaichnus*, a minute record of the

58 Ibid., p. 463 (6 and 16 April 1633).

59 Maximilian left Munich before the Swedes arrived (4 April 1632) and returned in mid-May 1635 after the pestilence had subsided. Much of the time he governed from fortified Braunau on the Inn, some 77 miles east of Munich.

60 Stahleder, *Chronik der Stadt München*, vol. 2, p. 503 (10 September 1638).

61 See, e.g., Hauptstaatsarchiv München, Dreißigjähriger Krieg – Akten 323, Acta Den Einfall der Schweden in Bayern, in specie deren Annäherung nach Dachau [...], Rat und Landrichter von Dachau an Maximilian, 19 April 1648, 22r.

convent's day-to-day happenings with occasional comments, provides a testimony to the volatile nature of military advances.[62] Situated on the Altmühl River about 40 miles south of Nuremberg outside Eichstätt's city walls, Mariastein was repeatedly assaulted and eventually destroyed by enemy troops. As the Swedes marched into Eichstätt's territory in April 1633, the sisters fled to fortified St. Willibaldsburg high above the city. The castle, however, proved no match for the enemy's onslaught, and after a ten-day bombardment, during which the frightened nuns watched from their post in the stronghold how the troops ransacked their cloister, Eichstätt capitulated and negotiated an accord with the Swedes.[63] Still, the settlement did nothing to reassure the nuns since, while the clergy in the city were promised to be spared and free to leave, the men and women who had sought refuge at the castle were to remain until their value was estimated for ransom.[64]

Religiosi tried to reclaim some measure of control over the shifting realities of war through one means in particular: the procurement of reliable information. As we saw, Prioress Kurz of Augsburg employed this strategy to guard the life of her sisters and the interests of her convent. In general, in these volatile times, contemporaries soaked up whatever news and information they could find about the war, which reveals a visceral desire to know what was going on.[65] Such impulse was only sensible since one's life might depend on making good judgments about one's next moves. However, not everyone had access to trustworthy information nor the means to obtain it. Rumors of advancing troops frequently set off large-scale flights of the peasant population to places they believed offered more protection than the open village. People like the Benedictine abbot, Veit Höser, on the other hand, tried to get confirmation of such news so that they could take appropriate rather than blind, panicked action.[66] Indeed, Höser's descriptions above of the chaotic situation among the fleeing villagers highlight a rather stark contrast in opportunities between

62 *Klara Staigers Tagebuch: Aufzeichnungen während des Dreißigjährigen Krieges im Kloster Mariastein bei Eichstätt*, ed. Ortrun Fina (Regensburg, 1981), p. 326. Staiger's entries start in earnest in 1632 with her election as prioress and with the attack of the Swedes. Her notes are interrupted from 1645 until August 1646, when she was gravely ill and could not continue her writing. Mariastein lay in close proximity to the seat of the Augustinian canons of Rebdorf, who functioned as the nuns' guardians.

63 *Klara Staigers Tagebuch*, p. 82.

64 *Klara Staigers Tagebuch*, p. 83 (13 May 1633). *Religiosi* and the clergy in general as well as secular leaders were regularly taken to extract money from governments or communities.

65 For an entry into the expansive literature on this topic, see Wolfgang Behringer, *Im Zeichen des Merkur: Reichspost und Kommunikationsrevolution in der Frühen Neuzeit* (Veröffentlichungen des Max-Planck-Instituts für Geschichte), vol. 189 (Göttingen, 2003).

66 Sigl, *Veit Hösers Kriegstagebuch*, p. 112.

different social levels. While many *Religiosi* and secular leaders had the chance to prepare for their flight, the broader populace was often caught off-guard. The former were generally better informed and sent out envoys to check the veracity of rumors, whereas commoners frequently lacked the means to secure military information and reacted to hearsay.

This thirst for information is connected to another intriguing feature: a proclivity to comment unreservedly on the military situation. Both male and female *Religiosi* regularly criticized the government for its bad management of the war. In Höser's estimation, 2,000 men quickly dispatched could well have been enough to save Regensburg and Straubing and to prevent the devastation of lower Bavaria.[67] The fault for all the misery, he charged, lay with the military and political commanders, especially Albrecht von Wallenstein (1583–1634), who had failed to come to Bavaria's aid:

> May those search their conscience and bear responsibility who have been entrusted with the protective sword and power! What now? Inclinate Capita Vestra, bow your heads, you poor abandoned and forsaken ones, you children of the Fatherland, who have been robbed of every help and – harsh are my words! – submit to the enemy or flee! Because the arch-enemy of the Roman faith and of our ancestral holy religion, Bernhard of Weimar, has already marched across the bridge and through the gates of Straubing toward us [...] without anyone's opposition – yes, not even a dog barked at him! [...] I, for my part, would rather flee and trust the steadfast assurance of the Lord: 'When they persecute you in one city, flee to the next!' (Matthew 10:23)[68]

Clara Staiger also frequently expressed criticism of the military. From early on she judged the Catholic troops (*Die unsrigen*) to be ineffectual at best and at

67 "Ich meine, der Feind hat sich mit der Einnahme von Regensburg das Tor zu uns aufgemacht. Er ist durch diese Pforte schon so weit ins Niederbayerische eingedrungen, daß er nun schon in Straubing einziehen konnte, ohne daß ihn jemand gehindert hätte, niemand sich auch nur entgegenstellte. Die Regensburger und genau so die belagerten Straubinger könnten [...] unter Beweis stellen, ob nicht schon 2000 Mann Infanterie jeder der beiden Städte zu ihrer Rettung genügt hätten, unter der Bedingung allerdings, daß sie schnell in Marsch gesetzt worden wären, und diesen so luderischen (ludricum= Luder=Luther) Feind aufgehalten und die Städte vor ihrem Fall bewahrt hätten. Ob nicht, meine ich, 2000 Mann Niederbayern vor einer so furchtbaren Verheerung, so viele Kirchen und heilige Stätten, so viele Menschen und Menschenseelen vor der Vernichtung beschützt hätten. Mögen nun jene mit sich ins Gewissen gehen und es verantworten, denen das schützende Schwert und die Macht anvertraut ist!" Sigl, *Veit Hösers Kriegstagebuch*, pp. 98–99. Höser indicated that he received his information through *Relationen* – news reports available at seasonal fairs.

68 Ibid., p. 99.

worst more harmful than the enemy troops.[69] These assessments suggest not so much a penchant for armchair military strategizing but rather frustration and pain – to the point of outrage – over the lack of political and military protection during this conflict. They also convey the sense that their deep-rooted understanding of how the world should work was crumbling: one's own soldiers behaved worse than the enemy's; and governments were no longer able to protect their subjects.[70]

To conclude, the testimonies above have highlighted the immensely volatile and destabilizing effects military operations of the Thirty Years' War had on every level of the community, from political leaders to commoners. Our discussion focused particularly on narratives related to sieges, and the movement and mayhem leading up to and following them. These sieges, it has become clear, were not long, drawn-out affairs but quite brief in character, usually lasting only a few days until an accord was reached or another party was able to turn the tables. From the contemporaries' point of view, the ground was continuously shifting. Agreements with the occupying party might provide some relief. Moreover, people in advantageous positions, such as many *Religiosi*, were able to counteract the chaos of events and movement by being "on top of things" with regard to military news and information, which enabled them to make shrewd decisions.

The "world of the siege" encompasses a broad spectrum of features. Some of these operations amounted to large-scale, long-term enterprises. More frequently, however, as the above discussion suggests, we encounter a rather complex mixture of short-term sieges, offensive pressure, and negotiations, with a populace trying to adjust to the ever-changing nature of the war.

Bibliography

Archival Sources

Diözösenarchiv Augsburg
HS 97

69 "Die unsrigen sein umb vil vil 1000 Mann störckher gewesen / Haben aber wenig außgericht / sunder nur land und leuth / wo sie hinkomen / verderbt / kirchen und clöster mehr als der feindt selbsten beraubt und sein biß in dritte monat / gegen ein ander vor Nürnberg gelegen." *Klara Staigers Tagebuch*, pp. 61–62 (30 August 1632). Staiger is referring here to the Battle at the Alte Veste.

70 For more information on these issues, see my forthcoming monograph.

Staatsbibliothek München (Munich)
Cgm 2943
Clm 1326

Hauptstaatsarchiv München (Munich)
Dreißigjähriger Krieg – Akten 323

Printed Sources

Deutsches Wörterbuch von Jacob und Wilhelm Grimm, vol. 3 (Munich, 1999 [1862]).
Hümmer, Friedrich K., ed., "Bamberg im Schweden-Kriege. Nach einem Manuscripte (Mittheilungen über die Jahre 1622–1634)," in *Bericht über Bestand und Wirken des historischen Vereins zu Bamberg* 52 (1890), 1–168 (part 1); and 53 (1891), 169–224 (part 2).
Klara Staigers Tagebuch. Aufzeichnungen während des Dreißigjährigen Krieges im Kloster Mariastein bei Eichstätt, ed. Ortrun Fina (Regensburg, 1981).
Mußinan, Joseph von, *Über das Schicksal Straubings und des baierischen Waldes während des dreyßig jährigen Krieges vom Oktober 1633 bis April 1634* (Straubing, 1813).
Sigl, Rupert, *"Wallensteins Rache an Bayern. Der Schwedenschreck." Veit Hösers Kriegstagebuch* (Grafenau, 1984).

Secondary Sources

Amelang, James S., "*Vox populi*: Popular Autobiographies as Sources for Early Modern Urban History," *Urban History* 20 (1993), 30–42
Asche, Matthias et al., eds., *Krieg, Militär und Migration in der Frühen Neuzeit* (Herrschaft und Soziale Systeme in der Fühen Neuzeit), vol. 9 (Berlin, 2008).
Bahlcke, Joachim, ed., *Glaubensflüchtlinge: Ursachen, Formen und Auswirkungen frühneuzeitlicher Konfessionsmigration in Europa* (Religions- und Kulturgeschichte in Ostmittel- und Südosteuropa), vol. 4 (Berlin, 2008).
Behringer, Wolfgang, *Im Zeichen des Merkur: Reichspost und Kommunikationsrevolution in der Frühen Neuzeit* (Veröffentlichungen des Max-Planck-Instituts für Geschichte), vol. 189 (Göttingen, 2003).
Brändle, Fabian et al., "Texte zwischen Erfahrung und Diskurs: Probleme der Selbstzeugnisforschung," in *Von der dargestellten Person zum erinnerten Ich: Europäische Selbstzeugnisse als historische Quelle (1500–1850)*, ed. Kaspar von Greyerz et al. (Selbstzeugnisse der Neuzeit), vol. 9 (Cologne, 2001), pp. 3–31.
Briegel, Manfred; Frühwald, Wolfgang, eds., *Die Erfahrung der Fremde: Kolloquium des Schwerpunktprogramms "Exilforschung" der Deutschen Forschungsgemeinschaft* (Weinheim, 1988).
Davies, Mererid Puw et al., eds., *Autobiography by Women in German* (Bern, 2000).

Eberhardt, Otto, "Exil im Mittelalter: Einige Streiflichter," in *Weltanschauliche Orientierungsversuche im Exil*, ed. Reinhard Andress (Amsterdamer Beiträge zur neueren Germanistik), vol. 76 (Amsterdam, 2010), pp. 13–36.

Evelein, Johannes F., ed., *Exiles Traveling: Exploring Displacement, Crossing Boundaries in German Exile Arts and Writings 1933–1945* (Amsterdamer Beiträge zur Neueren Germanistik), vol. 68 (Amsterdam, 2009).

Fulbrook, Mary; Rublack, Ulinka, "In Relation: The 'Social Self' and Ego-Documents," *German History* 28, no. 3 (2010), 263–272.

Greyerz, Kaspar von, "Ego-Documents: The Last Word?," *German History* 28, no. 3 (2010), 273–282.

Greyerz, Kaspar von, "Religion in the Life of German and Swiss Autobiographers (Sixteenth and Early Seventeenth Centuries)," in *Religion and Society in Early Modern Europe 1500–1800*, ed. Kaspar von Greyerz (Winchester, 1984), pp. 223–241.

Greyerz, Kaspar von, *Vorsehungsglaube und Kosmologie: Studien zu englischen Selbstzeugnissen des 17. Jahrhunderts* (Göttingen, 1990).

Hacke, Daniela, ed., *Frauen in der Stadt: Selbstzeugnisse des 16. – 18. Jahrhunderts: 39. Arbeitstagung in Heidelberg, 17.–19. November 2000* (Ostfildern, 2004).

Haude, Sigrun, "The Experience of War," in *Research Companion to the Thirty Years War*, ed. Olaf Asbach, Peter Schroeder (Farnham, 2014), pp. 257–268.

Heimers, Manfred P., *Krieg, Hunger, Pest und Glaubenszwist: München im Dreißigjährigen Krieg* (Veröffentlichungen des Stadtarchivs München) (Munich, 1998).

Hofmann, Hanns H., *Die Nürnberger Stadtmauer* (Nuremberg, 1967).

Jancke, Gabriele; Ulbrich, Claudia, eds., *Vom Individuum zur Person: Neue Konzepte im Spannungsfeld von Autobiographietheorie und Selbstzeugnisforschung* (Querelles: Jahrbuch für Frauen- und Geschlechterforschung 2005), vol. 10 (Göttingen, 2005).

Koopmann, Helmut, "Exil als geistige Lebensform," in *Exil: Transhistorische und transnationale Perspektiven*, ed. Helmut Koopmann, Klaus D. Post (Paderborn, 2001), pp. 1–19.

Kormann, Eva, *Ich, Welt und Gott: Autobiographik im 17. Jahrhundert* (Selbstzeugnisse der Neuzeit), vol. 13 (Cologne, 2004).

Krauss, Marita, "Heimat – Begriff und Erfahrung," in *Heimat, liebe Heimat: Exil und Innere Emigration (1933–1945)*, ed. Herrmann Haarmann (Berlin, 2004), pp. 11–27.

Krusenstjern, Benigna von, "Buchhalter ihres Lebens: Über Selbstzeugnisse aus dem 17. Jahrhundert," in *Das dargestellte Ich: Studien zu Selbstzeugnissen des späteren Mittelalters und der frühen Neuzeit*, ed. Klaus Arnold et al. (Selbstzeugnisse des Mittelalters und der beginnenden Neuzeit), vol. 1 (Bochum, 1999), pp. 139–146.

Krusenstjern, Benigna von, *Selbstzeugnisse der Zeit des Dreißigjährigen Krieges: Beschreibendes Verzeichnis* (Selbstzeugnisse der Neuzeit), vol. 6 (Berlin, 1997).

Krusenstjern, Benigna von, "Was sind Selbstzeugnisse? Begriffskritische und quellenkundliche Überlegungen anhand von Beispielen aus dem 17. Jahrhundert," *Historische Anthropologie* 2 (1994), 462–471.

Liebhart, Wilhelm, *Altbayerisches Klosterleben: Das Birgittenkloster Altomünster 1496–1841* (Münchener Theologische Studien), vol. 1/30 (St. Ottilien, 1987).
Medick, Hans; Krusenstjern, Benigna von, *Zwischen Alltag und Katastrophe: Der Dreißigjährige Krieg aus der Nähe* (Veröffentlichungen des Max-Planck-Instituts für Geschichte), vol. 148 (Göttingen, 1999).
Mortimer, Geoff, *Eyewitness Accounts of the Thirty Years War, 1618–1648* (New York, 2002).
Münch, Paul, *"Erfahrung" als Kategorie der Frühneuzeitgeschichte* (Munich, 2001).
Outram, Quentin, "The Socio-Economic Relations of Warfare and the Military Mortality Crises of the Thirty Years' War," *Medical History* 45 (2001), 151–184.
Pastenaci, Stephan, "Probleme der Edition und Kommentierung deutsch-sprachiger Autobiographien und Tagebücher der Frühen Neuzeit, dargestellt anhand dreier Beispiele," in *Edition von autobiographischen Schriften und Zeugnissen zur Biographie*, ed. Jochen Golz (Beihefte zu Editio) 7 (Tübingen, 1995), pp. 10–26.
Pfister, Christian, *Bevölkerungsgeschichte und Historische Demographie 1500–1800* (Munich, 1994).
Reicke, Emil, *Geschichte der Reichsstadt Nürnberg von dem ersten urkundlichen Nachweis ihres Bestehens bis zu ihrem Uebergang an das Königreich Bayern (1806)* (Nuremberg, 1896).
Roeck, Berndt, *Eine Stadt in Krieg und Frieden: Studien zur Geschichte der Reichsstadt Augsburg zwischen Kalenderstreit und Parität*, 2 vols (Göttingen, 1989).
Schiewek, Ingrid, "Zur Manifestation des Individuellen in der frühen deutschen Selbstdarstellung: Eine Studie zum Autobiographen Bartholomäus Sastrow (1520–1603)," *Weimarer Beiträge* 13 (1967), 885–915.
Schönauer, Tobias, *Ingolstadt in der Zeit des Dreißigjährigen Krieges: Soziale und wirtschaftliche Aspekte der Stadtgeschichte* (Beiträge zur Geschichte Ingolstadts), vol. 4 (Ingolstadt, 2007).
Schulze, Winfried, "Vorbemerkung," in Idem, *Ego-Dokumente: Annäherung an den Menschen in der Geschichte* (Berlin, 1996), pp. 9–30.
Scott, Joan, "The Evidence of Experience," *Critical Inquiry* 17 (1991), 773–797.
Stahleder, Helmuth, *Chronik der Stadt München: Belastungen und Bedrückungen*, vol. 2: *Die Jahre 1506–1705* (Hamburg, 2005).
Tallett, Frank, *War and Society in Early Modern Europe, 1495–1715* (London, 1992).
Theibault, John, "The Demography of the Thirty Years War Re-Revisited: Günther Franz and his Critics," *German History* 15 (1997), 1–21.
Willax, Franz, "Die Befestigungsanlagen Gustav Adolfs von Schweden um Nürnberg 1632," *Mitteilungen des Vereins für Geschichte der Stadt Nürnberg* 82 (1995), 185–235.
Wilson, Peter H., *The Thirty Years War: Europe's Tragedy* (Cambridge, MA, 2009).

CHAPTER 3

Colchester's Plight in European Perspective: Printed Representations of Seventeenth-Century Siege Warfare

Anke Fischer-Kattner

1 Siege in a War of Battles? Representing the History of the British Civil Wars

The online edition of the *Encyclopedia Britannica* has an up-to-date interactive interface, but it still presents the "English Civil Wars" in a rather traditional fashion: the set of conflicts that shook up the political system of the British Isles in the mid-seventeenth century is still labeled "English," and its course remains structured by a canonical succession of major pitched battles.[1] This is remarkable because the article in question was originally authored by Jane Ohlmeyer.[2] In her research, she has clearly demonstrated that an English conception of the Civil Wars in the Stuart kingdoms is misleading and much too limited.[3] This important conceptual shift is taken up in the text of the article, despite its conventional main title, as well as in the accompanying list of publications and alternate titles[4] for Ohlmeyer's entry. However, another important revision of the historiographical image of the Civil Wars is only mentioned in passing: while the well-known battles, such as that of Edgehill, are linked to their own separate entries in the online encyclopedia, the statement that "sieges and skirmishes – rather than pitched battles – dominated the military landscape in England during the first Civil War" stands without further details.[5]

1 See <http://www.britannica.com/event/English-Civil-Wars>, (accessed 23 October 2018).

2 The online format allows for feedback, which might lead to alterations by the editors. The "article history" of Ohlmeyer's text gives one major alteration, namely the replacement of an older, very differently organized article, on 11 July 2007, presumably by Jane Ohlmeyer's text. Unfortunately, it does not indicate who authored the replaced version.

3 E. g. John Kenyon, Jane Ohlmeyer, eds., *The Civil Wars: A Military History of England, Scotland, and Ireland 1638–1660* (Oxford, 1998); Jane Ohlmeyer, *Making Ireland English: The Irish Aristocracy in the Seventeenth Century* (New Haven, 2012); ead., *Civil War and Restoration in the Three Stuart Kingdoms: The Career of Randal MacDonnell, Marquis of Antrim, 1609–1683* (Cambridge, UK, New York, 1993).

4 Besides the "Great Rebellion," they cross-reference "Wars of the Three Kingdoms."

5 Quoted from paragraph one in the section on "The first English Civil War."

 | DOI:10.1163/9789004395695_004

Yet, seventeenth-century Britain did form part of the early modern world of the siege. Recent historiography has focused on siege warfare in the British Isles, also demonstrating its impact in England.[6] In an attempt to describe the historical experience of the Civil Wars, Charles Carlton has calculated that 31 per cent of all military actions in "England proper" were sieges. According to his reckoning, these accounted for 20,981 deaths (or 24 per cent of the war's total excluding uncounted victims of disease).[7] From a technological viewpoint within more traditional military history, the importance of sieges had been established even earlier. In 1979, Christopher Duffy had already pointed out how important fortifications and their conquest were for the British conflicts.[8]

Recent works of cultural history have taken up this crucial characteristic of the Civil Wars, placing it in a different light. Carlton, for example, emphasizes that the sieges which formed such an important part of the conflicts he examined were "intensely personal and vicious events."[9] His enquiry into the experience of siege warfare highlights the miseries of early modern war as they appeared to combatants below the level of the military leaders. This

6 While the sieges of Cromwell's campaign in Ireland had been accorded attention before, the history of operations in England had largely been presented as the history of the canonical succession of pitched battles as given by Britannica to this day. Yet, studies on sieges during the Civil Wars have multiplied in recent years. Many of the latest titles are written in the vein of local history. Publications of the last 15 years include e. g. James Dinn, "Nathaniel Nye and the siege of Worcester, 1646," *Transactions of the Worcestershire Archaeological Society* 3rd ser., 24 (2014), 173–180; Stuart B. Jennings, "The Third and Final Siege of Newark (1645–1646) and the Impact of the Scottish Army Upon Nottinghamshire and Adjacent Counties," *Midland History* 37, no. 2 (2012), 142–162; Elaine Murphy, "Siege of Duncannon Fort in 1641 and 1642," in *The 1641 depositions and the Irish Rebellion*, ed. Eamon Darcy, Annaleigh Margey, Elaine Murphy (Warfare, Society and Culture), vol. 6 (London, 2012), pp. 143–154; Rosie Serdiville, Douglas John Sadler, *The Great Siege of Newcastle, 1644* (Brimscombe Port, 2011); Mark Clinton, Linda Fibiger, Damien Shiels, "Archaeology of Massacre: The Carrickmines Mass Grave and the Siege of March 1642," in *Age of Atrocity: Violence and Political Conflict in Early Modern Ireland*, ed. David Edwards, Padraig Lenihan, Clodagh Tait (Dublin, 2007), pp. 192–203; C. Don Gilbert, "Two Sermons at the Siege of Worcester, 1646," *Transactions of the Worcestershire Archaeological Society* 20 (2006), 139–141; Kevin Downes, "The 'Great Lewis' and the Siege of Duncannon, 1645," *Decies: Journal of the Waterford Archaeological & Historical Society* 60 (2004), 115–116; Audrey Howes, "'Hull Must Absolutely Be Had:' the First Siege of Hull, July 1642," *East Yorkshire Historian* 4 (2003), 65–74; Maighréad Ní Murchadha, "From Kildare to Baldongan: Fr. Conor Donnogh and the Siege of Baldongan, 1642," *Journal of the County Kildare Archaeological Society* 19, no. 2 (2003), 269–288; John Barratt, *The Great Siege of Chester* (Stroud, 2003).

7 Charles Carlton, *Going to the Wars: The Experience of the British Civil Wars, 1638–1651* (London, 1992), pp. 154–155.

8 Christopher Duffy, *Siege Warfare: The Fortress in the Early Modern World, 1494–1660* (London, 1979), pp. 140–162.

9 Carlton, *Going to the Wars*, p. 156.

perspective is also drawing some attention to the suffering of the non-combatant, 'civilian' population,[10] particularly affected by siege warfare. Carlton's work thus complements classic military history, which still stresses technical and strategic aspects.[11]

Barbara Donagan has expanded the range of experiential history in the Civil Wars even further. She considers the legal and political implications of acts committed in the wars. For this purpose, she examines their representations in primary sources. In her cultural history of the wars, Donagan pays close attention to details of siege warfare. She introduces the 'Siege of Colchester' of 1648 as an extraordinary example of embittered conflict between Royalists and Parliamentarians, contrasting it with an earlier, relatively minor operation, the siege of Boarstall House, a royalist country seat, in 1645–6.[12] Donagan examines discursive practices in these sieges. She finds that in spite of the extreme acrimony of the conflict by the time of the siege of Colchester, 'traditional' codes of warfare were still in use. Because "moderating social links"[13] required certain "civilities"[14] between noble opponents, the conduct of war was regulated to a certain extent. In Donagan's view, this was due to cultural exceptionalism: she argues that "English legalism"[15] was a "distinctive"[16] characteristic of the society that comprised the warring parties. Wider cultural meanings of representations and practices of siege warfare are thus foregrounded.

The approach of 'new military history,'[17] taking into account experiential and cultural aspects of conflict, to the Civil Wars in Britain thus brings with it

---

10 The brutality of the "English Civil War," including its likeness to contemporary continental European conflicts, is established by Will Coster, "Massacre and Codes of Conduct in the English Civil War," in *The Massacre in History*, ed. Mark Leven, Penny Roberts (New York, 1999), pp. 98–105.

11 See e. g. the valuable overview by John Childs, *Warfare in the Seventeenth Century* (Smithsonian History of Warfare) (Washington, D.C., 2001). Yet, a first inquiry into the human experience of fortress warfare is already given in Duffy, *Siege Warfare*, 247–254.

12 Barbara Donagan, *War in England, 1642–1649* (Oxford, 2008). A less original and innovative study, written for the general public, on the siege and its context in the Civil Wars had been published only five years before: Phil Jones, *The Siege of Colchester, 1648* (Stroud, 2003). Before this, the event was merely treated in rather antiquarian fashion as an instance of local history, see e. g. David T. D. Clarke, *The Siege of Colchester, 1648* (Colchester, [1975]); yet with detailed biographical research: Daphne Woodward, Chloe Cockerill, *The Siege of Colchester: A History and Bibliography* ([Chelmsford], 1979).

13 Donagan, *War in England*, p. 346.

14 Ibid., p. 347.

15 Ibid., p. 394.

16 Ibid.

17 Cf. for a critical appraisal of this turn with its tendency to ignore actual operations: Peter Paret, "The New Military History," *Parameters* 21 (1991), 10–18. An inspiration for those researches whose work can be subsumed within the approach of new military history was

added values and new challenges. It takes representational and discursive aspects seriously as expressions of a war experience distinct from the commanders' tactical overview. It thereby opens up new insights into the complexity of war and helps to recover the importance of siege warfare from contemporary sources.

Yet, even pioneering works such as those by Carlton and Donagan remain wedded to a customary view of the Civil Wars in another respect. In contrast to revisionist scholarship, which since the 1980s has, among other things, called for an integration of the 'English Revolution' into a European context,[18] they keep to a distinctly British or even English perspective. In her invaluable interpretive work on siege representations, Donagan is taking a very quick glance at the rhetorical use of continental comparisons in English primary sources. She finds they are employed to describe the most horrific and apparently un-regulated aspects of war.[19] Her hint points to a more detailed examination of European comparisons. Indeed, as Will Coster has shown, such comparisons prove that, as regards the occurrence of massacres, "the English were not so different from their Celtic or European contemporaries."[20] Even Mark Fissel's study of Tudor and Stuart land warfare emphasizes that the fighters at Edgehill (23 October 1642), the first major engagement of Royalists and Parliamentarians, "assimilated the military revolution fully" as they were commanded by experienced "veterans of continental wars."[21] Although it is Fissel's avowed aim to delineate

---

John Keegan, *The Face of Battle* (London, 1976). In Germany, especially research on the letters sent through armed forces postal services in World Wars I and II spawned important methodological reflections on the possibilities of an experiential history of warfare: Klaus Latzel, "Vom Kriegserlebnis zur Kriegserfahrung: Theoretische und methodische Überlegungen zur erfahrungsgeschichtlichen Untersuchung von Feldpostbriefen," *Militärgeschichtliche Mitteilungen* 56, no. 1 (1997), 1–30.

18 See the historiographical debate on whether the wars were "A British Problem? A European Crisis?" in Ann Hughes, *The Causes of the English Civil War* (Basingstoke, 1991), p. 9. On the level of national historiography, a main concern of revisionist scholarship has been to question the revolutionary character of the uprisings in the course of the Civil Wars, yet this has not been unanimously accepted, cf. e. g. Gerald E. Aylmer, *Rebellion or Revolution? England, 1640–1660* (Oxford, 1986); and recently: Amanda Jane Whiting, *Women and Petitioning in the Seventeenth-Century English Revolution: Deference, Difference, and Dissent* (Turnhout, 2015); Nigel Smith, ed., *A Collection of Ranter Writings: Spiritual Liberty and Sexual Freedom in the English Revolution* (London, 2014); John Donoghue, *Fire under the Ashes: An Atlantic History of the English Revolution* (Chicago, 2013).

19 Donagan, *War in England*, p. 331. The huge interest which English readers took in accounts of the contemporary continental Thirty Years' War is exemplified by the conflict's coverage in more than 400 London periodicals identified by Jane E. E. Boys, *London's News Press and the Thirty Years War* (Woodbridge, Rochester, 2011), p. 9.

20 Coster, "Massacre," p. 103.

21 Mark Charles Fissel, *English Warfare 1511–1642* (London, New York, 2001), p. 303.

a distinctive "English" tradition of warfare, his means of dispelling older accusations of English inferiority to continental troops is to argue that "Englishmen had the ability to assimilate, adapt, and absorb remarkably."[22] Whether the result of these processes was a distinctive national military culture or not, they attest to the links between warfare in the British Isles and on the European mainland. A comparative perspective that takes into account exchange and entanglement thus promises to highlight crucial aspects for a reading of primary sources which were generated in and about siege warfare in the British Civil Wars. Such an analysis contributes decisively to a comprehensive early modern European history of siege representation.

2 Media Event and Existential Threat: Challenges of Siege Representations

Any historiography of siege warfare has to face a double challenge of representations. It offers a representation of past events, whether it foregrounds explanatory models or narrative.[23] Moreover, the primary sources with which historians work are historical representations of siege operations – with all the well-known pitfalls addressed through the craft of source criticism and with the help of some theoretical assumptions. These form a particular context for the historiographical representation of sieges such as that at Colchester. All in all, sieges are fundamentally ambiguous: they were constructed as media events by contemporary print publications, yet they were also physical events marked by existential dangers – violence, hunger, disease, causing trauma and suffering. The source material created in and about the siege of Colchester therefore has to be analyzed in the light of different approaches.

On the one hand, primary sources printed in mid-seventeenth-century England can be read as parts of an overarching media system.[24] Newsletters and early periodicals were covering foreign events, foreign publications were re-

22 Ibid., p. 1.

23 The challenges of postmodernity have instigated a self-reflexive movement in the history of historiography, in which the existence and location of the border between fact and fiction, discourses and materiality have been hotly debated. For a summary of the problem and the myriad of publications it spawned, see e. g. Ann Curthoys, John Docker, *Is History Fiction?* 2nd ed. (Sydney, 2010).

24 See for a summary of the development and use of the concept: Henrik G. Bastiansen, "Media History and the Study of Media Systems," *Media History* 14, no. 1 (2008), 95–112. Developed in the context of studies of the press during the early years of the Cold War, the concept has recently been recovered for media studies – on a more presentist, national level and not in the transnational and constructivist sense it takes on here – by Daniel

printed and translated, and visual materials reproduced. These activities connected the reading public of the British Isles to its continental European counterpart.[25] A systemic approach does take some cues from the field of media studies. Its practitioners, however, have mainly used the concept of a media system to describe twentieth-century developments. In this context, it has taken on very specific connotations; systems here seem to imply ideological coherence and central control, exercised in separate (national) societies. In contrast, historiographical usage of the concept has often simply emphasized pragmatic interconnections of different media types.[26] In this latter sense, early modern media communications can be said to have formed a system, consisting of diverse interrelated components.[27] The concept thus helps in highlighting relations of different material media, such as different types of prints considered below. Moreover, it draws attention to relations between the different consumers of printed content, who formed audiences craving for information on certain prominent topics – on the level of proto-national societies,[28] but also well below (on a local or regional level) and above this.

Within an overarching European media system, special 'media events' were thus creating clusters of existing and newly established connections between people and their conceptions of the world. Famous sieges were among the popular contents which circulated in this system.[29] Influenced by the traditions

---

C. Hallin and Paolo Mancini, *Comparing Media Systems: Three Models of Media and Politics* (Cambridge, UK, 2004).

25 The transmission of English news to the Habsburg Netherlands is e. g. traced by Paul Arblaster, "Posts, Newsletters, Newspapers: England in a European System of Communications," *Media History* 11 (2005), 21–36. Traditions of war reporting are described by Boys, *London's News Press*.

26 For such a use of the concept, see e. g. the introductory text book by Werner Faulstich, *Mediengeschichte: Von den Anfängen bis 1700* (Göttingen, 2006), pp. 8 (for an understanding of single media as systems) and 148; a specifically early modern perspective is combined with Parsons' and Luhmann's systems theory by Johannes Arndt, *Herrschaftskontrolle durch Öffentlichkeit: Die publizistische Darstellung politischer Konflikte im Heiligen Römischen Reich, 1648–1750* (Göttingen, 2013), pp. 30–35; a more pragmatic approach is chosen by Asa Briggs and Peter Burke, *Social History of the Media: From Gutenberg to the Internet* (Cambridge, UK, 2009), p. 19.

27 Important non-material media elements of this system, e. g. instances of orality and performed music, are often lost for historical research. Yet, cf. the interesting examples for the "Interactions Between Media," the connections between orality, manuscript writing and print, presented by Briggs and Burke, *Social History*, pp. 37–40.

28 For the emergence of the modern national state in the 19th century, such an idea has long been put forward by Benedict Anderson, *Imagined Communities: Reflections on the Origin and Spread of Nationalism* (London, 1983).

29 Databases of historic printed works can give an impression of the importance of the topic: a search for the title word "Belagerung" in VD17, the database of German-language

of military theory[30] and newly developing transnational conventions of (military) engineering,[31] visual and textual representations of these key events circulated between European centers of print culture with their respective regional markets.[32] In this way, English discourses of siege warfare were entangled with those in other European countries.

On the other hand, even with a focus on issues of representation, the subject of siege warfare poses a particular problem to cultural history. It can be regarded through a lens of aesthetics, as a spectacle for a refined audience and noble warriors.[33] Yet, experiential approaches, such as Carlton's, inspired by John Keegan's classic work,[34] have made clear that any history of war has to meet the challenge of existential dangers inherent in this kind of activity. With various vignettes from early modern Europe, Renaissance scholar Lauro Martines has recently painted a gruesome picture of armed conflict in this period.[35] The threat of violent death – whether in battle or by hunger and disease, which were regular components of siege warfare – was imminent for those who were willing or unwilling participants in a siege. This experience went beyond words and images. In the face of death and violence, all abstract considerations of discourses appear as highly questionable enterprises. Yet, death and violence

---

prints of the 17th century, and, respectively, "siege" in Early English Books Online (limited to texts printed between 1600 and 1700), results in 758 and 387 hits. Compared with the much broader title topic of "Krieg" (1504 hits, including authors by that name) or "war" (1674 records), these are quite impressive numbers. Siege was obviously a key word for advertising a publication.

30 The classics were taken up in famous theoretical works such as Niccolò Machiavelli, *The Art of War* (Cambridge, MA, 2001; originally 1521, transl. by Neal Wood in 1965), who treats fortification and siege in Book 7. Overviews of the theoretical discussions on war are presented by Martin van Creveld, *The Art of War: War and Military Thought* (Smithsonian History of Warfare) (Washington, D.C., New York, 2005; first ed. London, 2000); and Beatrice Heuser, *Den Krieg denken: Die Entwicklung der Strategie seit der Antike* (Paderborn, 2010).

31 Works published in this field ranged from rather abstract, mathematical, and complex aesthetic programs to very practical manuals, see Martha Pollak, *Cities at War in Early Modern Europe* (Cambridge, UK, 2010), pp. 61–95.

32 For the peddling of prints in the countryside, but also particularly in urban settings, see e. g. Jeroen Salman, *Pedlars and the Popular Press: Itinerant Distribution Networks in England and the Netherlands 1600–1850* (Library of the Written Word), vol. 29 (Leiden, 2013).

33 See for this innovative reading: Brian Sandberg, "'To Have the Pleasure of This Siege:' Envisioning Siege Warfare During the European Wars of Religion," in *Beholding Violence in Medieval and Early Modern Europe*, ed. Allie Terry Fritsch, Erin Felicia Labbie (Farnham, 2012), pp. 143–162.

34 Carlton, *Going to the Wars*, p. xi; Keegan, *Face of Battle*; the connection is confirmed by Keegan himself in a foreword to Carlton's work (Carlton, *Going to the Wars*, pp. ix–x).

35 Lauro Martines, *Furies: War in Europe, 1450–1700* (New York, 2013).

are embedded in and shape their own cultural contexts.[36] Thus, printed siege reports were part of contemporary efforts to re-/present the extreme experiences of warfare. Published texts and images continued individual and collective cognitive efforts of providing war events with a comprehensible structure and wider meaning.[37] The renewed presentations did not, however, simply elide the dangerous and destructive character of warfare. As will become clear from the examples examined below, early modern siege representations were drawing on shared visual and textual languages, in which documentary and propagandistic effects were entwined with the traumatic aspects of war experience.

Building on these considerations, a study of military operations and ensuing (published) communications in the British Civil Wars reveals how this conflict was connected to a wider world of war and its representation. A brief recapitulation of the events at Colchester prepares the ground for a close reading of printed sources about the 1648 siege. A broadsheet produced in London will be the key source for comparison. It (re-)presents the "Siege of Colchester" according to the conventions of European siege views and journals. This document becomes a prism shedding light onto contemporary contexts, onto media systems on local and (proto-)national levels, but also onto shared European concepts of siege warfare. Its contextualization with further contemporary representations of the siege then highlights the interplay of abstraction and concrete, strategy and suffering in armed conflict. As events at Colchester and the printed accounts that were produced around them are re-situated within a wider cultural context, they moreover shine a new light on the established connection between war and the imposition of authority. The Parliamentarian

36 The first-mentioned relation shapes studies like Mark S. Schantz, *Awaiting the Heavenly Country: The Civil War and America's Culture of Death* (Ithaca, 2008); or those collected in David M. Pritchard, ed., *War, Democracy and Culture in Classical Athens* (Cambridge, UK, New York, 2010). The sad climax of the second relation is described by Timothy Snyder, *Bloodlands: Europe between Hitler and Stalin* (New York, 2010). Alan Kramer, *Dynamic of Destruction: Culture and Mass Killing in the First World War* (Oxford, 2007), explores both as two sides of the same coin in the early 20th century.

37 Uniting the philosophy of history and psychology, Frank Ankersmit speculates that "the traumatic experience of certain historical events" enabled the emergence of "Western" history writing. This kind of historiographical activity then turns suffering into a mere "sign," by "dissociation" from the actual, emotional experience. Frank Ankersmit, "Trauma and Suffering: a Forgotten Source of Western Historical Consciousness," in *Western Historical Thinking: An Intercultural Debate*, ed. Jörn Rüsen (New York, 2002), pp. 72–85, quotes from p. 75 and 78. Yet, such a strict separation of experience and (written) representation is misleading in that it seems to suggest that representational activity is merely an *ex-post* activity. The cognitive efforts at grasping any event while it occurs already form part of the representational processes studied here.

besiegers' stern insistence on fatal personal consequences and political implications of the Royalists' earlier resisting the summons to surrender thus becomes less of a peculiar scandal or revolutionary act. It can rather be seen as an instance of the 'necessity' typical for siege warfare. This concept created an intersection of representation and action, of rhetorical and physical impositions of authority. Its use in siege warfare unites events at Colchester with the historiography of the continental wars during the period of early modern state formation.[38] While printed media were processing the trauma and suffering of siege warfare into contemporary history, they were exerting an influence on the events themselves and on their meanings over time.

3 A Siege? Events at Colchester in the Summer of 1648 and the Problem of Labeling

Approaching sieges through their representations brings about challenges for definition. What exactly is a siege? How is it to be distinguished from other forms of positional warfare?[39] A siege diary, reprinted as an antiquarian document to illustrate the history of Colchester in 1724 by Daniel Defoe, points to the problem. Defoe entitled it "A Diary: Or, an Account of the Siege and Blockade of Colchester, An. 1648,"[40] explicitly differentiating siege and blockade.

38 For English-speaking academia, a much-quoted origin of this connection lay in the debate on an early modern military revolution, initiated by Michael Roberts, *The Military Revolution, 1560–1660* (Belfast, 1956); taken up by Geoffrey Parker, *The Military Revolution: Military Innovations and the Rise of the West, 1500–1800* (New York, 1988); the debate is summarized and taken further in Clifford J. Rogers, ed., *The Military Revolution Debate: Readings on the Military Transformation of Early Modern Europe* (Boulder, 1995). In a German-language context, the bureaucratic effect of the necessity to raise money for early modern wars had already been pointed out by Otto Hintze, "Der Commissarius und seine Bedeutung in der allgemeinen Verfassungsgeschichte," in *Historische Aufsätze: Karl Zeumer zum 60. Geburtstag*, ed. Mario Krammer (Weimar, 1910), pp. 493–528; this idea was taken up for late-seventeenth-century England by John Brewer, *The Sinews of Power: War, Money and the English State, 1688–1783* (London, 1989).

39 See Peter Wilson's contribution below; a material definition on a Vaubanian basis (arguing against definitions merely based on time length and distinguishing sieges from "non-siege positional tactics", p. 348) is offered by Jamel Ostwald, *Vauban under Siege: Engineering Efficiency and Martial Vigor in the War of the Spanish Succession* (Leiden, 2006), Appendices B, pp. 326–339, and D, pp. 348–360.

40 Daniel Defoe, *A Tour Thro' the Whole Island of Great Britain, Divided Into Circuits or Journies,* [...] (London, 1724), 1:21. The document in question is a reprint from an eighteenth-century manuscript preserved in James Round's collection. It was considered an original piece by the Royal Historical Manuscripts Commission, which re-printed it in full: Historical Manuscripts Commission, ed., *Fourteenth Report, Appendix, Part IX: The*

The distinction appears confirmed when the manuscript mentions that Fairfax "resolv'd to turn his Siege into a Blockade"[41] by completely encircling the town with his troops. Yet, even after this, the royalist diarist still speaks of besiegers and besieged. He concludes his narrative (before a note listing the surviving 3,526 defenders), stating that "Thus ended the Siege of Colchester."[42] Despite the explicit terminological differentiation, the language of this representation ultimately makes the blockade appear as a tactical element of an overarching siege event.

Military historian Christopher Duffy found the actions of Thomas Fairfax' Parliamentary army against Royalists from Essex and Kent, who had sought refuge in Colchester, lacking in terms of vigor and tactical efficiency: "Active siege warfare hardly made an appearance."[43] His vantage point foregrounds technology, strategy, and formal definitions of early modern siege warfare. In that sense, Fairfax' use of siege artillery against the town "did little to hasten the capitulation,"[44] which was eventually effected by the successful blockade. With regard to participants' evaluations, Barbara Donagan, on the other hand, calls the parliamentarian circumvallations "the most sophisticated set of siege lines of either civil war, planned on the best modern principles."[45]

In the mid-twentieth century, local historian John Wellens, who was describing the operations at their tricentennial, found it "strange" that "the siege occurred at all."[46] He had, however, no doubt that this set of events in what came to be known as the "Second Civil War"[47] between King Charles I and the English parliament was a siege. Wellens' surprise is due to circumstances of the Second Civil War generally and in East Anglia in particular. He thus points to the link between terminology and wider military events. After the king had surrendered to a Scottish Army in 1646, thus ending the war he had formally

Manuscripts of the Earl of Buckinghamshire, the Earl of Lindset, the Earl of Onslow, Lord Emly, Theodore J. Hare, Esq., and James Round, Esq., M.P. (London, 1895), pp. 281–290. The identification of these documents is confirmed by Donagan, *War in England*, p. 374.

41 Historical Manuscripts Commission, ed., *Report 14, App. IX*, p. 284.

42 Ibid., p. 290.

43 Duffy, *Siege Warfare*, p. 156.

44 Ibid.

45 Donagan, *War in England*, p. 323.

46 John Wellens, "The Siege of Colchester," *East Anglian Magazine* 7, no. 9 (1948), 460–465, here: p. 460.

47 The anti-parliamentarian origins of this renewed conflict are emphasized by Robert Ashton, *Counter Revolution: The Second Civil War and its Origins, 1646–8* (New Haven, 1994).

declared in 1642,[48] there had been an "uneasy peace"[49] between the warring parties. In the winter of 1647/8, insurgents in a number of uprisings across the country expressed royalist sentiments. Although in captivity after his surrender to the Scots, the king tried to regain influence by political maneuvering between different factions. Parliament reacted by passing the Vote of No Addresses, ruling out negotiations with the monarch, in January 1648. Yet, this move did not find unanimous support in the country. Pressure for a reconciliation increased while new instances of violence made it ever less likely.[50] The new struggle embittered relations between the factions. A "sense of betrayal and desperation [...] characterized the second civil war."[51] It fueled the radicalization that would eventually lead to the abolishment of the monarchy and the beheading of Charles I in January 1649.

The origin of the operations at Colchester lay in a royalist revolt in Kent. It started in May 1648 and was led by George Goring, Earl of Norwich, who was planning to take London. Before he could join forces with many supporters, however, the parliamentary main army led by Thomas Fairfax managed to crush the movement in Kent at the beginning of June.[52] The parliamentarian London militia prevented Norwich and his remaining troops from entering the city, so the royalists moved into Essex instead. They were joined by sympathetic London apprentices and royalist officers, among them "seasoned soldiers"[53] – Lord Capel, Bernard Gascoigne, George Lisle and Charles Lucas. Meanwhile, the Essex trained bands, commanded by a former parliamentarian officer named Colonel Farr, had taken the parliamentarian county committee hostage at Chelmsford. Norwich moved to meet Farr and to integrate the trained bands into his royalist force. A parliamentarian detachment under Colonel Whalley followed Norwich's army as it moved from Chelmsford to Colchester, Lucas' home town, on June 10–12. By this time, the parliamentary main army had also arrived. In spite of unclear numerical relations between the military forces, Fairfax was determined to attack at once.[54]

48 Fighting in the British Isles had already begun with the Scottish Bishops' Wars in 1639. For a comprehensive survey of the conflict between the accession of Charles I to the English throne and the restitution of the monarchy, see e.g. Austin Woolrych, *Britain in Revolution, 1625–1660* (Oxford, 2002).

49 Ashton, *Counter Revolution*, p. 1.

50 Donagan, *War in England*, pp. 314–315.

51 Ibid., p. 312.

52 Woolrych, *Britain in Revolution*, pp. 411–412.

53 Ibid., p. 413.

54 Donagan, *War in England*, pp. 317–318. She supposes Fairfax' troops were numerically inferior to the Royalists. Woolrych, in contrast, counts 9,000 attackers against 4,000 defend-

On June 13, the two armies were preparing to fight a battle outside of Colchester. The city authorities had been convinced to surrender to the royalist forces, in spite of the divided sympathies of the townspeople.[55] Fending off the Parliamentarians' vigorous attacks, the Royalists retreated to the protection of the city walls and repelled repeated charges. As the battle had not brought about a decision, both sides settled for a siege. The Royalists began to fortify the old city walls while the Parliamentarians started to set up posts around the town in order to blockade it.[56] From June 13 to August 27, 1648, the defenders of Colchester withstood the enemies' efforts. The besiegers' forces were increasing, and Fairfax was eventually able to construct continuous lines of countervallation, closing the "noose around the Essex town"[57] at the beginning of July. Blockade, artillery fighting, sorties and storm attacks might not conform to a continental-style "siege in form" according to the soon-to-be dominant model of the Maréchal de Vauban.[58] Yet, for contemporary witnesses and in historiographical representations, it was clear that what was going on around Colchester was indeed a siege.

During a first phase of the operations, Fairfax' army was slowly starting to close in on the town. Its numbers were eventually increased by more parliamentarian troops and the local trained bands. Thus, the besiegers could construct and man a number of forts against sallies of the besieged, who had at first been able to roam the surrounding areas, attack parliamentarian stragglers and bring in provisions. While both sides were trying to strengthen their positions, a number of messages on issues such as the exchange of prisoners, options of surrender, accusations of unfair use of dangerous misshapen or even poisoned bullets, and the conditions of Colchester's trade were exchanged.[59] By mid-July, the besieged cavalrymen made some last attempts to break through the ring of countervallation and forts that were now cutting off the town from its hinterland. As they failed, morale inside the beleaguered city

---

ers, but his numbers include the "local forces from Essex and Suffolk" (Woolrych, *Britain in Revolution*, p. 413), whose loyalty was far from clear at the beginning of the siege.

55 The Puritans of Colchester had become known as a revolutionary crowd in the summer of 1642, when the local trained bands and an angry mob of possibly several thousand people had attacked the house of Sir Charles Lucas and other local Royalists as the latter were preparing to fight for the king. (Woolrych, *Britain in Revolution*, p. 230.)

56 Donagan, *War in England*, pp. 320–322.

57 Mark Kishlansky, *A Monarchy Transformed: Britain 1603–1714* (London, 1997), p. 180. Kishlansky has the same numbers as Woolrych for attack and defense, yet Donagan makes clear that this superiority was only coming about as the operations drew on: Donagan, *War in England*, p. 323.

58 Childs, *Warfare*, p. 148; Ostwald, *Vauban under Siege*, p. 12.

59 These formed an important aspect of the representational acts examined below.

was starting to fail as well. The much-valued horses of the cavalry had to be killed for food. Disease and desertions, the latter encouraged by Fairfax' public declarations, weakened the forces of the besieged. On July 26, the besiegers were beginning their bombardment of the town walls. The besieged were preparing for a storm attack in case of a breach.[60] By early August, victuals had become so scarce in the town that serious conflicts were breaking out between soldiers and civilian inhabitants. As all attempts by city authorities and inhabitants to obtain some relief failed,[61] pressure on the royalist commander Lord Norwich for surrender was greatly increasing. Moreover, any hopes of relief were dashed by the news of Cromwell's victory over Hamilton's Scottish royalist army at Preston on August 17, 1648.[62]

As the Royalists were defeated throughout the country and Colchester's situation was growing ever more desperate, Fairfax lost all willingness to grant privileged terms to the enemy officers. In late August, the latter tried to incite their soldiers to a last attempt at breaking through the besieging lines, but they could not convince their subordinates.[63] Threatened by mutiny, the Royalists had to surrender.[64] On August 28 the Lords and gentlemen became prisoners at mercy, while the common soldiers received passes to go home unarmed. The unfortunate town, hit hard by "Fire, Hunger, and Atrocity,"[65] had to pay a considerable sum in order to be spared from pillaging by the victorious army.[66] Lords Goring and Capel were sent to London in order to be judged by their peers in Parliament, but the royalist commanders of lower status, Lucas and Lisle, were executed instantly by parliamentarian firing squads.[67]

When Daniel Defoe visited Colchester in the early eighteenth century, the town still bore marks of the military operations, as "the batter'd Walls, the Breaches in the Turrets, and the ruin'd Churches still remain[ed.]"[68] Interestingly, he illustrated his reproduction of the royalist diary about the siege with

60 The date is given in Historical Manuscripts Commission, ed., *Report 14, App. IX*, p. 288.

61 See below for a detailed analysis of the representation of 'humanitarian' concerns.

62 Woolrych, *Britain in Revolution*, pp. 416–417.

63 See Historical Manuscripts Commission, ed., *Report 14, App. IX*, p. 290.

64 Donagan, *War in England*, pp. 328–329.

65 Ibid., p. 330 (chapter title).

66 The terms of surrender are summarized by Historical Manuscripts Commission, ed., *Report 14, App. IX*, p. 290.

67 Woolrych, *Britain in Revolution*, p. 418. For the literary afterlife of the men in contrast to the parliamentarian "martyr" Colonel Thomas Rainsborough, see Andrea Brady, "Dying With Honour: Literary Propaganda and the Second English Civil War," *The Journal of Military History* 70, no. 1 (2006), 9–30.

68 Defoe, *Tour*, p. 19.

an image that had been printed for the opposing side at the time of the events.[69] Defoe thus connected his account of Colchester to the media systems of the Civil Wars, harking back to European traditions of popular siege representation.

4 The Siege of Colchester: Contemporary Broadsheet Representation in a European Context

Siege warfare was depicted in a number of formats that attest to the widespread early modern interest in the subject. "Siege views" were taking up aesthetic and technical conventions of panoramic city views and topographical site plans, adding the exciting action of military operations.[70] Elaborate visual representations celebrated successful sieges for courtly elites. The famous paintings of Louis XIV at the sites of his victorious siege operations and his collection of three-dimensional *plans-reliefs* are cases in point. Other art forms that took up the subject were "intarsia, tapestry, wall painting, and low relief sculpture."[71] More widely available were engravings of sieges, often produced for propagandistic reasons as "part of a 'war of images'"[72] between the important European powers. The technical quality of engraved siege views varied significantly. Their creators ranged from the print artist (such as Frans Hogenberg, Jacques Callot and the surveyor-publisher Matthaeus Merian) or the fortification officer and expert military architect (e.g. Erik Dahlbergh)[73] to unnamed producers of cheap popular images for mass-consumption. The latter might have fabricated stock images with faulty perspective and clumsy arrangements of elements. Yet, even they were taking up the basic configuration and representational conventions current among the elite siege views.

Such visual delineations of siege warfare were accompanied by different versions of textual representation in the early modern European media system. These may e. g. be arranged according to their immediacy. The latest news about military operations were circulating as oral messages, often short,

---

69 Ibid., inserted image between p. 20 and p. 21. It is the same as the siege view analyzed below.

70 This is the way Martha Pollak uses the term in her respective chapter (Pollak, *Cities at War*, pp. 109–153). She distinguishes the more action-laden and three-dimensional views from the more map-like siege plans but at the same time points to the overlap between the two forms.

71 Pollak, *Cities at War*, p. 110.

72 Ibid., p. 122.

73 These examples are presented in respectful detail by Martha Pollak: ibid., pp. 121–142.

fragmentary, and incomplete. This information was taken up in personal letters and, often combining collected material from several sources, in handwritten newsletters for a wider audience.[74] Printed newsletters offered similar information, but could be produced in greater numbers, thus extending the reach of their information. Other prints, produced during the relevant events or afterwards, purported to give exact copies of documents circulated in the course of the military operations, e. g. of the letters exchanged between the officers of the opposing forces.[75] As the siege in question was ended, printers were quick in printing comprehensive accounts of the events, mostly in the form of siege journals – day-by-day narratives of the operation usually presented from the perspective of a participant-observer.[76]

Illustrated broadsheets combined textual accounts with siege view images. The broadsheet was a popular format in early modern Europe, usually printed on an unfolded folio sheet and frequently displaying diverse kinds of violence.[77] It presupposed both textual and visual literacy. Its European consumers of the 17th century fulfilled these requirements. From the era of the rise of siege warfare in the 16th century, when some cities had begun to respond to stronger artillery by 'modernizing' their fortifications in the *trace-italienne* style,[78] the reading public had been eager to obtain news about important sieges.[79]

At the beginning of the 17th century, the coverage of the Spanish siege of Ostend, a Calvinist stronghold in the Dutch War of Independence, for example raised international awareness. The town's story, presented to German readers

---

74 See for this format and its continuing importance in the age of print: Zsuzsa Barbarics, Renate Pieper, "Handwritten Newsletters as a Means of Communication in Early Modern Europe," in *Correspondence and Cultural Exchange in Early Modern Europe, 1400–1700*, ed. Francisco Bethencourt, Florike Egmond (Cambridge, UK, 2007), pp. 53–79.

75 See below, section 5, "Necessity and Extremities," for this segment of the media system.

76 Examples for this format from the 16th century, from the French Wars of Religion, are given by Michael Wolfe, "Walled towns During the French Wars of Religion (1560–1630)," in *City Walls: The Urban Enceinte in Global Perspective*, ed. James D. Tracy (Studies in Comparative Early Modern History), vol. 4 (Cambridge, UK, 2000), pp. 317–348, here: pp. 339–340. The continuing importance of this representational form is, for example, indicated by Hagen Haas, "'Denn die Bombe, wann sie fällt ...:' Zum Schicksal von Einwohnern belagerter Städte im absolutistischen Zeitalter," *Militär und Gesellschaft in der Frühen Neuzeit* 7, no. 1 (2003), pp. 41–59, relying heavily on the expertise of Daniel Hohrath.

77 Julius R. Ruff, *Violence in Early Modern Europe, 1500–1800* (Cambridge, UK, 2001), pp. 17–23.

78 See Parker, *Military Revolution*, pp. 10–13.

79 E. g. for the French Wars of Religion in the second half of the century: Wolfe, "Walled Towns"; the interest aroused by the "spectacle" of siege warfare in France and elsewhere is analyzed from a cultural history perspective by Sandberg, "To Have the Pleasure."

in three updated instalments, was translated into French and English versions.[80] The operations of Ostend came to be known as a "school" or even "university" for the military arts of the siege.[81] The printed representations of this siege event likewise set a standard: they offered a textual day-by-day account, a siege journal that gave participants' impressions from the course of events. This was combined with lavish visual materials, using representational cues of the map in order to provide an overview for the reader.

The same combination characterized siege representations on popular printed broadsheets. A German example of this media genre (figure 3.1) depicts another famous operation during the Dutch Revolt, the 1624/5 siege of Breda, likewise captured for the Spanish monarchy. The written notes on the siege plan, using German, French, and Latin, attest to the transnational use of images in the European media system. In spite of spatial limitations on the single page of a broadsheet, some narrative was obviously an indispensable element of the representation. This example gives a strongly condensed version, merely stating the conditions of surrender. It was necessary to read the richness of the image, which conferred a more detailed impression of the siege. It showed topographical conditions and the fortifications of besiegers and besieged, identified the besieging corps by their commanders' names, and depicted troops and named individuals moving around the militarily occupied landscape. The broadsheet thus presented the siege as a comprehensible whole, yet hinting at individual hardships and confusion.

The style and topic of the German broadsheet on Breda are also found in an English print produced in London (figure 3.2). Thomas Archer, who sold this broadsheet, had made sure that the text of the siege journal was adapted to the interest of his clients: his narrative, in diary format, emphasized the efforts of a relief army which included English troops. With only slight adaptations, Archer's print thus presented a shared European media event to its particular audience. This audience had been well prepared for this sort of representation. Soldiers returning from service on the continent had brought reports about

80 Anna E. C. Simoni, *The Ostend Story: Early Tales of the Great Siege and the Mediating Role of Henrick Van Haestens* ('t Goy-Houten, 2003). Within the last year of the siege, German, French, and English descriptions of events around Ostend appeared: *Belägerung der Statt Ostende: Journal: Tagregister und eigentliche beschreibung* [...], 3 vols. ([s. l.], 1604); *Histoire remarquable et véritable de ce qui s'est passé par chacun iour au siege de la ville d'Ostende,* [...] (Paris, 1604); Edward Grimeston, *A True Historie of the Memorable Siege of Ostend,* [...] (London, 1604). See for a study of the place-making characteristics of military and representational acts: Anke Fischer-Kattner, "Violent Encounters at Ostend, 1601–1604: Spatiality, Location, and Identity in Early Modern Siege Warfare," *Comparativ* 28, no. 2 (2018), 22-41.

81 E. g. *Belägerung der Statt Ostende*, vol. 1, fol. [35]v.

FIGURE 3.1 Eygentlicher Abriß und gestalt der Statt Breda [...] [Actual Outline and Form of the Town of Breda...]
BAYERISCHE STAATSBIBLIOTHEK MÜNCHEN, HBKS/G 39, URN:NBN:DE:BVB:12-BSB00064501-5

'modern' warfare.[82] Their texts were shaped by the representations of their

82 Henry Hexham, *A True and Briefe Relation of the Famous Seige of Breda Beseiged, and Taken in Vnder the Able and Victorious Conduct of his Highnesse the Prince of Orange* [...] (Delft, 1637); Francis Vere, *The Commentaries of Sr. Francis Vere Being Diverse Pieces of Service, Wherein He Had Command* [...] (Cambridge, UK, 1657); an English traveler-writer also promised an eye-witness account: William Lithgow, *A True and Experimentall Discourse, Upon the Beginning, Proceeding, and Victorious Event of This Last Siege of Breda* [...] (London, 1637).

FIGURE 3.2 A thirde and last mape, both of the Sedg of Breda by Spinola and how the Princ of Orage hath enguarterde his forces [...]

profession current in Europe. Thus, Britain's print culture, dominated by the workshops in the city of London, was integrated into the European media system of the 17th century. By the time of the Civil Wars, numerous publications had prepared the ground for an understanding of siege warfare that united the British Isles with the continental mainland.

Represented in the familiar siege-broadsheet format, the operations around Colchester were not so much depicted as a violent extreme in an environment of civilized English exceptionalism, but rather as an instance of – so to speak – the regular state of exception common to European siege warfare. By a printed broadsheet (figure 3.3), composed of two sheets of folio format, the interested reading public was informed about "The Siege of Colchester By the Lord Fairfax As it was with the Line and Outworks." The print was produced in London by "Tho[mas] Witham at the Golden Ball in Long Lane near West Smithfield."[83] It combined a siege view with textual elements in the left, bottom, and right border. The main part of the latter was "A Diary of the Siege of Colchester by the Forces under the Command of Generall Fairfax." The London print seller and the consumers of his product were presumably in agreement that the operations taking place around Colchester were a siege that could be understood within the range of common European representational conventions.

A commercial broadsheet raised the awareness of potential buyers through its striking visual aspects. The panoramic image of "The Siege of Colchester" provided spatial orientation – in a more abstract geographical sense, through an identification of the cardinal points, as well as in a rather practical social context, by situating the town within the network of roads that connected it to neighboring places such as London, Cambridge, or Malden. Moreover, the image presented a detailed 'human geography' of the siege. It positioned different troops of the besieging forces in the field and lines of countervallation, identifying the units, according to current European usage, by their commanders' names. In the city and its suburbs, some streets (e. g. East Street) and individual buildings (e. g. Lucas House) are highlighted with their names and indications of ownership. Some of this information can be found within the image,

83 Witham's imprint only appears with this particular piece in the database of Early English Books Online. A "Print Seller and Frame-Maker" of the same name and address produced a celebratory portrait engraving of George I as king in the early eighteenth century: Thomas Witham, GEORGIUS. D. G. MAG. BR. FR. ET. HIB. REX. F. D., [engraving], digitized version: <https://www.royalcollection.org.uk/collection/603756/georgius-d-g-mag-br-fr-et-hib-rex-f-d> (accessed 23 October 2018).

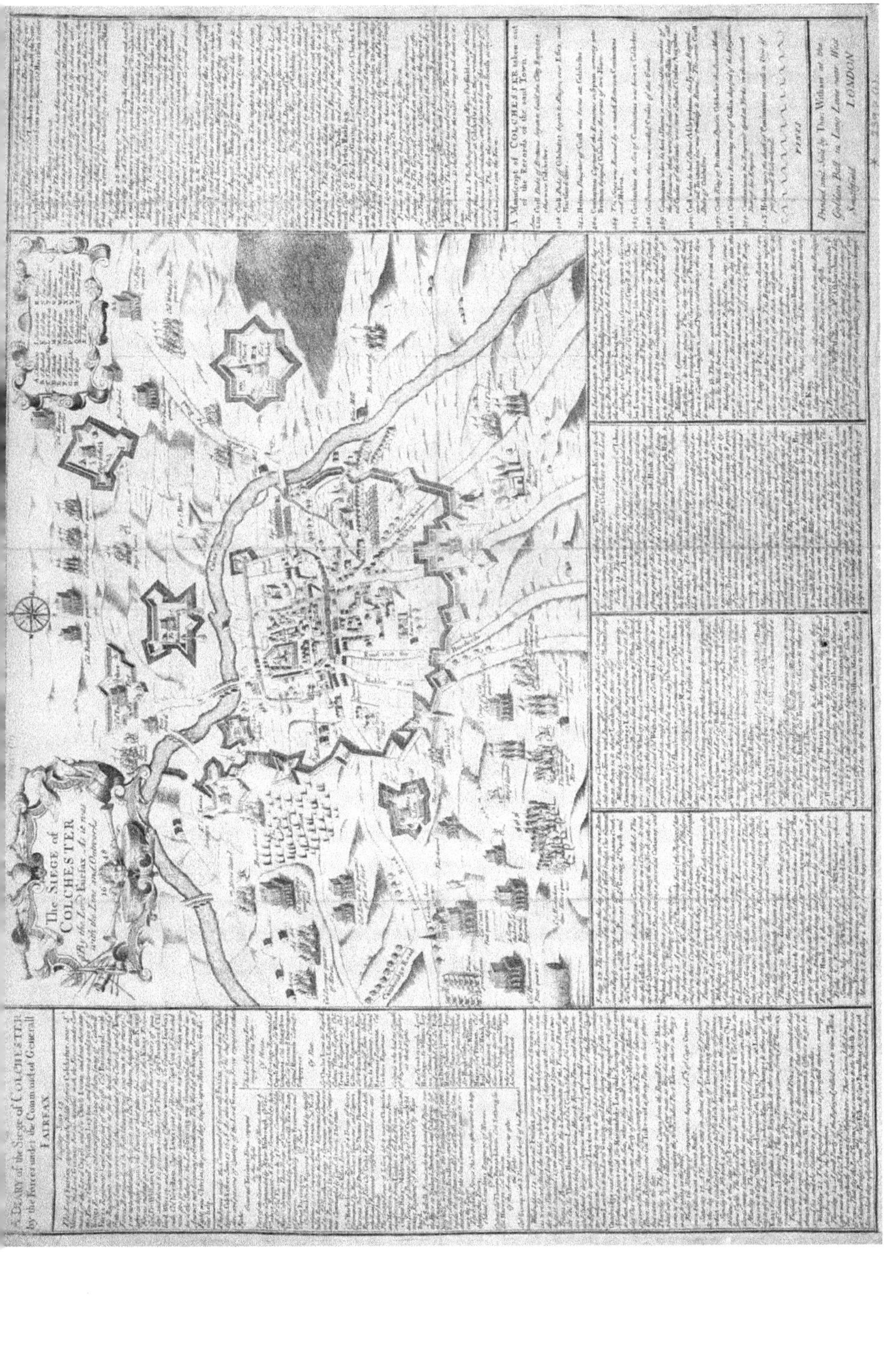

FIGURE 3.3 The Siege of Colchester by the Lord Fairfax, as it was with the Line and Outworks, 1648; [...]
© BRITISH LIBRARY BOARD, CARTOGRAPHIC ITEMS MAPS * 2390.(1.)

while explanations for the crowded town area are given by letters referring to a list in the upper right corner.

The siege view moreover depicts military action. The forces of horse and foot are shown as moving around, with the horses' hooves in the air and foot soldiers in lockstep. Guns are firing both from the town and from the siege lines (e. g. from Fort Needham), emitting large clouds of smoke. Musketeers are handling their weapons (e. g. near "Fort Ingo[l]desby" in the north-west corner of the siege lines). Other prints, like the abovementioned English version of the siege of Breda with its representation of dead bodies in a melee to the north-west of the besieged city, depict the consequences of fighting. In contrast, the siege view of Colchester shows potentially destructive action without results. Apart from a hint at consequences through the comment "burnt," which accompanies the depiction of flames and smoke coming out of "Grimsted house," the image represents an open-ended event.

Nonetheless, the printed image gives some hints as to the eventual outcome (or at least the one for which the besiegers, the side responsible for this broadsheet, were fighting): the siege lines are marked by intense shading, suggesting the materiality of massive works in the 'modern' style, whereas the ancient city walls occupied by the defenders are presented as thin and fragile. This does to some extent merely illustrate the difficulties of defending perpendicular medieval walls against ever stronger early modern artillery.[84] Yet, as Christopher Duffy has pointed out in his critique of the operations at Colchester, the besiegers did not primarily concentrate on attacking the walls by artillery fire or mining in order to force a breach.[85] The representational conventions thus visualized the power and imminent success of the besiegers. Their works appear to control the space around the city. Although their blockade was far from complete in the first weeks of the siege, the generalizing cartographic characteristics of the siege view make the lines appear continuous and strong all along.

All in all, the land- and cityscape in the image is subordinated to the ongoing military conflict. Fields, villages, and houses are occupied by the armed

84 See e. g. Geoffrey Parker, "The 'Military Revolution,' 1560–1660 – a Myth?", *The Journal of Modern History* 48, no. 2 (1976), 195–214, here: p. 203; for medieval origins of this development: Kelly DeVries, "Gunpowder Weaponry and the Rise of the Early Modern State," *War in History* 5 (1998), 127–145; the importance of other factors, especially manpower and material resources, is emphasized by the lavishly illustrated, somewhat popular work of Peter Harrington, *English Civil War Fortifications 1642–51* (Oxford, 2003), p. 55. Archaeological remains of the conflict (and of the settlement since pre-historic times) are described by Adrian Gascoyne and David Radford, *Colchester – Fortress of the War God: An Archaeological Assessment* (Oxford, 2013).

85 Duffy, *Siege Warfare*, p. 156.

forces. The roads, the main commercial connections of the town, are severed by the siege lines. The siege view thus complements the accompanying text, in which some of the citizens' desperate attempts to evade the worst effects of warfare are mentioned. The siege view condenses time, the chronological evolution of the siege, into space. It locates various events and the figures of military actors in one synoptic setting.

The text works, so to speak, the other way around. The siege journal gives a day-by-day account of the operations. Narrating a chronologically ordered story, it emphasizes the linearity of the siege time over the spatial dimension. Its organizational principles as well as its participant-observer perspective (reporting from the camp of the parliamentarian besiegers) are well known from continental examples. In this case, the diary records events between June 13, when Norwich's soldiers retreated into Colchester after their battle with Fairfax's army, and August 22, 1648, a few days before the surrender of the town. It includes a list of the units engaged in the fight on both sides, again identifying them by the names of their commanders according to shared European convention.

After the siege journal, the printer has added "A Manuscript of Colchester taken out of the Records of the said Town." This curious piece connects the siege of the present to ancient history and medieval legend: it is a table of historical events from the third century A. D., when Duke Coell, here identified as the founder of Colchester, expanded his dominion over all of Britain. His position was attacked by the Roman commander Constantius, who purportedly besieged Colchester from 260 AD. Yet, the besieger became family in 264, when Constantius married Coell's daughter Helena – her name alluding to another famous siege, that of Troy. Yet, this Helena, according to medieval English legend, was the mother of Emperor Constantine, who converted himself and his empire to Christianity.[86] According to this text, Constantine was born in Colchester in 265. The striking addition of this chronological table was more than just a concession to the rising fashion of antiquarianism and local history.[87] It revealed timeless values in the frightening current events: the ancient siege of Colchester by Constantius demonstrated the possibility of reconciliation. It moreover promised the growth of a powerful Christian state out of the miseries of siege warfare.

86 See for the legendary figures of King Cole and Helena: Antonina Harbus, *Helena of Britain in Medieval Legend* (Cambridge, UK, 2002), 64–71.

87 Graham Parry, *The Trophies of Time: English Antiquarians of the Seventeenth Century* (Oxford, 1995).

As the "Manuscript of Colchester" foreshadows a positive outcome for the violent military operations of the present, it complements the "Diary of the Siege of Colchester." This text narrates the struggle from the perspective of the besiegers. They identify themselves by the plural personal pronoun 'we,' for example in the account of the battle turning into a siege, stating "when we enter'd the Suburbs, the Lord Goring was Summoned but returned an Answer not becoming a Gentleman."

The journal closes with an entry for August 22, so the story is open-ended, for the Royalists only surrendered on August 28.[88] Nonetheless, just like the siege view described above gave visual clues for the besiegers' superiority, the journal indicated that time itself was working against the besieged. Their distress was obvious, and they were said, on the last day described in the "Diary," to have initiated a parley. Yet, the Parliamentarians, whose forces had meanwhile beaten the King's Scottish allies, were obviously no longer in the mood to grant any concessions. The journal indicates that surrender upon "mercie" would not be the same as the "honourable conditions" Norwich's officers were hoping for. Indeed, the Parliamentarians would execute two royalist officers, Sir Charles Lucas and Sir George Lisle, on the spot. Their defiant last words to the parliamentary firing squads and alleged gallantry in the face of death would become a stock element of royalist propaganda.[89] This outcome is not represented in the printed broadsheet. Yet, the journal does make clear that relations between the warring factions had become thoroughly embittered. Thus, the broadsheet was to prepare the reading public for a parliamentarian interpretation of the succeeding events.

The text of the journal highlighted the losses incurred in the operations. In the beginning of the siege, the besiegers were under dangerous gun fire from the city. For Thursday, June 15, the diary states that "The Besiegeds Canon from the Royall Fort at St. Maries, plaid very hard, killed severall of our men, as they did the day before; some, as they were raising the first Work, called Fort Essex, others as they were stragling in the field." As the siege was dragging on, perpetual danger and the conditions caused by 1648 being "the wettest [year]

88 The document itself does not give a date of publication, but its incompleteness suggests that it was published while the siege was still going on. It was definitely part of a parliamentarian propaganda effort against the Royalist upsurge of the summer of 1648. (See below, section 5, "Necessity and Extremities," for further elements of the pamphlet war that accompanied the military conflict and its aftermath.) Yet, the abovementioned inclusion of the antique legend of Coell indicated a conciliatory stance. This would definitely have been discredited by the full story of the surrender and execution of the Royalist leaders, so the Parliamentarians did not necessarily have an interest in updating the story through a new pamphlet after the conclusion of the siege.

89 E. g. Wellens, "Colchester," pp. 464–465; Brady, "Dying With Honour."

of the century"[90] were wearing on the nerves of the soldiers. When a party of the besieged sallied forth from the city on July 5 to rout the attackers at East Bridge and conquer two cannons, they were beaten back by the Parliamentarians. The siege journal records "29 slain on the ground, the Drakes recovered, and our former ground also." Besides that, "Lieut. Col. Weston, Lieut. Col. Weeks, and 80 & odd prisoners were taken, most of them sore cut for shooting poysoned bullets (20 of them died the next day)." The diary next lists the victims on the parliamentary side. In this way, the mutilation and killing of the prisoners are put into a context of well-deserved retribution. The whole affair was certainly raising mutual distrust and hatred. The accusation of using "chewed or poisoned bullets"[91] – misshapen and dirt-rolled projectiles that would cause more damage and gangrenous wounds – was serious. Rumors about such practices circulated on both sides and embittered the relations between the opposing forces and with the civilian population.[92]

Moreover, the printed siege journal indicates communicative and propagandistic aspects that increased the hatred between the parties. While the opponents still followed the conventional script for a siege from summons to parley (and ultimately surrender),[93] their interpretations – as presented in printed accounts – were often contradictory. They emphasized irreconcilable differences over shared fundamental values. On Sunday, July 16, Fairfax "sent a Summons again to Surrender the Town: The Lord Goring, Lord Capell, & Sir Charles Lucas jointly returned Answer (in writing,) under their hands to our Generall, That if the Trumpeter came any more with such a Summons, they would hang him up." The printed siege journal thus shows the Royalists haughtily disregarding their deteriorating position – confirmed a few days later, when deserters coming over from the town into the besiegers' camp on July 22 were "much complaining of their Diet in horse-flesh." A rhetorical attack on the immunity of a messenger went beyond the usual defiance expected from defenders, even when they were still hoping for relief. The printed journal explained that "The Conditions then offered to the Souldiers, was, Liberty, and Passes to go to their severall homes, submitting to the Authority of Parliament."

90 Wellens, "Colchester," p. 463.
91 Donagan, *War in England*, p. 342.
92 See ibid., 342–343; Ian Gentles, *The English Revolution and the Wars in the Three Kingdoms, 1638–1652* (London, 2007), p. 340. The broadsheet informed its readers that such bullets had been discovered with prisoners on June 28, who claimed that they were used by Norwich's orders. Fairfax was said to have threatened the refusal of quarter in case of further use of such material. Responsibility for the extremities was thus shifted to the enemy.
93 A protocol for siege warfare is for example presented by John Childs, "Surrender and the Laws of War in Western Europe, c. 1650–1783," in *How Fighting Ends: A History of Surrender*, ed. Holger Afflerbach, Hew Strachan (Oxford, 2012), pp. 153–168.

In the eyes of the parliamentarian narrator, the overbearing royalist commanders were playing with their subordinates' fair chance of escape. On July 25, the Parliamentarians therefore made a striking propaganda effort. During a nighttime attack, they "shot 20 Arrows, (with papers of advertisement offered) into the Town, to undeceive the Souldiers; acquainting them with what Conditions were offered them, and shall still be made good unto them, if they come out: which coming to some of their knowledge above 200 came out by that day 7. night." In the eyes of the royalist officers, such an act of inciting their men to disloyalty and desertion must have been a vile move. Though written from the perspective of one side, the text thus hinted at the difficulties that were arising in the course of the conflict (and of the Second Civil War generally). The 'atrocity' of the killings of Lucas and Lisle after their surrender clearly had a longer story behind it.[94] It was constructed by the interplay of military events with their contrasting representations in the media systems reporting on the Civil Wars.

For its representation of space and time of the siege, the printed pamphlet made use of the conventions of siege views and journals in order to point out peculiarities of the operation at Colchester. Reading the parliamentarian broadsheet as part of a system of printed media of the Civil Wars and, more generally, mid-seventeenth-century Europe, thus promises a more profound understanding of the conflict and the historical sources spawned by it.

5 Necessity and Extremities: Propaganda Efforts in the Civil Wars' Media System

The text of the Colchester broadsheet positioned the space of the besieged city (represented in the illustration) in the dimension of directional time. As the room for action of the besieged was shown to contract continuously, the story, although unfinished, pointed to a seemingly inevitable ending. Surrender and relatively harsh conditions were to be expected in the eyes of the parliamentarian publisher of the pamphlet. Royalist publications did their best to contradict such claims. A periodical leaflet entitled "The Colchester Spie" declared in mid-August that there was absolutely no question of the besieged entering into negotiations with the enemy. In poetic form, the timeless strength of the Royalists was affirmed: "Though the Saints hope to starve them out/ Alas it

94 See for the contested legal interpretations of their case, Brady, "Dying With Honour," pp. 15–19.

nere will bee."[95] The moral and physical endurance of the besieged are defended against parliamentarian "lies" with great pathos: "There is as much probability that the incomparable Royalists in Colchester will come to an agreement with the perjured Traytor Fairfax, [while they have a horse, a dog, or cat to feed upon] as it is likely, that the distant poles of heaven shall meet and conjoyne in one." This extravagant simile (including the bracketed insertion in the original text) was based on a claim that the besieged, against the statements of parliamentarian reporting, still had abundant provisions. The royalist pamphlet took much care to make this point. It reported that parliamentarian deserters were still trying to get into the town and that a stratagem employed by Fairfax, who tried to lure the defenders out of the town by letting cattle run loose under the walls, had failed: Charles Lucas had allegedly managed to capture the animals for the defenders.[96] This catch would supplement the supplies on which the defenders of Colchester were usually feasting. The horseflesh mentioned by enemy propaganda had only been consumed in order to get accustomed to it "in case of a long siege and need compelling them to feed frequently thereon."[97] Both sides reported similar acts and events, yet gave them a thoroughly different meaning by contextualizing them in different narratives and media formats.

Printed siege representations, particularly broadsheets and multi-page publications that combined text and image, can be read as attempts to impose cognitive order on a situation which entailed immense insecurity and existential dangers for participants. The synoptic character of siege views and journals gave their consumers an impression of overview and control.[98] Concluding from the broadsheet's contents, the audience of a piece like Witham's, probably mostly London-based, with a little spare money for buying a cheap broadsheet or at least with access to it being presented, had important political and ideological stakes in the event.[99] In a way, these readers were even given a

95 *The Colchester Spie: Truly Informing the Kingdome of the Estate of That Gallant Town, and the Attempts of Fairfax Against It: With Some Other Remarkable Intelligence From the English and Scots Army, the Last News From the Navie, Also From Westminster and London*, No. 2 (August 10–17, 1648), [Title page].

96 Ibid., [2].

97 Ibid.

98 In this sense they could be said to initiate for the siege what John Keegan has called the "rhetoric of battle history." (Keegan, *Face of Battle*, p. 36.) My thanks goes to Jamel Ostwald for pointing out this parallel. However, in contrast to Keegan's unbroken historiographical tradition, the early modern sources on siege warfare, as will become evident in this section, did immediately undermine the image of orderly war they were presenting at first sight.

99 Colchester is only about 50 miles from London, and both sides were soliciting more active support from there – Fairfax relying on Parliament as his employer, the Royalists hoping

more privileged vantage point than that of the military leaders, who were still on the ground during the fighting. The panoramic overview might even offer some sort of reassurance to Royalists who saw Witham's broadsheet. Yet, in another respect, any printed representation was very close to the action and dangers of the siege. Such prints were part of a media system that was formed by and itself influenced the current conflict. Literary scholars and historians have pointed out the importance of the moral and political debate that was emerging in such a situation.[100]

New Military History already expanded its interests to the wider socio-cultural contexts of war and soldiering, but Barbara Donagan adds further important aspects. She focuses on acts of symbolic communication in war, taking into account instances of demonstrative chivalry as well as violent extremities. Using an expanded concept of political acts recalling the approach of New Political History,[101] she moreover demonstrates that such actions and their representation had deep political implications. Donagan thus takes a fresh look at the rising tensions between Parliament and its army, as both were claiming authority over the revolutionary movement. Her analysis integrates siege warfare and its symbolic actions into the politics of the Civil Wars and thus presents a valuable starting point for a comparative look at the prism of local[102] media representations of sieges.

Reading the broadsheet "The Siege of Colchester" with Donagan's approach in mind highlights specific aspects that undermine the illusion of remoteness and control. While the general perspective of the print is that of the parliamentarian military, it does hint at the opponents' and civilian suffering from the conflict. The immediate threats and fears of siege warfare thus appear as pinpoints piercing through the canvas of the overview picture. The journal for

---

for the support of the apprentices, who were rebelling against the parliamentary government. (Woolrych, *Britain in Revolution*, pp. 413–414.) Even though the siege operations did not immediately affect the City, they were closely connected to its and the whole country's fate.

100 A literary studies perspective on the events at Colchester is e. g. presented by Brady, "Dying With Honour," 12–19.

101 The AHA magazine, *Perspectives on History* 49, no. 5 (2011), had a thematic focus on "Political History Today," see particularly the contribution by Steven Pincus and William Novak, "Political History after the Cultural Turn;" online edition: <https://www.historians.org/publications-and-directories/perspectives-on-history/may-2011/political-history-today/political-history-after-the-cultural-turn> (accessed 23 October 2018); see for this concept in German-language research: Ute Frevert, Heinz-Gerhard Haupt, eds., *Neue Politikgeschichte: Perspektiven einer historischen Politikforschung* (Frankfurt am Main, 2005).

102 In this term, I am including the London printing presses that might be considered 'regional' or 'national' as well; yet, for my purposes, these materials, centered on the locality of the siege and connecting it to Civil War politics, can be regarded as 'local.'

example reported from the last weeks of the siege, as the situation in the town was worsening drastically: "Tuesday 15 [of August]. Many men came over this day from the Besieged, & the poorer sort of people began to rise for want of bread." While this short message might still be interpreted as a simple demonstration of parliamentarian superiority, the detail given in the description of the next day reaches further:

> Wednesday 16. They rise in great Numbers, and come to the Lord-Gorings Quarters, some bringing their Children starved to death, they crying out, so long as Horseflesh, Dogs, and Cats were to be had, they did not complain. This day the Mayor of Colchester sent a Letter to the Gen. That the Inhabitants might come out, for that they had no provision, it being all seised by the Souldiers. Our Generall returned answer, He pittied their condition, but to grant that, was to make the Town hold out longer, and did not stand with his trust to permit it.

By giving the wording of the civilians' protestations before the defending commander's quarters and by mentioning starved children, the text went into an emotional register. It connected overview representation to the immediate experience of hunger and death. It did this through explicit references to the oral and written communication produced in the situation of conflict.

Thereby, the broadsheet text pointed to a local media system spawned by the siege. This system included a host of printed publications that claimed to make the communications of the situation itself available. It included e. g. letters from Fairfax' 'leaguer' (army camp) and messages exchanged between the officers of the warring parties or between them and the town representatives petitioning for relief. Sometimes such documents merely claimed news value, appearing in supposedly 'raw' letter form, sometimes they were blended in with political tracts and proclamations.[103]

103 Titles available in the database of Early English Books Online are e. g. *A Copy of Some Papers Lately Passed Betwixt the Lord Fairfax on the One Side and the Earle of Norwich, Lord Capel, and Sir Charles Lucas, on the Other, at Colchester* (London, 1648); *The Demands and Proposals of The Earle of Norwich, and Sr. Charles Lucas (in the Name of Themselves, and the Rest of His Majesties Officers and Soulders in the City of Colchester) to Generall Fairfax* [...] (London, 1648); *The Copy of a Letter Sent From a Person of Much Honour and Reason, Accidentally The Resolution of Sr Marmaduke Langdale Concerning the Relieving of the Lord Goring. The Lord of Loughborough. The Lord Capel. Sir Charles Lucas, and the Rest of the Officers and Souldiers Now Blocked Up in the Town of Colchester, by the Lord Generall Fairfax. And Their Further Proceedings Touching Major Generall Lambert; Agreed Upon by a Councel of War, and Assented to by Most of the Colonels and Officers in the Northern Army.* [...] (London, 1648); *A Letter, From a Gentleman in Colchester, to His Friend in London*

A contemporary publication collected correspondence between the different military and civilian leaders involved in the fights and negotiations around Colchester. This work claimed to present an exchange between Fairfax and the "Bay and Say makers" of the town in the form of original letters. The artisans and their supporters in the royalist council of war had petitioned the general of the besieging forces to allow trade to go on in spite of the blockade. Yet, Fairfax was not in the mood for concessions. His answer on June 24, 1648, made clear that he would not compromise his military position:

> It had bin good that the unavoydable consequence of War, of that of restraining trade to a town beseiged, is one had bin considered of by the Inhabitants of your Town before Their admittance of those forces, which have necessarily drawen it upon them, and which indeed first began the new disturbances upon this county and the Kingdome, and that interruption to your Trade which is complayned off.[104]

(London, 1648); *15 Junii, 1648. The Particulars of the Fight at Colchester (Sent in a Letter to the Honorable William Lenthal Esq; Speaker of the Honorable House of Commons)* [...] (London, 1648); *5 Iulii, 11 at Night. A Letter From the Leaguer Before Colchester, Sent to the Honorable Committee at Derby-House, of the Great Fight Between his Excellency the Lord Fairfax, and the Forces in Colchester* [...] (London, 1648); *6. Julii, 1648. From the Leaguer at Colchester, More Certain News of the Fight on Wednesday Last; and of Their Present Condition* (London, 1648); *The Lord Goring, the Lord Capel, and Sir Charles Lucas Their Letter, Directed to the Lord Generall, and the Lord Fairfax His Letter Directed Unto Them, Concerning the City of Colchester* (London, 1648); *Joyfull Nevves from Colchester* [...] (London, 1648); *A Message Sent From His Highnesse the Prince of Wales, to the Major of Yarmouth, Concerning the Landing of His Forces There for the Relief of Colchester.* [...] (London, 1648); *A True and Exact Relation of the Taking of Colchester, Sent in a Letter From an Officer of the Army* [...] (London, 1648); *Another Bloudy Fight at Colchester in Essex, Between the Forces Commanded by Sir Charles Lucas, Col. Washington, and Col. Lunsford, and the Lord Generall Fairfax's With the Generals' Message to the L. Goring, and His Lordships Answer Thereunto Concerning His Majesties Pardon.* [...] (London, 1648); *The Lord Gorings Message to the Lord Generall, Concerning the Surrendring of the Town of Colchester, With All the Ordnance, Armes, and Ammunition. And the Severall Attempts of Generall Hastings, and Sir Charles Lucas to Escape Away With their Horse* [...] (London, 1648); *A True Relation of the Surrendring of Colchester to his Excellency the Lord Generall Fairfax. As It Was Sent in a Letter to the Honourable William Lenthal, Esquire, Speaker of the Honourable House of Commons* (London, 1648); *A Letter From His Excellency the Lord Fairfax Generall of the Parliaments Forces: Concerning the Surrender of Colchester, the Grounds and Reasons of Putting to Death Sir Charles Lucas and Sir George Lysle; With the Articles and Explanation of the same.* [...] (London, 1648).

104 *Severall Papers and Letters Betwixt His Excellency the Lord Fairfax, The Earle of Norwich, Lord Capell, Sir Charles Lucas, About the Surrender of Colchester* [...] (London, June 27, 1648), p. 6.

Fairfax used the stern rhetoric of an instructor to remind the town about the practices of war. He pointed out that practical and moral mistakes had been made both by the city's inhabitants, who let the Royalists take up their defense within the town walls, and by the latter, who were responsible for the disruptions caused by renewed conflict in the first place. Hardships for the townspeople were part and parcel of serious siege warfare, a necessity rather than a choice for the military commander.

Fairfax' statement indicates that early modern siege warfare might be considered a precursor to modern ideas of a 'state of exception' or 'emergency.'[105] It can be connected to early modern conceptions of the 'necessities' of (siege) warfare. The wording is frequently recurring in early modern descriptions of sieges, both with a specific meaning of a lack of important items and in a more general sense, e. g. as a justification of unpopular administrative measures.[106] The situation of a siege was thus presented as a moment for exceptional acts, legitimized and – in spite of their severity – to a certain extent regulated by rules of behavior in war. Under such circumstances, symbolic communication and physical pressure could forcibly align local interests and conditions with a politics of wider responsibility: Fairfax sternly reminded the artisans and merchants of Colchester that their own act of letting the Royalists into town had had negative consequences for the whole commonwealth. Their disadvantages were merely part of a wider set of effects of their own political behavior. They had strengthened the evil side in the eyes of the Parliamentarians.

Symbolic communication, physical pressure, and public representations of the siege of Colchester united to inculcate an important political message: no siege was merely of local importance. The impact of war necessarily entangled local interests with larger political circumstances. Any act leading up to and during the siege thus carried a load of far-reaching responsibilities. The specific situation of the siege created conditions under which morally questionable acts of violence – such as the maltreatment of prisoners of war for their alleged use of poisoned bullets, the intentional destruction of the suburbs by

---

105 This concept was theorized in the 1920s by the German right-wing jurist Carl Schmitt, *Die Diktatur: Von den Anfängen des modernen Souveränitätsgedankens bis zum proletarischen Klassenkampf* (6th ed. Berlin, 1994; original 1921); id., *Politische Theologie: Vier Kapitel zur Lehre von der Souveränität* (Berlin, 1928). His ideas are critically taken up by Giorgio Agamben, *Ausnahmezustand*, Homo sacer II.1 (Frankfurt am Main, 2004).

106 For "necessitas" as a legitimating principle of territorial lords' interferences with city freedom: Heinz Schilling, *Die Stadt in der Frühen Neuzeit* (2nd ed., Munich, 2004), p. 45; see also a city official's description of the famous second Turkish siege of Vienna in 1683, where all kinds of unpopular measures (forced work on fortifications, requisitioning, fire safety measures) are declared highly necessary ("hochnothwendig"): Nikolaus Hocke, *Kurtze Beschreibung* [...] (Vienna, 1685).

fire, or the refusal to relieve the suffering of the civilians[107] – became necessities of warfare. One of the last desperate attempts of the besieged to improve their situation illustrates this in a shocking manner.

The royalist officers had realized that their plea for honorable surrender had come too late as Fairfax' troops were already preparing to storm the town. Their general was unwilling to grant any concessions to the enemy leaders, who tried to improve their strategic position and make some further resistance possible by turning out civilians in order to save provisions for the soldiers: "Munday [August] 21. The Besieged turned out of the Town in the night, many men, women, & children, but the next morning took them in again." The terror of the victims caught between the lines for a night can hardly be imagined. Neither side felt able to show any mercy to them.[108] Violence, inhumanity even, appeared necessary in order to impress a politico-military message on the enemy and the public. Acts and publications were closely linked in this situation.

An important part of the nationally oriented media discussion spawned by the events around Colchester were publications of correspondences between the main actors. By printing and publishing these, both sides were trying to publicly legitimize their positions and actions while at the same time confirming that threats were not empty and promises would be kept. Thus, some communications during the parley for surrender in August 1648 were instantly published by the Speaker of the revolutionary House of Commons in London.[109] The harsh fate of the civilian population of Colchester was treated in this publication. The Mayor and Aldermen of the town had done their best in one of their petitions to Fairfax "humbly to pray the continuance of [his] good affections towards [them], and this miserable decayed and wasted Town."[110] In spite of previous severity, Fairfax now openly accepted this plea for protection from the townspeople to some extent: he signed his answer as "Your assured friend," and declared that he and his council of war had resolved to include town representatives in the negotiations for surrender.[111] Yet, while the besiegers assured the town's administrators of their continuing favor, the situation of

107 Donagan even speaks of atrocities in the sense of modern international law: Barbara Donagan, "Atrocity, War Crime, and Treason in the English Civil War," *The American Historical Review* 99, no. 4 (1994), 1137–1166.

108 See Donagan, *War in England*, pp. 339–341.

109 [John Rushworth], *A Letter Sent to the Honorable William Lenthall, Esq. Speaker of the Honorable House of Commons. With Severall Letters From the Lord Norwich, Lord Capel, Sir Charles Lucas; and Their Agreement for the Delivery of the Town of Colchester. The Petition of the Major* [*sic*] *and Aldermen of the Town; and the Generals Answer. With the Result of the Counsell of War* (London, 1648), [p. 8, no pagination].

110 Ibid., [p. 7].

111 Ibid.

necessity was clearly not yet past. Fairfax' council indicated in its resolution that it would consider to distinguish between the town and "the Authors of its suffering and destruction,"[112] but the threat of excessive violence in storm and plunder was still hovering over the civilian populace.

After the surrender, contemporary witnesses were shocked by the lamentable state of the town. It had suffered from fire, hunger, and atrocities which had accompanied the deteriorating military situation of the defenders.[113] Nonetheless, Colchester was spared the worst possible fate. In the midst of the extremities of siege warfare, the town leaders had managed to avert a complete breakdown of communications with the besiegers. At least, they were permitted to sit at the table in the negotiation of their lot. The reproduction of their letters to Fairfax in parliamentarian print media assured their success in this very limited, yet crucial sense.

Barbara Donagan's work has shown the immense value inherent in a fine-grained analysis of the communicative function of such acts and their representations within a local media system. Yet, it begs further questions when a comparative perspective expands the analytical framework. The way space and time of a siege were depicted in the contemporary sources fed back into the operations and their contemporary understanding – in England, the British Isles, and beyond. The (printed) representations provided a totalizing understanding of events and contributed to legitimize action within larger strategic, political, and socio-cultural contexts.

6 Making Sense of Siege Representations in 17th-Century Europe

The popular format of the illustrated broadsheet appealed to consumers through rich visual material and a text promising the 'highlights' of the operations. The image foregrounded the spatial dimension of siege warfare in more or less masterfully executed siege views from an overview perspective. The accompanying text added chronological depth and a narrative dimension in the typical day-by-day account of the siege journal. A comparison of the broadsheet representation of the operations around Colchester with that of other European sieges thus points to overarching cultural conventions of siege warfare and its depiction in early modern print media. This perspective comple-

112 Ibid., [p. 8].

113 For a contemporary pamphlet eliciting commiseration with "Colchester's Teares," see Donagan, *War in England*, 312; for the destruction wrought by the siege, ibid., pp. 330–346.

ments the experiential and communication-oriented approach of New Military History.

As producers and consumers of printed news and curiosities in 17th-century Europe had to make sense of the frequently recurring siege events around them,[114] they were relying on preexisting representational conventions and developing them further at the same time. The prints were creating a cultural framework, in which some aspects of the traumatic experiences of siege warfare could be contained.[115] The published representations of the collective suffering inherent in siege warfare rendered unique events comparable. They stressed the subordination of time and space of violence to familiar formal rules such as the 'laws of war.'[116] Yet, such rules and conventions were not just mitigating factors. Inherent in the norms of siege and warfare more generally was the threat of violent severity. Comparing siege representations across early modern Europe makes clear that a stern interpretation, without slackness or mercy, was always part of the threatening game in military conflict. If such an interpretation was applied, it was not simply an abuse or false understanding of the conventions of war – although the opponents might accuse each other of this. Thus, even the subordination of events to conventional scripts in (print) representations involved elements of insecurity, which showed that threats to life and property were still dangerously close.

Moreover, the example of Colchester has demonstrated that the abstracted strategic viewpoints of printed broadsheets only seemed to be removing the consumers in space and time from a siege at first sight. On closer inspection, publications could become starting points for a rapprochement to the suffering of soldiers and civilians in the siege situation. The horrors of destruction, violence, hunger, and disease, which always accompanied warfare, could in this way be converted into shared experience. They could become collectively owned by the members of 'public spheres'[117] of changing extension. Some of

114 For the importance of siege warfare, see e. g. Frank Tallett, *War and Society in Early-Modern Europe, 1495–1715* (War in context) (London, 1992), p. 3.

115 Frank Ankersmit suggests that this activity can be regarded as part of the larger European development of "Western historical consciousness," which allowed 'Westerners' to observe disturbing events from a scientific viewpoint at a distance. (Ankersmit, "Trauma and Suffering," pp. 8–9.)

116 See for a distillation of formal principles from medieval sources: Maurice H. Keen, *The Laws of War in the Middle Ages* (London/Toronto, 1965); an early modern perspective is taken by Childs, "Surrender and the Laws of War."

117 For this concept: Jürgen Habermas, *The Structural Transformation of the Public Sphere: An Inquiry Into a Category of Bourgeois Society* (Cambridge, MA, 1989; German original 1962). While Habermas' work has been read critically (e. g. Craig Calhoun, ed., *Habermas and the Public Sphere* (Cambridge, MA, 1992)), it has been adopted for historical analysis of

these were formed on local and regional levels immediately touched by the events. In the British Civil Wars, circulating prints were moreover meant to unite a proto-national public. Beyond that, printed media of siege warfare were uniting 'transnational' groups. They could strengthen existing links of confessional solidarity or contribute to their constitution by circulating across borders – reminding readers of the successes and sufferings of their own people (whether political allies, co-religionists or simply other civilian citizens) in current siege events. For all of the groups addressed, the traumatic side of the events did shine through the visual and textual depictions. It was a defining element of practices and representations during the 'state of emergency' of a siege that fixed rules of conduct and the 'necessities' of violent excess coexisted.

Barbara Donagan's historiographical work has convincingly established the importance of the siege of Colchester for the course of English and British history. A comparative perspective on the common European format of the siege broadsheet points to crucial additional aspects. This form of representation relates the specific military event with its terrifying experiential character to an overarching culture that informed practices and understandings of warfare. The conduct of a siege was depicted according to shared spatial and temporal conventions. Military leaders interpreted such conventions in specific ways, according to general values as well as the particular situation they found themselves in. The media system, which comprised printed publications and word-of-mouth relations of events, thus mirrored and actively shaped a shared European imagination of (siege) warfare. This cultural image in turn fed back into the strategic, spatio-temporal 'games' played by the actors of a siege. It exerted an influence on local experiences of violence and suffering – and turned them into much more widely shared collective history.

Realities and representations of siege warfare in early modern Europe contributed to the emergence of the authority and ideas of the (national) state. The printed media about events at Colchester made clear that local violence and conflict had causes and consequences in a larger political context. Other famous sieges in seventeenth-century Europe, such as in the war for Dutch independence (Eighty Years' War, 1568–1648), in the diverse conflicts which shook central Europe in the Thirty Years' War (1618–1648) or in the wars fought by and against Louis XIV of France (r. 1643–1715), were described in comparable

early modern (mass) communication by cultural historian Peter Burke in order to structure his overview of the fifteenth to eighteenth centuries: Briggs and Burke, *Social History of the Media*, pp. 61–90. Habermas' underlying idea of a public constituted by collective communication is valuable for an understanding of the formation of (early) modern states in connection with media.

fashion. It will take further investigation into the details of siege representation and its influence on ideas of authority and state control, but this initial probing of the connection between the emergencies of siege warfare and the cultural element of state building promises important insights into the complexities subsumed in the familiar statement that 'War makes States.'[118]

Bibliography

Printed Sources

15 Junii, 1648. The Particulars of the Fight at Colchester (Sent in a Letter to the Honorable William Lenthal Esq; Speaker of the Honorable House of Commons) [...] (London, 1648).

5 Iulii, 11 at Night. A Letter From the Leaguer Before Colchester, Sent to the Honorable Committee at Derby-House, of the Great Fight Between His Excellency the Lord Fairfax, and the Forces in Colchester [...] (London, 1648).

6. Julii, 1648. From the Leaguer at Colchester, More Certain News of the Fight on Wednesday Last; and of Their Present Condition (London, 1648).

A Copy of Some Papers Lately Passed Betwixt the Lord Fairfax on the One Side and the Earle of Norwich, Lord Capel, and Sir Charles Lucas, on the Other, at Colchester (London, 1648).

A Letter, From a Gentleman in Colchester, to His Friend in London (London, 1648).

A Letter From his Excellency the Lord Fairfax Generall of the Parliaments Forces: Concerning the Surrender of Colchester, the Grounds and Reasons of Putting to Death Sir Charles Lucas and Sir George Lysle; With the Articles and Explanation of the Same. [...] (London, 1648).

A Message Sent From His Highnesse the Prince of Wales, to the Major of Yarmouth, Concerning the Landing of his Forces There for the Relief of Colchester. [...] (London, 1648).

Another Bloudy Fight at Colchester in Essex, Between the Forces Commanded by Sir Charles Lucas, Col. Washington, and Col. Lunsford, and the Lord Generall Fairfax's With the Generals' Message to the L. Goring, and his Lordships Answer Thereunto Concerning his Majesties Pardon. [...] (London, 1648).

A True and Exact Relation of the Taking of Colchester, Sent in a Letter From an Officer of the Army [...] (London, 1648).

---

118 See for a revival of this debate in political science: Charles Tilly, "War Making and State Making as Organized Crime," in *Bringing the State Back In*, ed. Peter Evans, Dietrich Rueschemeyer, Theda Skocpol (Cambridge, UK, 1985), pp. 170–187.

A True Relation of the Surrendring of Colchester to His Excellency the Lord Generall Fairfax. As It Was Sent in a Letter to the Honourable William Lenthal, Esquire, Speaker of the Honourable House of Commons (London, 1648).

Belägerung der Statt Ostende: Journal: Tagregister und eigentliche beschreibung [...], 3 vols ([s. l.], 1604).

Defoe, Daniel, *A Tour Thro' the Whole Island of Great Britain, Divided Into Circuits or Journies,* [...], vol. 1 (London, 1724).

Grimeston, Edward, *A True Historie of the Memorable Siege of Ostend,* [...] (London, 1604).

Hexham, Henry, *A True and Briefe Relation of the Famous Seige of Breda Beseiged, and Taken in Vnder the Able and Victorious Conduct of his Highnesse the Prince of Orange* [...] (Delft, 1637).

Histoire remarquable et véritable de ce qui s'est passé par chacun iour au siege de la ville d'Ostende, [...] (Paris, 1604).

Historical Manuscripts Commission, ed., *Fourteenth Report, Appendix, Part IX: The Manuscripts of the Earl of Buckinghamshire, the Earl of Lindset, the Earl of Onslow, Lord Emly, Theodore J. Hare, Esq., and James Round, Esq., M.P.* (London, 1895).

Hocke, Nikolaus, *Kurtze Beschreibung* [...] (Vienna, 1685).

Joyfull Nevves from Colchester [...] (London, 1648).

Lithgow, William, *A True and Experimentall Discourse, Upon the Beginning, Proceeding, and Victorious Event of this last Siege of Breda* [...] (London, 1637).

Machiavelli, Niccolò, *The Art of War* (Cambridge, MA, 2001; originally 1521, transl. by Neal Wood in 1965).

[Rushworth, John], *A Letter Sent to the Honorable William Lenthall, Esq. Speaker of the Honorable House of Commons. With Severall Letters From the Lord Norwich, Lord Capel, Sir Charles Lucas; and Their Agreement for the Delivery of the Town of Colchester. The Petition of the Major* [sic] *and Aldermen of the Town; and the Generals Answer. With the Result of the Counsell of War* (London, 1648).

Severall Papers and Letters Betwixt His Excellency the Lord Fairfax, The Earle of Norwich, Lord Capell, Sir Charles Lucas, About the Surrender of Colchester [...] (London, June 27, 1648).

The Colchester Spie: Truly Informing the Kingdome of the Estate of That Gallant Town, and the Attempts of Fairfax Against It: With Some Other Remarkable Intelligence From the English and Scots Army, the Last News From the Navie, Also From Westminster and London, No. 2 (August 10–17, 1648).

*The Copy of a Letter Sent From a Person of Much Honour and Reason, Accidentally The Resolution of Sr Marmaduke Langdale Concerning the Relieving of the Lord Goring. The Lord of Loughborough. The Lord Capel. Sir Charles Lucas, and the Rest of the Officers and Souldiers Now Blocked Up in the Town of Colchester, by the Lord Generall Fairfax. And Their Further Proceedings Touching Major Generall Lambert; Agreed

Upon by a Councel of War, and Assented to by Most of the Colonels and Officers in the Northern Army. [...] (London, 1648).

The Demands and Proposals of the Earle of Norwich, and Sr. Charles Lucas (in the Name of Themselves, and the Rest of his Majesties Officers and Soulders in the City of Colchester) to Generall Fairfax [...] (London, 1648).

The Lord Gorings Message to the Lord Generall, Concerning the Surrendring of the Town of Colchester, With All the Ordnance, Armes, and Ammunition. And the Severall Attempts of Generall Hastings, and Sir Charles Lucas to Escape Away With Their Horse [...] (London, 1648).

The Lord Goring, the Lord Capel, and Sir Charles Lucas Their Letter, Directed to the Lord Generall, and the Lord Fairfax His Letter Directed Unto Them, Concerning the City of Colchester (London, 1648).

Vere, Francis, *The Commentaries of Sr. Francis Vere Being Diverse Pieces of Service, Wherein He Had Command* [...] (Cambridge, UK, 1657).

Witham, Thomas, GEORGIUS. D. G. MAG. BR. FR. ET. HIB. REX. F. D., [engraving], digitized version: <https://www.royalcollection.org.uk/collection/603756/georgius-d-g-mag-br-fr-et-hib-rex-f-d> (accessed 23 October 2018).

Secondary Sources

Agamben, Giorgio, *Ausnahmezustand*, Homo sacer II.1 (Frankfurt am Main, 2004).

Ankersmit, Frank, "Trauma and Suffering: a Forgotten Source of Western Historical Consciousness," in *Western Historical Thinking. An Intercultural Debate*, ed. Jörn Rüsen (New York, 2002), pp. 72–85.

Arblaster, Paul, "Posts, Newsletters, Newspapers: England in a European System of Communications," *Media History* 11 (2005), 21–36.

Arndt, Johannes, *Herrschaftskontrolle durch Öffentlichkeit: Die publizistische Darstellung politischer Konflikte im Heiligen Römischen Reich, 1648–1750* (Göttingen, 2013).

Ashton, Robert, *Counter Revolution: The Second Civil War and its Origins, 1646–8* (New Haven, 1994).

Aylmer, Gerald E., *Rebellion or Revolution? England, 1640–1660* (Oxford, 1986).

Barbarics, Zsuzsa; Pieper, Renate, "Handwritten Newsletters as a Means of Communication in Early Modern Europe," in *Correspondence and Cultural Exchange in Early Modern Europe, 1400–1700*, ed. Francisco Bethencourt, Florike Egmond (Cambridge, UK, 2007), pp. 53–79.

Barratt, John, *The Great Siege of Chester* (Stroud, 2003).

Bastiansen, Henrik G., "Media History and the Study of Media Systems," *Media History* 14, no. 1 (2008), 95–112.

Boys, Jane E. E., *London's News Press and the Thirty Years War* (Woodbridge, Rochester, 2011).

Brady, Andrea, "Dying With Honour: Literary Propaganda and the Second English Civil War," *The Journal of Military History* 70, no. 1 (2006), 9–30.

Brewer, John, *The Sinews of Power: War, Money and the English State, 1688–1783* (London, 1989).

Briggs, Asa; Burke, Peter, *Social History of the Media: From Gutenberg to the Internet* (Cambridge, UK, 2009).

Calhoun, Craig, ed., *Habermas and the Public Sphere* (Cambridge, MA, 1992).

Carlton, Charles, *Going to the Wars: The Experience of the British Civil Wars, 1638–1651* (London, 1992).

Childs, John, *Warfare in the Seventeenth Century* (Smithsonian History of Warfare) (Washington, D.C., 2001).

Childs, John, "Surrender and the Laws of War in Western Europe, c. 1650–1783," in *How Fighting Ends: A History of Surrender*, ed. Holger Afflerbach, Hew Strachan (Oxford, 2012), pp. 153–168.

Clarke, David T. D., *The Siege of Colchester, 1648* (Colchester, 1975).

Clinton, Mark; Fibiger, Linda; Shiels, Damien, "Archaeology of Massacre: The Carrickmines Mass Grave and the Siege of March 1642," in *Age of Atrocity: Violence and Political Conflict in Early Modern Ireland*, ed. David Edwards, Pádraig Lenihan, Clodagh Tait (Dublin, 2007), pp. 192–203.

Coster, Will, "Massacre and Codes of Conduct in the English Civil War," in *The Massacre in History*, ed. Mark Leven, Penny Roberts (New York, 1999), pp. 98–105.

Creveld, Martin van, *The Art of War: War and Military Thought* (Smithsonian History of Warfare) (Washington, D.C., New York, 2005; first ed. London, 2000).

Curthoys, Ann; Docker, John, *Is History Fiction?* (2nd ed., Sydney, 2010).

DeVries, Kelly, "Gunpowder Weaponry and the Rise of the Early Modern State," *War in History* 5 (1998), 127–145.

Dinn, James, "Nathaniel Nye and the Siege of Worcester, 1646," *Transactions of the Worcestershire Archaeological Society* 3rd ser., 24 (2014), 173–180.

Donagan, Barbara, *War in England, 1642–1649* (Oxford, 2008).

Donagan, Barbara, "Atrocity, War Crime, and Treason in the English Civil War," *The American Historical Review* 99, no. 4 (1994), 1137–1166.

Donoghue, John, *Fire under the Ashes: An Atlantic History of the English Revolution* (Chicago, 2013).

Downes, Kevin, "The 'Great Lewis' and the Siege of Duncannon, 1645," *Decies: Journal of the Waterford Archaeological & Historical Society* 60 (2004), 115–116.

Duffy, Christopher, *Siege Warfare: The Fortress in the Early Modern World, 1494–1660* (London, 1979).

Faulstich, Werner, *Mediengeschichte: Von den Anfängen bis 1700* (Göttingen, 2006).

Fischer-Kattner, Anke, "Violent Encounters at Ostend, 1601-1604: Spatiality, Location, and Identity in Early Modern Siege Warfare," *Comparativ* 28, no. 2 (2018), 22-41.

Fissel, Mark Charles, *English Warfare 1511–1642* (London, New York, 2001).
Frevert, Ute; Haupt, Heinz-Gerhard, eds., *Neue Politikgeschichte. Perspektiven einer historischen Politikforschung* (Frankfurt am Main, 2005).
Gascoyne, Adrian; Radford, David, *Colchester – Fortress of the War God: An Archaeological Assessment* (Oxford, 2013).
Gentles, Ian, *The English Revolution and the Wars in the Three Kingdoms, 1638–1652* (London, 2007).
Gilbert, C. Don, "Two Sermons at the Siege of Worcester, 1646," *Transactions of the Worcestershire Archaeological Society* 20 (2006), 139–141.
Haas, Hagen, "'Denn die Bombe, wann sie fällt…:' Zum Schicksal von Einwohnern belagerter Städte im absolutistischen Zeitalter," *Militär und Gesellschaft in der Frühen Neuzeit* 7, no. 1 (2003), 41–59.
Habermas, Jürgen, *The Structural Transformation of the Public Sphere: An Inquiry Into a Category of Bourgeois Society* (Cambridge, MA, 1989; German original 1962).
Hallin, Daniel C.; Mancini, Paolo, *Comparing Media Systems: Three Models of Media and Politics* (Cambridge, UK, 2004).
Harbus, Antonina, *Helena of Britain in Medieval Legend* (Cambridge, UK, 2002).
Harrington, Peter, *English Civil War Fortifications 1642–51* (Oxford, 2003).
Heuser, Beatrice, *Den Krieg denken: Die Entwicklung der Strategie seit der Antike* (Paderborn, 2010).
Hintze, Otto, "Der Commissarius und seine Bedeutung in der allgemeinen Verfassungsgeschichte," in *Historische Aufsätze: Karl Zeumer zum 60. Geburtstag*, ed. Mario Krammer (Weimar, 1910), pp. 493–528.
Howes, Audrey, "'Hull Must Absolutely Be Had:' the First Siege of Hull, July 1642," *East Yorkshire Historian* 4 (2003), 65–74.
Hughes, Ann, *The Causes of the English Civil War* (Basingstoke, 1991).
Jennings, Stuart B., "The Third and Final Siege of Newark (1645–1646) and the Impact of the Scottish Army Upon Nottinghamshire and Adjacent Counties," *Midland History* 37, no. 2 (2012), 142–162.
Jones, Phil, *The Siege of Colchester, 1648* (Stroud, 2003).
Keegan, John, *The Face of Battle* (London, 1976).
Keen, Maurice H., *The Laws of War in the Middle Ages* (London, Toronto, 1965).
Kenyon, John; Ohlmeyer, Jane, eds., *The Civil Wars: A Military History of England, Scotland, and Ireland 1638–1660* (Oxford, 1998).
Kishlansky, Mark, *A Monarchy Transformed: Britain 1603–1714* (London, 1997).
Kramer, Alan, *Dynamic of Destruction: Culture and Mass Killing in the First World War* (Oxford, 2007).
Latzel, Klaus, "Vom Kriegserlebnis zur Kriegserfahrung: Theoretische und methodische Überlegungen zur erfahrungsgeschichtlichen Untersuchung von Feldpostbriefen," *Militärgeschichtliche Mitteilungen* 56, no. 1 (1997), 1–30.

Martines, Lauro, *Furies: War in Europe, 1450–1700* (New York, 2013).

Murchadha, Maighréad Ní, "From Kildare to Baldongan: Fr. Conor Donnogh and the Siege of Baldongan, 1642," *Journal of the County Kildare Archaeological Society* 19, no. 2 (2003), 269–288.

Murphy, Elaine, "Siege of Duncannon Fort in 1641 and 1642," in *The 1641 depositions and the Irish Rebellion*, ed. Eamon Darcy, Annaleigh Margey, Elaine Murphy (Warfare, Society and Culture), vol. 6 (London, 2012), pp. 143–154.

Ohlmeyer, Jane, *Civil War and Restoration in the Three Stuart Kingdoms: The Career of Randal MacDonnell, Marquis of Antrim, 1609–1683* (Cambridge, UK, New York, 1993).

Ohlmeyer, Jane, *Making Ireland English: The Irish Aristocracy in the Seventeenth Century* (New Haven, 2012).

Ostwald, Jamel, *Vauban under Siege: Engineering Efficiency and Martial Vigor in the War of the Spanish Succession* (Leiden, 2006).

Paret, Peter, "The New Military History," *Parameters* 21 (1991), 10–18.

Parker, Geoffrey, *The Military Revolution: Military Innovations and the Rise of the West, 1500–1800* (New York, 1988).

Parker, Geoffrey, "The 'Military Revolution,' 1560–1660 – a Myth?", *The Journal of Modern History* 48, no. 2 (1976), 195–214.

Parry, Graham, *The Trophies of Time: English Antiquarians of the Seventeenth Century* (Oxford, 1995).

Pincus, Steven; Novak, William, "Political History after the Cultural Turn," online edition: <https://www.historians.org/publications-and-directories/perspectives-on-history/may-2011/political-history-today/political-history-after-the-cultural-turn> (accessed 23 October 2018).

Pollak, Martha, *Cities at War in Early Modern Europe* (Cambridge, UK, 2010).

Pritchard, David M., ed., *War, Democracy and Culture in Classical Athens* (Cambridge, UK, New York, 2010).

Roberts, Michael, *The Military Revolution, 1560–1660* (Belfast, 1956).

Rogers, Clifford J., ed., *The Military Revolution Debate: Readings on the Military Transformation of Early Modern Europe* (Boulder, 1995).

Ruff, Julius R., *Violence in Early Modern Europe, 1500–1800* (Cambridge, UK, 2001)

Salman, Jeroen, *Pedlars and the Popular Press: Itinerant Distribution Networks in England and the Netherlands 1600–1850* (Library of the Written Word), vol. 29 (Leiden, 2013).

Sandberg, Brian, "'To Have the Pleasure of This Siege:' Envisioning Siege Warfare During the European Wars of Religion," in *Beholding Violence in Medieval and Early Modern Culture*, ed. Allie Terry-Fritsch, Erin Felicia Labbie (Farnham, 2012), pp. 143–162.

Schantz, Mark S., *Awaiting the Heavenly Country: The Civil War and America's Culture of Death* (Ithaca, 2008).

Schmitt, Carl, *Die Diktatur: Von den Anfängen des modernen Souveränitätsgedankens bis zum proletarischen Klassenkampf* (6th ed. Berlin, 1994; original 1921).
Schmitt, Carl, *Politische Theologie: Vier Kapitel zur Lehre von der Souveränität* (Berlin, 1928).
Schilling, Heinz, *Die Stadt in der Frühen Neuzeit* (2nd ed., Munich, 2004).
Serdiville, Rosie; Sadler, Douglas John, *The Great Siege of Newcastle, 1644* (Brimscombe Port, 2011).
Simoni, Anna E. C., *The Ostend Story: Early Tales of the Great Siege and the Mediating Role of Henrick Van Haestens* ('t Goy-Houten, 2003).
Smith, Nigel, ed., *A Collection of Ranter Writings: Spiritual Liberty and Sexual Freedom in the English Revolution* (London, 2014).
Snyder, Timothy, *Bloodlands: Europe between Hitler and Stalin* (New York, 2010).
Tallett, Frank, *War and Society in Early-Modern Europe, 1495–1715* (War in context) (London, 1992).
Tilly, Charles, "War Making and State Making as Organized Crime," in *Bringing the State Back In*, ed. Peter Evans, Dietrich Rueschemeyer, Theda Skocpol (Cambridge, UK, 1985), pp. 170–187.
Wellens, John, "The Siege of Colchester," *East Anglian Magazine* 7, no. 9 (1948), 460–465.
Whiting, Amanda Jane, *Women and Petitioning in the Seventeenth-Century English Revolution: Deference, Difference, and Dissent* (Turnhout, 2015).
Wolfe, Michael, "Walled Towns During the French Wars of Religion (1560–1630)," in *City Walls: The Urban Enceinte in Global Perspective*, ed. James D. Tracy (Studies in Comparative Early Modern History), vol. 4 (Cambridge, UK, 2000), pp. 317–348.
Woodward, Daphne; Cockerill, Chloe, *The Siege of Colchester: A History and Bibliography* [Chelmsford], 1979).
Woolrych, Austin, *Britain in Revolution, 1625–1660* (Oxford, 2002).

CHAPTER 4

More Honored in the Breach? Representations of Honor in Louisquatorzian Sieges

Jamel Ostwald

> … [That the entire garrison] will leave the place [of Menin] the [date left blank] by the breach, with their arms, baggage and horses, without being searched, drum beating, musket balls in mouth, matches lit, flags flying, [with] twelve large cannon of various calibers, four mortars and their carriages, and arms and munitions to fire twenty rounds, to march together by the shortest route, in one day, to Lille….
>
> SURRENDER TERMS REQUESTED BY THE FRENCH COMMANDER OF MENIN, 22 August 1706.

∵

The above excerpt describes a common martial ceremony of the early modern period – a garrison's peaceful evacuation from a fortress at the end of a siege. The theatricality of such surrender ceremonies has long intrigued historians. The minute details of the defender's evacuation – flags flying, beating drums, arms at the ready, musket balls ready to be spit into the gun barrel, evacuating through the breached wall they were threatening to defend if denied their terms – incongruously represented the defeated garrison's *symbolic* willingness to continue the fight even after it refused to resist *in reality*. What are modern historians to make of these genteel siege manners of an earlier aristocratic age, steeped as we are in the battle-centric mindset of the Napoleonic era, and the tenaciously self-sacrificial combat of World War II and modern suicide bombers? What do such ceremonies and their "marks of honor" tell us about how contemporaries viewed surrender?

The dominant historical interpretation has been to incorporate these martial ceremonies into a broader conceptualization of siege warfare as an artificially restrained, formulaic, and ritualized form of combat. In this view, the potential for the destruction of entire cities, as famously occurred in the Old Testament, or Magdeburg in 1631, was avoided through ritual de-escalation. It

 | DOI:10.1163/9789004395695_005

is argued that the Old Regime – from the end of the 'age of religious wars' (conventionally dated c. 1648) up to the revolutionary age of the 1790s – saw few such atrocities thanks to limitations on war facilitated by the application of Baroque formalism and Enlightenment rationalism to early modern siegecraft.[1]

Honor is seen as a key element in this 'regularized' system of de-escalation: to preclude further resistance, besiegers allowed garrisons to abandon their fortresses with ritual "marks of honor." Even the tune being played and the position of the surrendering soldier's musket are declared of the utmost importance in John Wright's eighty-year-old article on the customs of siege warfare. Still the most cited author on the subject, his cataloging of the "fine distinctions" of these ceremonial details reinforces the view of Baroque rituals as precisely calibrated performances, or "calculated pomp."[2] Such tokens helped both attacker and defender choreograph an end to their siege.[3] In this interpretation, both sides used the evacuation performance to facilitate a rational, mutually-advantageous exchange: the besiegers won the town without further bloodshed or lost time, while the besieged could console themselves with having done their duty, and be recognized as honorable opponents in a formal evacuation ceremony.

Among many traditional military historians this moderation carries the taint of the artificial niceties of "limited" war, witness the frequency with which

---

1 Elsewhere I have questioned the "scientific" aspect of this rhetoric of siege history, see Jamel Ostwald, *Vauban under Siege: Engineering Efficiency and Martial Vigor in the War of the Spanish Succession* (Leiden, 2007).

2 For urban rituals of possession generally: Edward Muir, *Ritual in Early Modern Europe* (New York, 1997), pp. 239–246. On the theatricality of Louis XIV's royal sieges, see Joël Cornette, *Le roi de guerre: Essai sur la souveraineté dans la France du Grand Siècle* (Paris, 1993), pp. 255–260.

3 John Wright, "Sieges and Customs of War at the Opening of the Eighteenth Century," *American Historical Review* 39 (1934), 640–643; David Chandler, *The Art of Warfare in the Age of Marlborough* (1976; repr. New York, 1995), pp. 267–269; Christopher Duffy, *Fire and Stone: The Science of Fortress Warfare 1660–1860* (1976; repr. Edison, NJ, 2006), pp. 152–153; Simon Pepper, "Siege Law, Siege Ritual, and the Symbolism of City Walls in Renaissance Europe," in *City Walls: The Urban Enceinte in Global Perspective*, ed. James D. Tracy (Cambridge, 2000), pp. 573–604; Barbara Donagan, "The Web of Honour: Soldiers, Christians, and Gentlemen in the English Civil War," *The Historical Journal* 44, no. 2 (2001), 384–385; Randall Lesaffer, "Siege Warfare and the Early Modern Law of War," SSRN Scholarly Paper (Rochester, 2006), <http://papers.ssrn.com/abstract=926312>; John A. Lynn, "Introduction: Honourable Surrender in Early Modern European History, 1500–1789," in *How Fighting Ends: A History of Surrender*, ed. Holger Afflerbach, Hew Strachan (New York, 2012), pp. 106–107; John Childs, "Surrender and the Laws of War in Western Europe, c. 1650–1783," in *How Fighting Ends: A History of Surrender*, ed. Holger Afflerbach, Hew Strachan (New York, 2012), pp. 153–68.

these deadly sieges are compared to formal dances and staged plays.[4] Recent scholars have examined capitulations with more respect for ritual. These scholars, inspired by anthropological studies of ritual performances, argue that such ceremonies defined status and honor, that such evacuation details were significant, every deviation a potential slight requiring protest. The most familiar parallels in Louis XIV's court were calibrated rituals at Versailles, such as the calculated distance his representative would descend down the Ambassador's staircase to greet a visiting delegation, or the seating arrangement (and chair height) at a royal dinner.[5] Whether one mocks ceremonial evacuations or uses them to gain insight into contemporary mentalities, all assume their overarching importance.

This chapter acknowledges the symbolism within these liminal ceremonies, but questions whether evacuation rituals were formalized, whether contemporaries truly cared about such details, and whether various audiences accepted such marks as a straightforward proxy for the garrison's honor. It compares the rhetoric of siege surrenders with the contradictory ways in which contemporaries discussed honor in surrender during the War of the Spanish Succession (1701–1714). It explores how garrisons' claims to honor were interpreted and contested by several layers of observers, rather than assume that contemporaries either passively equated evacuation marks with honor or negotiated the minute details of evacuation marks. Despite explicit references to an "honorable" surrender and "marks of honor," there was little consensus that gaining such evacuation marks was an important marker of an honorable defense, or that garrisons denied these marks had served *dis*honorably. Instead, observers, with their own honor at stake, debated garrisons' conduct using a broader range of measures.

1 "More Honored in the Breach…"

To understand the role of honor in early eighteenth century siegecraft, we must first understand the broader context in which these capitulation documents were created. The War of the Spanish Succession provides an excellent case study of the surrender process. It began after the death of King Carlos II, when the Austrian Habsburg claimant challenged Carlos' choice of a French

---

4 For a recent example, see John Childs, "The Laws of War in Seventeenth-Century Europe and Their Application during the Jacobite War in Ireland, 1688–1691," in *Age of Atrocity: Violence and Political Conflict in Early Modern Ireland*, ed. David Edwards et al. (Dublin, 2007), p. 296.

5 For a recent discussion of Baroque court rituals and its historiography, see Giora Sternberg, *Status Interaction during the Reign of Louis XIV* (New York, 2014).

Bourbon as successor. France quickly occupied Spanish territories in the name of the new king Philippe V. The Habsburg Archduke Charles was thus forced to capture the disputed territories by force, backed by a coalition of anti-Bourbon princes; the resulting war was fought across four European theaters, as well as in the Americas and at sea. Over the course of this thirteen-year war, the Bourbons (mainly France and Spain) and the Allies (primarily Austria and the Holy Roman Empire, the Netherlands and England) conducted approximately 125 sieges in western and central Europe, ranging from attacks lasting barely a day to defenders holding out for months.[6] Ninety percent of the 36 towns besieged in the Low Countries theater, the focus of this chapter, surrendered.[7]

The practice of negotiating formal surrender agreements was centuries old when the War of the Spanish Succession began in 1701 – in this sense, the purported artificial niceties of negotiated surrender were hardly limited to the 'age of limited war.'[8] By definition, defenders had already refused the attacker's pre-siege summons to surrender. Assuming the defenders were unwilling to fight to the bitter end, they usually initiated the surrender process, presenting their terms as the starting point for the negotiation.[9] The garrison commander, sensing his defense would soon become untenable, called his general officers to a council of war. Their precarious situation was summarized for the attendees: the parlous state of the place's works, the exhaustion of the men, the lack of food and munitions, as well as the proximity of the attackers to the final bulwark, the curtain wall. After deliberation ended, the attendees added their signatures to the council's minutes. The commander then drew up articles of surrender.[10] Prescriptive manuals not only warned governors to word their demands as precisely as possible, but further recommended they request

6 All these figures drawn from Ostwald, *Vauban under Siege*.

7 In theory, each of the 49 distinct Flanders forts, citadels and towns under attack could receive its own terms. In the war as a whole, 85% surrendered.

8 Generally, see the pre-modern chapters in Afflerbach and Strachan, eds., *How Fighting Ends*.

9 This section is drawn largely from contemporary treatises, including: Antoine de Ville, *De la Charge des gouverneurs des places* (Paris, 1639), pp. 349–351, 435–440; Alain Manesson Mallet, *Les Travaux de Mars*, 3 vols (Paris, 1672), 3:307–313; Louis de Gaya, *Science militaire* (The Hague, 1689), pp. 102–109; Pierre-Claude Guignard, *L'école de Mars*, 2 vols (Paris, 1725), 1:315–321, 2:451–456; Charles Sevin, marquis de Quincy, *L'art de la guerre ou Maximes et Instructions sur l'art Militaire*, 2 vols (The Hague, 1728), 1:450–455; Alvaro Navia Osorio, marqués de Santa Cruz de Marcenado, *Réflexions militaires et politiques*, 12 vols (Paris, 1738), 9: 198–249. On occasion, the besiegers would summon the garrison one last time before it launched a general storm.

10 Besiegers expected the defenders to provide concrete written terms. J. A. C. Hugill, *No Peace Without Spain* (Oxford, 1991), p. 96.

inflated terms in case they would be negotiated downward.[11] A white flag of truce was displayed at each sector under attack while drummers beat the "chamade," the universal signal to initiate a parley. Both sides ceased their fire and hostages of equal number and rank were exchanged in order to ensure good faith in the negotiations. Treatises cautioned defending commanders to send competent negotiators for this task, preferably those who could hold their wine; they were also warned that they were too important to personally negotiate in the besieger's camp.[12] The besieged representatives debated the proposed terms with the siege commanders. Rarely, a garrison might revive the medieval practice of offering to surrender in several days if not relieved by then – such old-fashioned requests were rejected out of hand. Discussions would then last hours; significant revisions of the garrison's requests might require a day or more. Talks might occasionally break down, leading to a renewal of hostilities, which usually resulted in yet another chamade and further negotiations within a day or two. After the final responses were hammered out and recorded, copies would be transcribed and signed. The garrison negotiators returned to deliver a copy to their commander for final approval.

The resulting capitulation document, named after the *capita* or headings of the individual articles, followed the format of other multilateral contracts, such as mercenary troop contracts, also termed capitulations. The larger column on the right side of the page listed the garrison's initial demands. In the left-hand margin was space to record the besieger's responses to each point. These documents were thus a contractual agreement between the two sides, though the implementation of most terms was cemented by good faith and the threat of reciprocity. Public perception also played a role: most of the surrender details were disseminated to the public at large. Signed manuscript copies of the document were sent to the heads of the signatory powers, full copies might be published in the town, publishers in other countries circulated printed copies of their own, and the outcome would be further publicized throughout Europe in correspondence, newsprint, pamphlets, and published histories.[13]

---

11 Antoine de Pas, marquis de Feuquières, *Memoirs Historical and Military*, 2 vols (London, 1735), 2:317; Guignard, *L'école de Mars*, 1:316.

12 Michel de Montaigne had already dedicated an essay to the subject in the 16th century: *Si le Chef d'une Place Assiegée Doit Sortir pour Parlementer*. The Montaigne Project, at <http://artflsrv02.uchicago.edu/cgi-bin/philologic/getobject.pl?c.0:2:5.montaigne.58263>.

13 Copies of capitulations can be found most readily in the appendices of François Eugène de Vault and Jean-Jacques-Germain Pelet, eds., *Mémoires militaires relatifs à la succession d'Espagne sous Louis XIV*, 11 vols (Paris, 1835); and Guillaume de Lamberty, *Mémoires pour servir à l'histoire du XVIII siècle*, 14 vols (The Hague, 1724).

The signatories' publicly-given word of honor served as an additional incentive to fulfill the agreement's terms.[14]

The primary purpose of this agreement was to regulate the transition of a town from the besieged to the besieger. This included the exact process by which the garrison would hand over the fortress, its future combatant status, as well as specification of the "honors" granted during the evacuation. Once the capitulation had been signed by both parties, the first concrete step was for the garrison to turn over one of the town's gates to the attackers, with guards posted to prevent pillage and unauthorized fraternization. Any garrison mines were to be revealed, as well as their magazines. A few days later, the defenders would march out according to the terms. This procession was both symbolic and practical: its formal closure of hostilities relieved the garrison of its anxiety about their fate, and was possibly the last time the townspeople would see their former defenders. In this idealized view, the evacuation performance declared to those present that even though the defenders were surrendering, they had nonetheless forced the enemy to acknowledge their valorous defense.

After this cathartic ceremony of dispossession put a formal and visible end to the previous sovereign's rule, the conquering general would perform his own ceremony of possession. This triumph was full of celebration, notable speeches and presentation of the keys to the city. These, then, were the general outlines of how a besieged place changed hands.

2 "... Than in the Observance"

The surrender process described above is well known. But our title's allusion to Hamlet's famously-ambiguous line prompts a necessary complication. The full phrase from the fictional Danish prince, "a custom, More honor'd in the breach, than [in] the observance," encourages us to ask two questions. First, it prompts us to enquire whether the honorable siege surrender was the norm or the exception. Second, we must enquire whether evacuating through the breach, or the acquisition of other marks of honor, was deemed honorable, and by whom. Was it a simple matter of either an honorable or a dishonorable surrender, or perhaps it was a matter of the quantity or quality of symbolic marks of honor you could acquire – an evacuation by the breach truly being "more honor'd?" What is the relationship between the phrase "an honorable surrender," the

14 As when a Dutch field deputy warned that violating the terms would besmirch the commander's honor. A. J. Veenendaal, Jr., ed., *De briefwisseling van Anthonie Heinsius 1702–1720*, 15 vols (The Hague, 1976), 5:527 #1019.

"marks of honor" given during the evacuation, contemporary references to a garrison's "honor," and how different participants and observers interpreted such details? Only by examining the context of actual siege surrenders can we understand the complicated ways in which contemporaries represented a surrendering garrison's honor and dishonor.

2.1 *Capitulations in Context: Conquest and Garrison Fate*

As a preliminary matter, capitulation documents spent very little space discussing the marks of honor which dominate modern historical attention. To give one example, Béthune's French defenders dedicated 3% of its length to prescribing the evacuation marks, a portion of one article out of twenty-eight.[15] Much more care was taken to specify *who* exactly could leave with the garrison: wives, children, domestics, and military administrators were all carefully identified as being covered by the agreement. Béthune's fifth article spent 130 words identifying all those to be evacuated. Attention was also paid to the garrison's destination, march route, and who was to pay – half of a capitulation's content spelled out the military transition. Just as important were all of the financial and political terms. A third of the average garrison's demands carefully accounted for their personal property and made sure they were not detained for debts they owed the townspeople – a few hostages would be kept as surety. Capitulations also promised a framework for future interactions between the townspeople and their new sovereign, though this was increasingly specified in a separate agreement negotiated directly between the victors and the town magistrates. Almost all the language in almost every capitulation, in other words, was on practical, mundane issues rather than symbolic evacuation marks.

Nevertheless, our focus here is on that small sliver of the capitulations that dealt with the garrisons' evacuation and fate, since this has been the focus of almost every historian to date.[16] The principal context for all these capitulations is that of undisputed conquest: contemporaries acknowledged that marks of honor were a sop for the losers, a declaration that they had lost with

15 Béthune's capitulation in Vault and Pelet, eds., *Mémoires militaires*, 10:298–306.

16 Additional scholarship on the subject has appeared after this chapter was written. The French scholar Émilie Dosquet is completing her research on this subject for the Nine Years' War, while Paul Vo-Ha's dissertation has just been published as *Rendre les armes – Le sort des vaincus XVI–XVIIe siècles* (Champ Vallon, 2017). Two other works have also recently appeared: Erik Swart, "Defeat, Honour and the News: The Case of the Fall of Breda (1625) and the Dutch Republic," *European History Quarterly* 46, no. 1 (2016), 6–26; and Domagoj Madunić, "Taming Mars: Customs, Rituals and Ceremonies in the Siege Operations in Dalmatia during the War for Crete (1645–69)," *Hungarian Historical Review* 4, no. 2 (2016), 445–470.

honor. By 1700, garrisons were generally expected to lose, so contemporaries used an additional indicator of how complete their defeat was: their fate. The large number of Spanish Succession sieges allows us to survey six possible garrison fates.[17] Though every defender preferred to preserve their fortress as well as their honor, only four Flanders garrisons managed to do both. In these cases, possession indisputably marked a clear victory for the defenders and defeat for the besiegers. The Bourbon defenders of Fort Saint-Donas took a second, more ambiguous, tack, abandoning their fortifications in the night rather than submit as prisoners, forcing observers to decide whether being denied honorable terms excused sneaking away in the middle of the night.

The remaining forty-four Flanders garrisons were forced, with "inexpressible sadness," to capitulate and perform the "dolorous" evacuation ceremony.[18] The most lenient besiegers would grant a free evacuation – in contemporary shorthand, "honorable terms."[19] This meant that the garrison would receive a variety of the symbolic marks of honor during its exit. More importantly, such a garrison was free to march to rejoin their field army. In fact, the primary dichotomy was between an "honorable" surrender and one where the garrison was taken prisoner – as we shall see, far more attention was paid to the context of the surrender than the performance of the surrender ceremony. Less generously, an unlucky garrison might be allowed to evacuate freely, but had to promise to not fight for a specified period of time. Of the forty-nine Flanders fortifications, just a quarter managed to procure their freedom in a capitulation – only a quarter won an "honorable" surrender in this crucial sense.

Most garrisons denied a free evacuation obtained, as contemporaries wrote, "no other conditions" than to surrender as prisoners. Imprisoned garrisons marched out with few of the marks reserved for an honorable evacuation, sometimes none – almost all contemporary treatises were mute on the marks of honor allowed to prisoners.[20] Such defenders had uncertain futures awaiting them. Some capitulation documents promised a prisoner exchange without delay. Other agreements were silent on the issue, requiring later negotiations. Most imprisoned garrisons were given the familiar two-column capitulation

17 For a brief English summary, see Samuel Brewster, *Jus Feciale Anglicum* (London, 1725), p. 36.

18 As described by Guignard, *L'école de mars*, 1:321.

19 For one example, 1708.12.28–30 *Post Man*. All English newspapers are dated in Old Style, eleven days behind (most of) the Continent's New Style calendar.

20 Santa Cruz provided one of the most detailed discussions of capitulation terms, drawn from both his study of history and his experience in Spain during the War of the Spanish Succession (*Réflexions*, 9:203–249). He claimed that prisoners were allowed to evacuate through the breach with their arms, and then lay them down, though many imprisoned garrisons were *not* given this honor.

listing garrison demands and besieger responses, but six garrisons' voices were silenced completely, with the capitulation dictated to them by the attackers. All told, half of the 49 Flanders garrisons were denied an "honorable" surrender and instead taken prisoner.

Contemporaries distinguished garrisons taken prisoner from those taken "at discretion." Even though such soldiers became, practically speaking, prisoners of war as well, their evacuation terms were harsher. Their property might be forfeit and even their physical well-being tenuous – this was equivalent to "unconditional surrender." A few taken at discretion were given a capitulation; contemporaries here referred to the "honor" of even being granted a capitulation or surrendering "upon articles." Such documents were very brief and contained almost no promises beyond protection of their lives, and sometimes their baggage and the officers' swords. Often this was the fate of defenders who resisted until a general storm on the breach. One of Liège's small hilltop forts delayed too long, as one English attacker explained: "whielst wee were attacquing the breach thay beat a parle, but our men and officers thought it too laite and would not hear them."[21] Sometimes, however, garrisons who surrendered before such a last-ditch defense were also consigned to this ignominious fate. Four Flanders fortifications were taken at discretion.

The harshest terms a garrison might face was to have its appeals for clemency ignored altogether, and to be 'put to the sword.' Although a few Flanders garrisons defended a storm on the breach and had their initial appeals for surrender ignored, there were no cases of rampant bloodletting within any Flanders places, though it did occur on several occasions in the Iberian theater. Honorable surrenders were, then, as honored in the breach as in the observance: every capitulating garrison demanded to be given a free evacuation; yet fewer than half of our Flanders defenders received the "honorable" capitulation they requested. Certainly not the massacres of an earlier age, but not quite the stereotyped gentility portrayed in much of the literature.[22] After possession had been determined, contemporaries then distinguished between free garrisons, those imprisoned, and those taken at discretion.

21 John Churchill, Duke of Marlborough and Sidney Godolphin, *The Marlborough-Godolphin Correspondence*, ed. H. Snyder, 3 vols (Oxford, 1975), 1:126.

22 Some of this contrast between the two periods may result from an exaggerated perception of earlier brutality. See, for example, J.-L. Charles, "Le sac des villes dans les Pays-Bas au XVIe siècle: Étude critique des règles de la guerre," *Revue internationale d'histoire militaire* 24 (1965), 288–301; and Olaf van Nimwegen, *The Dutch Army and the Military Revolutions, 1588–1688* (Rochester, 2010), pp. 144–145.

2.2 *Random Evacuation Marks?*

The two most important outcomes of a siege were possession of the fortress and the garrison's fate. As a French general reported of their recapture of Le Quesnoy, "I won't say too much about the details, the essential thing is that the place is ours."[23] After possession and garrison fate were determined, participants then considered how to mark the honors the garrison acquired. It is difficult, however, to ascertain how garrisons viewed such performative evacuation marks. Beyond their demands preserved in the capitulation, they did not record their thoughts on the marks, rarely even recorded their actions throughout the siege. Siege journals from the War of Spanish Succession focused almost exclusively on the enemy's attacks, while the commanders' post-siege justifications cared only about justifying their decision to surrender, its timing, and their fate.[24] They obviously desired recognition, but too many historians have gone beyond this truism to equating evacuation marks as a mirror for the garrison's honor itself. Such rhetoric represents them as highly stylized markers of honor, their combination allowing fine distinctions between one garrison and another. Analysis of the details of the forty-nine capitulations (really only fourteen, since most marks were reserved for free garrisons) illustrates, on the contrary, how inconsistently such marks were requested. For most of the war, the symbolic marks of honor remained unstandardized.

Specific evacuation marks clearly had precise symbolic meanings within European military culture, representing a theoretical readiness to continue fighting. But *which* evacuation marks any garrison requested appears almost random for most of the war, though drawn from the limited menu of possibilities discussed by Wright.[25] The marks were formulaic in isolation, but no precise script, simple or ornate, was followed in combination. This was true despite how easy it would have been to copy (or inflate) past requests, given the ritualistic nature of the marks and the availability of so many printed models at hand. And it was also the case despite the apparent need to do so, considering the warnings to avoid any omissions in one's demands lest a close-reading foe take advantage of them. The order in which specific marks of honor were

---

23 Paris, Service historique de la Défense (hereafter SHD), GR A^4 carton 8, Flandres 1712 folder, *maréchal de camp* d'Alègre letter 1712.10.06.

24 The same is true for the very few sources from the burghers' perspectives I have discovered.

25 For examples of the standard "honors" due to French officers of various ranks as they entered and left towns, see Briquet's collections of ordinances, issued during the war and collected in *Code militaire ou Compilation des ordonnances des roys de France concernant les gens de guerre*, 2 vols (Paris, 1708), 2:297. From an Allied perspective, see Humphrey Bland, *A Treatise of Military Discipline* (2nd ed., London, 1727).

requested, how exactly they were phrased, where in the capitulation they appeared (generally they appeared in the first half), whether they were contained within a single article or not, and even which marks were requested, varied from garrison to garrison. Requested marks did not match the patterns suggested in prescriptive manuals such as Antoine de Ville's *De la Charge des Gouverneurs*; nor, for that matter, did De Ville mention the possibility that a garrison's honor was determined by adding up all the marks of honor granted them. But the French engineer was not even consistent within his own work – the marks he encouraged a cautious garrison to request were actually fewer than those he recommended the besiegers allow. Nor did other treatises follow his precise recommendations.[26]

Marks that historians identify as especially meaningful were often ignored. Arms being "at the ready," for example, were a symbol of the garrison's combat readiness. Such significance might have mattered in practice, but most garrisons did not even attempt to regulate this supposedly-critical detail: they simply requested "arms and baggage." A rarity, Aire's French garrison requested their fusils and muskets held high (*haut*). Presumably this was an indication of greater honor, but years earlier Antwerp's Spanish defenders received a similar right to shoulder its arms (article 2: *fusils sur l'épaule*), even though they meekly surrendered without any defense. Nor did garrisons bother to abandon anachronistic terms: they continued, for example, to insist on evacuating with "matches lit at both ends," despite their armies having abandoned matchlock muskets for flintlocks years earlier.

In the rare cases where we can compare multiple capitulations of the same town, there is no clear pattern. In the one case where we might distinguish a personal signature, the same officer surrendering two different fortresses after epics defenses, French marshal Boufflers requested different marks for Namur's citadel (1695) than those he demanded thirteen years later for Lille. A few capitulations even made a blanket request for "all" or the "usual" marks, but still prescribed specific ones. After Gelders and Rheinberg surrendered to Allied blockade in 1703, both requested that they evacuate "with all the marks of honor that can be given in such a case," yet then proceeded immediately to specify only a handful of them: "drum beating, flags [*enseignes*, rather than the more common *drapeaux*] flying, match lit at both ends, ball in mouth."[27] In 1710 Saint-Venant's French defenders requested a slightly different order (drums,

26 De Ville's *Charge* was one of the only works focusing on the defense (until Vauban's 1706 treatise), and was published multiple times, first in 1639 and as late as 1674.

27 Rheinberg article 14 in Lamberty, *Mémoires*, 2:417. Gelders article 5 in *Points et articles* at <http://vd18.de/de-ulbsa-vd18/content/titleinfo/45556286>. It is not clear if blockaded garrisons expected different terms from those taken by siege.

flags, balls in mouth, matches lit at both ends), but also appended "and all the honors of war."[28] The Imperial garrison of Freiburg (1713) in Germany asked for "all the marks of honor" in the first article, and then specified in article three "with sack and pack, drums beating, flags flying, arms on the shoulder, as is due in this sort of evacuation."[29] As Freiburg's "sack and pack" suggests, garrisons might add their own flourishes. Venlo's Franco-Spanish garrison was particularly dramatic in 1702, demanding that it evacuate with "kettledrums and drums beating, trumpets blaring, flags and standards flying [...]" (article 3). In short, garrisons haphazardly requested specific marks of honor – they were not yet fixed in a set sequence, and many garrisons failed to request all the individual marks theoretically available to them.

The one Flanders exception to free garrisons gaining some marks of honor was Oostende. Its Bourbon defenders were allowed a free evacuation, but pledged to only return to military service after six months. Thus, their fate was better than a garrison taken prisoner (no exchange was necessary), yet worse than an unconditional evacuation. Perhaps as an indication of its limbo status, the Dutch siege commander heer van Ouwerkerk only allowed them to evacuate with *swords* and baggage – terms, it should be noted, that were worse than the marks given to many prisoners. Though the attacking generals were said to be unanimous on the decision, one Swedish lieutenant-general in Dutch service sympathized with the commander's "great mortification" at his officers marching out with their swords at their sides. Since the garrison was not prisoner, he reasoned to the Allied army commander the English Duke of Marlborough, the Allies could at least have given them some useless muskets for the ceremony.[30] Ouwerkerk's terms were unusual, but elicited little attention beyond this bruised Swedish conscience. Other accounts (public and private) repeated the atypical stipulations, but none of the other participants or observers complained of the matter, or even remarked upon it, much less explained the decision.[31] This exception suggests, then, that the denial of arms to a free garrison was unusual, or perhaps that they distinguished a fully free

28 Lamberty, *Mémoires*, 6:130.

29 Vault and Pelet, eds., *Mémoires militaires*, 11: 645.

30 British Library, Additional Manuscripts 61309, ff. 119–120 (hereafter BL, Add. MSS). The cheap fusils, he noted, wouldn't even cost four *sous* – addressing a constant concern about the costs of surrender. The French commander, the comte de La Mothe Houdancourt, did not mention this evacuation detail in his apologetic letter back to Court, though it was evident in the capitulation copy he sent along. SHD, GR A[1] 1938 #91–#92; see also Gisbert Cuper, *Het Dagboek van Gisbert Cuper, gedeputeerde te velde, gehouden in de Zuidlijke Nederlanden in 1706*, ed. A. J. Veenendaal, Sr. (The Hague, 1950), p. 45.

31 Marlborough approved of their refusal of the garrison's terms. BL, Add. MSS, 61397 f. 169.

evacuation from one with conditions attached. With the exception of the aggrieved commander, few found it troubling.

Another scenario in which evacuation marks confused the distinction between free and imprisoned was the request to evacuate by the breach. In an age when garrisons pledged, were expected, and even ordered, to resist until they had repulsed at least one assault on a breach in the main walls, such an evacuation would serve as undeniable proof of following orders – a willingness to stand firm at the last ditch, to hold out "to the last extremity."[32] And yet, Flanders garrisons rarely requested this quintessential symbol of a stout defense. A handful in the war's early years did indeed demand, and were granted, this mark of honor.[33] But in the Low Countries, it was mostly *bicoques*, small weak Spanish places, whose resistance was brief and whose garrisons were often taken prisoner.[34] Yet these garrisons, unable to acquire an "honorable surrender," were, if the later author Guignard is to be believed, given the "most honorable" mark![35] More confusing, many later French garrisons who were widely acknowledged as having conducted good defenses and who were granted all their requested marks did not even *request* a breach evacuation, despite theoretical entreaties to inflate one's initial terms. None of the Flanders sieges from Lille 1708 on requested the term, a few times because, as with Lille's citadel, there was no breach. Surrendering before a breach theoretically violated Louis' public 1705 order to withstand at least one storm on the breach, and should have undermined the garrison's claims of a stout defense. Yet it was often ignored by French apologists and even the King himself.[36] Thus, a breach evacuation was in theory an honorable mark, but in reality it was more important for those garrisons needing some mark to offset their otherwise "dishonorable" fate. Marching out through the breach gave little indication, on its own, of whether the garrison had conducted a good defense, or whether it even acquired a free evacuation.

One might conclude that the order of particular evacuation marks was of little importance, since the marks were executed simultaneously. Yet a clear sequence did in fact develop late in the war. In 1706, Menin was the first garrison to request the full panoply of arms, drums, balls in mouth, matches lit, flags, with four cannon by the breach. In December 1708 Boufflers evacuated the citadel of Lille with an almost-as-comprehensive list: "arms and baggage and horses, drum beating, ball in mouth, match lit at both ends, and munitions

32 1706.04.25 *Daily Courant*; Veenendaal, Jr., ed., *Briefwisseling Heinsius*, 4:166.
33 Venlo, Stevensweert, Roermond.
34 For example, *Le Mercure galant*, 1702.10, p. 332, on Roermond and Stevensweert.
35 Guignard, *L'Ecole de mars*, 1:321.
36 SHD, GR A^1 1988 #24.

to fire twenty shots each, and flags flying...." Despite surrendering before a breach, Lille was the longest French defense of the war, lasting four months. Boufflers was an experienced *maréchal de France*, seconded by French subordinates, defending a French town. His troops were overwhelmingly French, and unlike previous French commanders defending Spanish fortresses, he did not have to share command with a Spanish governor. The conclusion of the siege received more coverage than previous surrenders, though Court memorialists noted every surrender. These facts may explain why Lille's citadel provided the model evacuation for French garrisons during the last four years of the war, with the exceptions of Saint-Venant and Bouchain (taken POW).[37] Thus we finally have a formalized sequence for garrisons requesting marks of honor expected by historians of Court rituals – in the eighth year of the last of Louis XIV's numerous wars. Even then, there were exceptions, and the wording of these marks was still individualized, though the general sequence of marks was consistent. Post-war treatises sought to record such details for future warriors, though they spent far more time on the marching order of an evacuation. Notably, these later authors did not repeat the sequence set by Lille in 1708. Even with their honor at stake, garrisons were as vague in their requests for specific evacuation marks as theorists.

2.3 *The Surrender as Viewed by Other Participants*

What little authorial intent we can infer from garrison demands only tells us a small part of the story, and only from the perspective of the petitioner. The signed capitulation proved that the defenders had relinquished possession and the document further specified the defenders' fate. Its format further clarified that its co-author, the besiegers, determined the final draft, regardless of how carefully demands were crafted. The power imbalance was particularly striking when attackers responded to long, carefully-crafted requests with an imperious "Refused."[38] Among these arbiters we find even less concern with performative evacuation marks, and the same obsessive focus on more practical matters.

Garrisons might have been haphazard with their requests for specific evacuation marks because, as a rule, besiegers never bothered to modify or reject such demands. Assuming the victors allowed a fully free evacuation, they accepted whichever combination of flags, drums, matches and the like requested,

37 This pattern, it should be noted, did not hold for the non-Flanders theaters, or Allied garrisons. Alicante's English garrison (1709) still asked for drums beating, matches lit, ball in mouth and "all the other marks given in such a case."

38 E.g. Béthune article 27.

in whatever order, in whichever article they appeared. Except for Oostende, which failed to receive a fully free evacuation, besiegers did not view the specific marks given to an "honorable" garrison as worth contesting. The Spanish veteran Santa Cruz gave attackers this same advice to grant a free garrison whatever marks it requested in his 1720s treatise.[39] Flanders garrisons taken prisoner or at discretion also requested these same marks, but were generally denied.

While besiegers cared little about the performative marks of honor, they carefully monitored more practical, physical, markers. Evacuation marks were intertwined with practical military concerns, morale foremost among them, but the victors ensured that the vanquished did not inflate symbolic marks into a practical military advantage. The number of 'trophy' cannon allowed an evacuating garrison was a precise, quantitative measure that was both symbolic and functional. Because these were expensive items useful for military operations, everything from newspaper accounts to playing cards used the number of captured cannon as an indication of the scale of a victory, whether battle or siege. Louis de Gaya's 1689 treatise, for example, did not volunteer a single performative evacuation mark in his twelve-page guide on how to surrender, but warned governors to be precise when requesting the number and caliber of cannon.[40] Besiegers, for their part, downgraded the requested cannon to no more than a handful, six cannon at most, and a few mortars.[41] In the case of Aire, the marquis of Goesbriand requested six cannon, four mortars, and fifty rounds each. The Allied besiegers responded to this fourth article by granting four cannon, two mortars, and twelve rounds, before adding in a postscript two more cannon "for the consideration of M. Goesbriand."[42] From both a tactical and symbolic perspective, larger calibers were better, witness Venlo desiring (in article 3) to evacuate with ten twenty-four pounders – the besiegers reduced it to two six-pounders, or twelve-pounders if no six-pounders were available. The number of rounds was a less contested, and less important, measure – besiegers usually accepted one to two dozen rounds without comment, though garrisons occasionally forgot to request precise numbers. More rounds might be requested but were usually reduced, though at Freiburg, the garrison was given fifty shots per gun instead of twenty-five "to give a mark of esteem to General Harrsch, the Baron Wachtendonck and their garrison."[43]

39 Santa Cruz, *Réflexions*, 9:230.

40 Gaya, *Science militaire*, p. 108. Also Feuquières, *Mémoires sur la guerre*, 1:406–408.

41 E.g. Veenendaal, Jr., ed., *Briefwisseling Heinsius*, 5:547 #1055.

42 This revision back up to six pieces was prompted by a letter from Goesbriand to Marlborough, requesting the additional two cannon as a "mark of distinction." BL, Add. MSS 61314, f. 178, 1710.11.01.

43 Vault and Pelet, eds., *Mémoires militaires*, 11:645.

A specified number of artillery pieces, often twelve, was the *only* consistent evacuation request made by garrison commanders. It was also the only evacuation term that the besiegers bothered modifying, usually to decrease, but on rare occasions these trophies were used as a marker for greater recognition.

Besiegers only modified or rejected practical requests: those that might let a garrison escape, or that hindered future operations, or that might cost them money. Besieging commanders insisted the defenders reveal any mines and their magazines, and moved forward the date of the evacuation when garrisons attempted to delay. They made sure to divert an evacuating garrison away from their next intended target: Venlo's defenders requested to march up the Maas to Roermond, but as this was the next Allied target, they were diverted westward to Antwerp (article 5). In contrast, the symbolic terms were of little consequence to the conqueror. For the besiegers, fortress possession and garrison fate, not evacuation marks, were their clearest markers of victory.

Nor did the garrison's side spend much time discussing marks of honor, though we have almost no detailed accounts of the surrender process from the garrison's perspective.[44] The orders Louis XIV sent to his beleaguered commanders emphasized only that the garrisons avoid capture, and delay as long as possible – no expectation of particular marks. The commanders' post-siege self-justificatory accounts were full of explanations for why the garrison had to surrender when it did, and apologized profusely if they were imprisoned, but did not bother justifying the evacuation marks they'd received, if they even specified them. They might even understate them: as Menin's tearful commander reported to the field commander the duc de Vendôme, his defense was short but he acquired the "ordinary honors of war" and, more importantly for future planning, would march with four cannon and two mortars to Douai within three days.[45] These sources avoided discussion of which symbolic marks were important, or why they requested one mark instead of another. Rather, the defenders sought to avoid dishonor by representing their unique mixture of marks as "usual" and "ordinary," regardless of how they compared with other capitulations. Overall, both defenders and attackers in the War of Spanish Succession paid surprisingly little sustained attention to the evacua-

44 For a detailed example, see Jean Martin de La Colonie, *Mémoires de Monsieur de La Colonie: maréchal de camp des armées de l'Electeur de Bavière*, ed. Anne-Marie Cocula (Paris, 1992), pp. 286–287, on Rain (Germany) in 1704. The instructions from Douai's commander to his negotiator similarly focused on getting a free evacuation and generic "honor:" 1710.06.26 SHD, GR A1 2215 #247 Albergotti to Valory.

45 SHD, GR A1 1939 #189. This despite the fact that he was the first Flanders defender to request and acquire practically all of the performative marks of honor, including evacuating by the breach.

tion performance, and ignored potential distinctions within the marks given to honorable garrisons.

2.4 *Audience Reception*

How did those beyond the siege interpret these distant performances? Audience reception is particularly important with honor claims because honor, the respect of one's peers, could not simply be appropriated by a garrison – there was too much competition and self-interest for claims of honor to be accepted at face value. External honor, as scholars have emphasized, has to be validated by others, by one's peers or 'honor group.'[46] With capitulations, however, the key questions are: which audience, and which markers? Performative marks of honor made a statement for those witnessing the evacuation. But the vacating garrison's ephemeral ceremonial behavior left few other traces. The capitulation texts therefore served as one important way for a garrison to justify its honor claims to a broader audience, by referring to the text for evidence of the terms it was granted. For those far from the siege site, however, commentary on the garrison's conduct drew upon a much broader range of indicators disseminated through multiple media. Word-of-mouth reports and rumors spread widely as troops marched from post to post; some of these verbal reports were written down in letters, recorded as court rumors (especially from the officer sent to report back to the King), and printed as raw reports from the front in the dozens of public papers. Yet unlike many other ritual processions, the details of the evacuation ceremonies themselves were ignored by practically all of these hundreds of written accounts, excepting an imprecise and often-inaccurate repetition of the terms specified in the surrender document.[47] Printed newsletters with space to spare might publish a transcription of the full capitulation document, but most summarized the terms in a cursory and idiosyncratic fashion.[48] Accounts throughout the war referred to the "usual" or

46 For discussions of early modern martial honor, see Brian Sandberg, *Warrior Pursuits: Noble Culture and Civil Conflict in Early Modern France*, (Baltimore, 2010); Roger B. Manning, *Swordsmen: The Martial Ethos in the Three Kingdoms* (Oxford, 2004); John Lynn, *Giant of the Grand Siècle* (Cambridge, 1997), pp. 251–254. More generally, see Faramerz Dabhoiwala, "The Construction of Honour, Reputation and Status in Late Seventeenth- and Early Eighteenth-Century England," *Transactions of the Royal Historical Society* 6 (1996), 201–213; and Hervé Drévillon, Diego Venturino, eds., *Penser et vivre l'honneur à l'époque moderne* (Rennes, 2011).

47 Contrast with the attention to detail in Théodore Godefroy, *Le cérémonial François*, 2 vols (Paris, 1649).

48 The main French, Dutch and English newspapers consulted include: *Gazette de France; Mercure galant; Clef du Cabinet; Lettres historiques; Esprit des Cours de l'Europe; Mercure historique et politique; Nouvelles extraordinaires; Gazette d'Amsterdam; Gazette de*

"ordinary" or "accustomed" honors, occasionally "all the marks of honor", with flags flying and drums beating receiving the most mention. An abbreviated "flags flying, &c." was used as well, obscuring details. On a few occasions observers off-handedly ignored a garrison's marks altogether. The *Daily Courant*'s detailed transcription of the thirty articles granted Venlo, faithfully translated from the *Oprechte Haerlemsche Courant*, deleted the evacuation marks (excepting cannon) the garrison had so carefully drafted.[49] More curt, and just as inaccurate, was a French paper's bored dismissal of the terms: "there is nothing in its capitulation that isn't in any of the others."[50] In this otherwise-laudatory account, the evacuation marks were completely ignored. Such marks could hardly be finely calibrated when no one bothered to accurately report them, nor when they were almost universally summarized as "usual" and "ordinary."

Visual sources provide even less evidence of the importance of marks of honor, or of the evacuation ceremony. Of the innumerable siege paintings, tapestries and engravings from the period, only a handful represent the surrender ceremony (*la reddition*). All of these graphic representations portray the event from the victor's perspective, those with the most to celebrate. Overwhelmingly they follow the visual syntax of ritual submission, notably the bowing presentation of the keys to the city iconically portrayed in Velázquez's *Las Lanzas*. A variation depicted a foregrounded equestrian conqueror greeting the head of a long column of the conquered receding back to the town gate in the distance. None provide enough detail to verify such minute marks – assuming they were even faithful to the real event – beyond undifferentiated soldiers with the occasional flag unfurled, shouldered guns and an absence of smoldering matches lit at both ends. In other words, a garrison's evacuation performance was witnessed by a small audience of besiegers and townspeople, almost none of whom bothered recording any more than the barest outlines of the ceremony.

For the precise details of these evacuation marks to persist beyond the performance, media coverage of the surrender was key: it was the primary way most people would be made aware of the ceremony and its terms. As we have just seen, however, garrisons had little control over how their words and deeds were represented to the rest of the world beyond the capitulation document. Whatever their requests and whatever their performance, they were usually flattened into a generic "usual honors of war" with a few cannon. Perhaps the

---

Rotterdam; The Daily Courant; The London Gazette; The Post Man; The Post Boy; The Supplement; The Present State of Europe; The Observator; The Review.

49 1702.09.24 *Daily Courant*, p. 1.

50 *Mercure galant*, septembre 1702, p. 448.

fact that the rest of the world paid so little attention to the details also explains why most garrisons appear almost indifferent when detailing such marks.

Many observers assumed that a garrison's honor was more accurately assessed by other indicators: measured not through fleeting performances, but through physical 'trophies,' through comparison with similar defenses and with expected conduct, and through testimonials. Unfortunately for garrisons, these other markers often undermined the symbolic honors performed in the evacuation ceremony. One such indicator was the number and condition of the evacuating troops – a small number of haggard troops illustrated the desperation of the garrison, while a large number implied that the defenders should have resisted for longer. Most prominently reported was the number of cannon allowed. But the allowance of a handful of cannon, sometimes explicitly associated with the garrison commander's honor, was easy to undermine with further context. Venlo's commander, for example, might brag that he received two cannon for his honor. Yet that was only a fifth of the ten guns he had requested. Worse, his failure to win even ten guns was further underscored by widespread newspaper reports that he left behind 188 cannon and 32 mortars.[51] Cannon, individually and collectively, had long been trophies, making such an imbalance hard to miss.[52]

The French besieging Ath in the previous war undermined the significance of cannon, and the stability of evacuation honors more generally, with a different argument. Initially, the Spanish defenders were reportedly embarrassed to evacuate through the breach without cannon and mortars. The French supposedly convinced them with an inversion of the traditional interpretation: being given an "honorable capitulation" was actually dishonorable because it meant you could have defended yourself further, whereas surrendering without arms and baggage was a "glorious testimony" of the extremity to which you defended the place.[53] It is unclear whether this anecdote, reported in a French account of the siege, was mocking the gullibility of the Spanish governor (earlier, a throw-away line made fun of the defenders for having so many capitulation models on which to base their own), or whether the Spanish were actually convinced of the argument. It is hard to avoid concluding that these besiegers

51 1702.09.25 *Daily Courant*. News of surrenders usually included an inventory of captured stores.

52 An observer noted that four pieces with the Arms of England (captured in battle thirteen years earlier) had been found in Menin, and would be returned to England. BL, Add. MSS 4742, f. 76.

53 Ernest Matthieu, ed., *Relation du siège de la ville d'Ath en 1697* (Mons, 1910), pp. 45–46. The source confusingly shifts from an honorable capitulation, to evacuating without artillery, to the glory of evacuating without arms and baggage (*sortie faitte le bâton blanc à la main*).

recognized the artificiality of this game, and cynically manipulated the term "honor" in order to end the siege and minimize the artillery leaving the town.

Evacuating by the breach had its own interpretive problems. As garrison commanders undoubtedly hoped, a few accounts explicitly described a breach evacuation as a mark of special recognition.[54] In practice, however, numerous detained garrisons, denied freedom and other marks of honor, were given a breach evacuation. An English summary of the desperate surrender of Huy's château illustrates the layers, or perhaps confusion, of contemporary use of "honor" regarding evacuation: "the Governour beat the Chamade, but could obtain no other Conditions than that the Garrison should have the Honour to march out at the Breach, with their Arms, and then lay them down, and surrender themselves Prisoner of War."[55] The artificiality of only allowing the defenders to evacuate through the breach with their arms and then lay them down and march off into captivity was hard to hide, particularly from the townspeople and enemies observing the ceremony, no matter how much closure it gave the garrison. The winning side could also undermine the significance of this specific marker in more subtle ways. On a few occasions, reports recounted how desperate garrisons frantically worked to make the breach slope passable the night before the evacuation.[56] Such garrisons, it was implied, had not really surrendered at the last possible moment, despite their departure through the breach: the true measure of good conduct, in other words, was the stage at surrender, rather than the evacuation terms granted. Few seem to have been impressed by these desperate attempts to acquire some kind of mark of honor.

Overall, outside observers paid surprisingly little heed to any of the specific evacuation marks. Consistently, they privileged different contextual details: who dictated the surrender, and at what stage of the attack. Garrisons obviously initiated the negotiation process by beating the chamade, but others considered the context of that decision a telling indicator of garrison honor. Did the defenders, with their backs against the wall, compel recognition of their honorable conduct, or had the attackers instead intimidated them into

54 Veenendaal, Jr., ed., *Briefwisseling Heinsius*, 4:166, regarding Albuquerque (Spain) in 1705. Note that the Allied besiegers intended this mark to be seen as a notable concession, in order to convince the Spanish of their good intentions after they had massacred a previous Bourbon garrison.

55 Francis Hare, *The Conduct of the Duke of Marlborough during the Present War* (London, 1712), p. 66. Indicative of the confusion, other accounts omitted the word Honour in their, otherwise verbatim, summaries.

56 1702.09.24 *Post Boy*. Spanish Succession besiegers rarely sought to embarrass them in that fashion, allowing them the opportunity to evacuate by a specified gate instead, as desired.

prematurely surrendering, or even bribed them to surrender early in exchange for 'honorable' terms? Public French accounts emphasized how the Allies had only taken the weak and undermanned place of Venlo with overwhelming firepower rather than courage. Even then, the attackers had been forced to "acquiesce" to the proposed terms, with the siege commander offering many honors and graces to various members of the garrison.[57] Allied accounts, beyond the reach of the Sun King's censure, cared less about such complimentary markers and purported testimonies of a job well done. They insisted instead on the interpretive precedence of possession over post-siege pleasantries. They further explained that the besiegers had forced the town to surrender in a fit of fright, a departure from the courage expected of true warriors. The besieging camp had fired off a celebratory *feu de joie* for the capture of Landau in Germany, and the Venlo garrison, fearing it signaled a storm and recalling how its Fort Saint-Michel had been taken by sword days earlier, frantically sought to have its chamade heard over the roar of the salvoes. Several Allied accounts mirthfully reported the besieged frantically beating their drums in order to be heard, of terrified burghers sprinting to the ramparts to wave their white cloths of surrender.[58] Venlo's embellished ceremonial marks, cursorily noted, could hardly make everyone forget how precipitously and prematurely the fortress had surrendered. Even some Frenchmen remained unconvinced of the garrison's desired-for association between performative marks and earned honor. The French artillery officer Saint-Hilaire recounted that Venlo's large, well-supplied garrison was given "all the honors of war that they demanded and which they assuredly did not merit."[59] An English sergeant, equally unimpressed with their "ordinary" defense, recorded in his journal that they marched out "with Baggage and small Arms only," not being granted "any other Conditions."[60] This Sergeant Millner was most likely mistaken, but, clearly, Venlo's evacuation honors had not definitively dictated how others viewed their conduct.[61]

57 1702.10 *Mercure Galant*, p. 317. Quote on p. 330.

58 Townspeople's white cloths: David Chandler, ed., *Military Memoirs of Marlborough's Campaigns, 1702–1712* (1968; repr. London, 1998), pp. 23–24.

59 Armand de Mormes de Saint-Hilaire, *Mémoires de Saint-Hilaire*, ed. L. Lecestre, 6 vols (Paris, 1903), 3:128.

60 J. A. Millner, *A Compendious Journal of All the Marches, Famous Battles, Sieges …* (1733; repr. East Sussex, 2005), pp. 29, 34–35. Less explicably, the English Sergeant Wilson incorrectly stated that Kaisersweert in 1702 surrendered as prisoners of war. David Chandler, ed., *Military Miscellany II: Manuscripts from Marlborough's Wars, the American War of Independence and the Boer War* (Gloucestershire, 2005), p. 34.

61 For another detailed case, see my discussion of media coverage of Stevensweert's surrender at <https://jostwald.wordpress.com/2012/07/27/i-read-the-news-today-oy-vay-pt-iii/> (accessed 23 October 2018).

Media coverage of Oostende illustrates how little an enemy might respect a garrison, even one that avoided prisoner status. As with most other sieges, accounts differed on the specifics of its 1706 evacuation, but consistently mentioned the lack of cannon and the six months of enforced idleness. The French, always searching for a silver lining, put a positive spin on the result by emphasizing a counterfactual comparison, without mentioning the denial of arms: the Allies should have taken Oostende at discretion, but its brave garrison was able to force a free evacuation nonetheless.[62] Others argued that acquiring eventual freedom was hardly impressive, for La Mothe's conduct still fell far short of expectations.[63] Allied narratives contrasted initial French boasts of a long, difficult siege with its brief, almost bloodless capture: eight days of open trenches and three of bombardment. Even more embarrassing than upended French grandstanding was the unflattering historical comparison drawn between La Mothe's paltry defense and the town's heroic three-year resistance a century earlier.[64] Reports of the first inconclusive parley, accompanied by the stated certainty that it would soon be forced to skulk back and submit to Allied terms, made clear who surrendered to whom. Foreign publications ridiculed the garrison's pretensions to honor. The most vituperative came from *The Observator*'s Country-man character. An entire issue in July 1706 mocked the numerous French defeats of that year, focusing most on Oostende:

> As soon as my Lord Overkirk began to fling his Bombs on one side, and the English Fleet did the like on the other side of the Town, the French and Spaniards began to Squeak like so many Rats and Wessels between two Fires. Ah, Master [i.e. the Observator character], 'tis a sad thing to be Roasted at that rate; and while a Body is turning upon the Spit to be Basted with huge ugly Bombs and stinking Carcasses [incendiary bombs]; 'tis enough to Fright any Body. I'll warrant it the poor Frenchmen Drip'd more T[urd] than Tallow; the heat of the Fire shrivel'd their poor thin hunger-starv'd Carcasses.
>
> But there is one thing I observ'd upon the Papers that seems very Chomical, I cou'dn't forbear Laughing at it: Master Mothe, the French Governor in the Town, when it was Surrender'd, excused the Bravery of his Men,

62 As reported at Court on 10 July. Louis François du Bouchet, marquis de Sourches, *Mémoires du marquis de Sourches sur le règne de Louis XIV*, ed. G. J. Cosnac, É. Pontal, 13 vols (Paris, 1882–1893), 10:118–119.

63 Dangeau's court journal recorded no mention of denied arms, only that the stipulation of no service for six months was "fort mauvaise." Philippe de Courcillon, marquis de Dangeau, *Journal du marquis de Dangeau*, 19 vols (Paris, 1854–1860), 11:151.

64 1706.07.13 *Gazette d'Amsterdam*, p. 5, among others.

> which he said Was quite lost in Defending a Ravelin; but he did not attribute it to a Natural Cause, but to Witchery and Devildom, and said All his Men were Bewitch'd. Aye, thought I, and so was thy Master Bewitch'd too, when he sent a Mothe [i.e. moth] to take Care of the Cloathing of such a Town as Ostend. The Notion of Witchery is a poor excuse for Cowardice, and being over-match'd in Bravery and Skill in Martial Affairs.[65]

Military officers in the siege camp likely resisted insulting their defeated foe to their face, but nonetheless expressed their disdain at the defense's conduct.

Bourbon defenders were forced to yield the Spanish town of Menin six weeks later. As our introductory epigraph suggests, the garrison was granted the right to evacuate through the breach with all the marks of honor, though the requested twelve cannon were reduced to four.[66] According to the evacuation marks, it had achieved all that surrendering defenders could hope. Some were willing to accept this judgment, comparing it favorably with Oostende. John Tutchin's Observator character parried Country-man's jab that Menin's governor and garrison had been bewitched like Oostende by referencing testimonials of his good conduct:

> that Governor [the comte de Caraman] is a Brave Man, made a good Defence, and tho' an Enemy, is worthy of Praise: He has been Faithful in his Trust, wanted neither Courage nor Conduct in his Post, and has had the Honour to Defend a Town against the best Troops in the World much longer than could be expected, considering what Vigorous Attacks were at first made upon that Place.[67]

Observator's comments illustrate yet another common rhetorical strategy, to highlight the superiority of your own side's glorious achievement by complimenting the enemy for honorably resisting such an unstoppable force.

Nevertheless, others emphasized Caraman's dishonorable conduct by comparing it with what should have been. The siege commander the Earl of Albemarle noted Caraman's poor defense of a strong place, making one wonder

65 1706.07.10 *Observator*.

66 The 1706.08.27 *Post Boy* contrasted the four cannon and two mortars the French retained with what the Allies gained: "118 Pieces of Brass Cannon, of several sizes; 59 Mortars; 4668 Muskets; 93500 weight of Gun-powder; 8753 Cannon-Balls; 11517 Hand-Granadoes; several Tun of Musket-shot; great quantities of Match; abundance of Swords; 1251 Sacks of meal, and other Provisions in proportion."

67 1706.08.24 *Observator*.

what exactly all those marks of honor he granted really signified.[68] The French gunner Saint-Hilaire, perhaps smarting at the loss of so many of his guns and munitions, concluded that while Menin had a good reputation, it was poorly defended and should have resisted longer.[69] France's chief engineer the *maréchal* Vauban considered the town one of the best fortresses in the theater. He was, therefore, quite annoyed when Caraman justified his short defense by complaining of the quality of its works. He rued that Caraman would not have been in a position to lose the town so quickly if they had had a protocol in place to weed out incompetent commanders.[70] French technicians challenged Caraman's technical conduct of the defense, dismissing the garrison's performative marks in order to defend the conduct of their own branches.

Public commentary on the best defense of the theater, Lille, also shows how begrudgingly the enemy offered recognition to a foe, and illustrates the markers they believed indicated true honor. The French were extremely pleased with Boufflers' four-month defense – he and his subordinates were toasted at Court, promoted, and well-compensated. The French version of events again inverted the normal dynamic, portraying the besieger's veritable surrender to the defender's demands. The *Clef du Cabinet* newsletter reported that Austrian commander Prince Eugene of Savoy was so impressed with the "glorious" defense of the town that he allowed Boufflers to dictate the terms, and granted all the advantageous conditions that "honor and duty would allow."[71] After the citadel finally fell, French papers enthusiastically declared Boufflers the "admiration of all Europe." They also absolved him of dereliction of duty by noting that he only followed orders to surrender before a breach in order to save the garrison.[72] The account circulating around Versailles, from one of Boufflers' subordinates, sought to salve French wounds with testimonials provided by the enemy, rather than evacuating marks:

> After the capitulation had been signed, and the day before the garrison evacuated, Prince Eugene asked if he might dine with Boufflers, to which the marshal consented. The visit passed with great honors [*honnêtetés*]

68 Veenendaal, Jr., ed., *Briefwisseling Heinsius*, 5:467 #895. See as well the verse and illustrations mocking both Vendôme and Caraman in *Byvoegsel tot de Koninglyke Almanach, genaamd het Gulde-Jaar der Bondgenooten, 1706* [...] (Brussels: 1706), p. 19.

69 Saint-Hilaire, *Mémoires*, 4:352.

70 Sébastien Le Prestre de Vauban, *Oeuvres militaires*, 3 vols (Paris, 1795), 2:294–296.

71 1708.12 *Clef du cabinet*, p. 454.

72 1708.12 *Mercure galant*, p. 337 admiration quote. Also 1708.12 *Clef du cabinet*, p. 420. The official *Gazette de France* was curiously silent, devoting only a single postscript sentence on the "honorable" capitulation it won. 1708.12.15 issue, p. 600.

> exchanged. Prince Eugene told him that while he had personally acquired great glory capturing Lille, the marshal had gained even more glory defending it. He begged the marshal to dine with him the next day at his own tent after the garrison had evacuated, and he rendered him all sorts of honors.[73]

For elite French society where social interactions were moderated by honorifics and courtesy, such flattering speech acts from the enemy were far more meaningful than positions of muskets. The duc de Saint-Simon, famous for his attention to gradations of Court etiquette and points of honor, focused exclusively, and at length, on the compliments Boufflers personally received.[74] How Eugene viewed such professional courtesies offered in such situations is unclear. Other Allied officers expressed their indifference about such complimentary markers, witness Dutch general Wassenaar-Obdam's *pro forma* report, during the evacuation of Kaisersweert, that he "made [the governor] a compliment appropriate to these sorts of occasions."[75]

France's enemies relied on very different evidence when judging this "glorious" defense. At most, they considered the defense honorable rather than glorious and used their recognition of the enemy as evidence of their own gentility and ability. Allied frustration with the unusual length and difficulty of this siege admittedly tempered their joy. *The Observator* again highlighted courteous English recognition of a worthy foe, concluding that Lille surrendered "upon Terms, which, all Circumstances consider'd, seem fair enough on both Sides; [...]. [Boufflers] has been treated honourably, as his brave Defence seems to merit; for we must praise Vertue, even in an Enemy."[76] This small dollop of grudging respect for his honorable conduct, however, was as generous as the English press would get.

Other accounts questioned whether the garrison had truly done its duty by criticizing its failure to, once again, meet expectations. Some pointed out that the citadel, the mightiest in Europe, should have extracted much more time and blood, particularly given Boufflers' reported boasts to defend the place to

---

73 Dangeau, *Journal*, 12:282–283.

74 Louis de Rouvroy, duc de Saint-Simon, *Mémoires de Saint-Simon*, ed. A. de Boislisle, 41 vols (Paris, 1879), 16:479–481. Generally, Emmanuel Le Roy Ladurie and Jean-François Fitou, *Saint-Simon and the Court of Louis XIV*, trans. A. Goldhammer (University of Chicago Press, 2001).

75 Veenendaal, Jr., ed., *Briefwisseling Heinsius*, 1:297 #524.

76 1708.12.04 *Observator*, p. 2.

the last extremity.[77] For the Allies "constrain'd [him] to yield up the [citadel], without the firing of one Cannon-shot, or Storming; only by Sapping."[78] Referencing one's "surprise" that a garrison beat the chamade as early as it did was another way to undercut a garrison's claim that it dictated the final terms, and that it fought as long, or as hard, as it could. English songs and poems handled Boufflers only a bit more gently than Louis' other marshals, one noting that the governor had "bravely" withstood attack, for "a while" at least.[79] Another poem returned to historical comparisons to minimize his defense:

> But Anna [Queen Anne] conquers, and her Chiefs survive,
> Low are Lisle's Walls, and Churchill's still alive.
> Ten rolling Years the Trojan Bullwarks won,
> But now, Three Months a greater Act have done.
> To Silver Saints, in vain Great Bourbon cry'd,
> On haught Boufflers he in vain rely'd.[80]

It was just as easy to drown Boufflers' honorable conduct in the glory won by the conquerors.[81] As *The Post Man*'s summary illustrates, glory, the natural reward of conquest, was enhanced under difficult conditions against a worthy foe:

> If the Difficulties of an Enterprize increase the Glory thereof, the taking of Lille must be one of the most Glorious Performances there ever was. The Place was in it self as strong as Art could make it, defended by a numerous Garrison, under the Command of a Mareschal of France, and other experienc'd Generals' provided with all manner of necessaries, and incouraged to a vigorous defence by the approach of a powerful [relief] Army, which was as numerous, if not stronger than the Forces of the Allies; and yet the la[t]ter, without having any Communication with their own Territories, […] found means to subsist about it, till the Reduction of the Cittadel[…]. The French thought this beyond the Power of Man, and

77 *The History of the Campaign in Flanders, in the Year 1708* (London, 1709), p. 23. For other critiques, see Maurice Sautai, *Le siège de la ville et de la citadelle de Lille* (Lille, 1899), p. 270.

78 1708.12.08 *Daily Courant*, p. 1. Deane emphasized that the citadel only resisted for as long as it did because the Allies had run out of munitions. *Journal*, p. 74.

79 *An Epistle from the Duke of Burgundy to the French King* (London, 1709), p. 3.

80 "Upon the Siege and Surrender of Lisle" in *Poetical Reflexions Moral, Comical, Satyrical* (London, 1708), p. 6.

81 Eugene's reported admission that Boufflers had gained more honor was not repeated in English accounts.

> therefore boasted, that without striking a Blow, they would oblige the Allies to abandon the Siege. And indeed the Difficulties the Allies had to struggle with, were so many and so great in themselves, that the Confidence of the Enemy did not appear altogether unreasonable; but what would have been impossible to the Duke of Vandome and other French Generals, was happily accomplished by Prince Eugene of Savoy, and the Duke of Marlborough, two Heroes who seem design'd by Providence for humbling the Pride of France, and reducing it within its ancient Bounds.[82]

Multiple markers of honor and shame, glory and disgrace, were trotted out at the end of every difficult siege: the garrison's subpar performance; the besiegers exceeding expectations; the superior conduct of the attacker vis-à-vis an experienced and well-prepared enemy; the emptiness of enemy boasts; the impotence of relief attempts; and the historical scale of the victory. Boufflers may have retained several cannon and mortars, but the Allies could point out that the "key to France" was now in their hands, along with all its resources. Reports from Gallic centers further informed Anglophone readers that they were in great consternation about its fall, regardless of their public boasts.[83] In this telling, the entire siege performance, not just the garrison's small evacuation scene, proved the siege a glorious Allied success and an embarrassing French loss. Claims of glory, in other words, had to be justified with reasoning and evidence, and contemporaries chose markers that redounded to their own credit.

Successive French garrisons would copy Boufflers' evacuation terms, and were given similar courtesies, but observers once again highlighted the gulf between such marks and the conduct of the defense measured by other criteria. La Mothe's defense of Ghent at the end of 1708 was as underwhelming as his conduct at Oostende. His massive garrison of 12,000 men resisted for less than a week of open trenches. As usual, the French papers put on a good face, mentioning the garrison's free evacuation and some (but not all) of the marks of honor it was granted.[84] Underneath this public façade, however, was French outrage at La Mothe's conduct. The provincial intendant of inland Flanders expressed his shock at the brevity of the defense, while his peer in charge of the maritime province described La Mothe's chaotic and timorous governance inside the town.[85] Saint-Simon excoriated La Mothe for his stupidity and rejected his 'honorable' surrender, noting that the Allies themselves were mocking

---

82 *Post Man*, 1709.01.01. Also: 1708.12.08 *Daily Courant*; 1708.12.10 *Supplement*, p. 1.

83 1708.12.08 *Daily Courant*, for example.

84 1709.01 *Lettres historiques contenant ce qui se passe de plus important en Europe*, p. 108.

85 1709.01.01 SHD, GR A¹ 2154 #1; Joseph E. Nève, *Gand sous l'occupation de Louis XIV, 1678–1679, 1701–1706, 1708* (Ghent, 1929), p. 330.

him – he weighed enemy testimony more heavily than performative marks.[86] France's grand master of artillery expressed his indignation to Boufflers, the exemplary defender of Lille:

> the bizarre adventure of Ghent will augment even further your efforts and your glory […]; it must seem very bizarre to you to see a place surrender without facing either cannonfire or *coup de main*, and to see the recent admirable example that you gave being so poorly followed. Your heart, being what it is, must undoubtedly be in a violent state over all this; knowing me like you do, you can also guess that I suffer as well.[87]

Evacuation marks, even the garrison's fate, were once again undermined with markers of incompetence, particularly when compared with other defenses. Nor did such counterfeit marks fool the King; a free evacuation did not protect La Mothe from being exiled from Court for his uninspired defense.[88]

For his part, Marlborough's opinion of Ghent's defense radically diverged from the public honors he granted: "I believe Monsieur de la Mott will not be able to give a good reason for what he has done."[89] Elsewhere he highlighted the garrison's shortcomings by expressing relief that the town surrendered so quickly, for "they might have given us much more trouble."[90] Allied public accounts undermined its honor by highlighting the disappointing defense and carelessly described the "usual" marks of honor.[91] One paper impugned the commander's honor when it reproduced a Parisian letter stating that La Mothe was "frighten'd" into surrendering.[92] The *Post Man* reported on the "surprising" ease with which the town surrendered, and reproduced what was purported to be an intercepted letter wherein French Secretary of State of War Daniel Voysin warned the garrison commander to "dispute the ground Inch by Inch." Voysin's letter further emphasized that La Mothe's honor depended on a stout defense, given his lackluster resistance at Oostende and his battlefield defeat

86 Saint-Simon, *Mémoires*, 17:2–5.

87 Saint-Simon, *Mémoires*, 18:494. Hervé Drévillon, *L'impôt du sang: Le métier des armes sous Louis XIV* (Paris, 2006), p. 330.

88 Saint-Simon, *Mémoires*, 17:3–5.

89 *Marlborough-Godolphin Correspondence*, 3:1188 #1200.

90 John Churchill, Duke of Marlborough, *The Letters and Dispatches of John Churchill, First Duke of Marlborough from 1702 to 1712*, ed. G. Murray, 5 vols (London, 1845), 4:389. For surprise at the surrender, see Landsberg, *Nouvelle manière de fortifier les places*, p. 39.

91 E.g. 1708.12.30 *Post Man*, p. 2

92 1709.01.03 *English Post*, p. 1. Deane described the townspeople as "running up and down like men distracted." Deane, *Journal*, p. 75.

at Wijnendaal earlier that year.[93] Another courant bluntly stated that the Allies were generous to give Ghent "more favourable terms… than they deserved."[94] Later accounts reminded readers of La Mothe's continued travails – *The Supplement* reported on the factional warfare waged at court between La Mothe's circle and his detractors over his early surrender.[95] All this criticism was leveled against a garrison that had received an "honorable surrender" with all possible evacuation marks of honor.

Tournai's surrender the next year illustrates how irrelevant a garrison's honor claims could be – or how controversial, depending on the audience. The French abandoned the strongly-fortified town for the citadel after a month of attack, and were forced to subsequently surrender that a month later, due to a lack of supplies. The observation army commander Eugene flattered the marquis de Hautfort-Surville that further, futile, resistance could not increase the "estime" he had already earned with such a "longue et belle" defense.[96] An English private was less charitable in noting the defenders' collective loss of nerve, who "would stand noe longer to be knocked in the head, but threw doun there arms and would act noe more."[97] Cognizant of the garrison's dire straits, the besiegers insisted on prisoner status "for the Honour of the Arms of the Allies." In addition to the greater glory gained by forcing an enemy to yield without reserve, reports circulated that Eugene insisted the garrison be punished for French outrages in Italy, and for the French king's refusal to accept a negotiated surrender earlier in the siege.[98] The resulting capitulation document explicitly refused Surville's request to evacuate with arms and flags flying (response to article 5). Yet when the garrison evacuated, one novice spectator noted that "Instead of their being prisoners of war, as reported, [they] march'd out with drumes beating, colours flying, trumpets sounding, the governour lolling very stately in his coach received by our officers with all imaginable tokens of honour."[99] According to the garrison's second-in-command, Surville received a verbal promise on the citadel esplanade that his troops would in fact be

93 1708.12.28 *Post Man*.

94 1709.01.08 *Post Man*.

95 1709.02.11 *The Supplement*.

96 Undated letter in Joseph Rechberger Ritter von Rechcron, ed., *Spanischer Successions-Krieg. Feldzug 1709* (Vienna, 1886), 2:300 – volume 11 of *Feldzüge des Prinzen Eugen von Savoyen*, ed. Abtheilung für Kriegsgeschichte des k. k. Kriegs-Archivs.

97 Deane, *Journal*, p. 88.

98 Quote in 1709.08 *Present State of Europe*, p. 314.

99 1709 letter of surgeon Thomas Armstrong in University of Michigan Libraries, Special Collections, D281.A2 A74. A report from the 1709.09.06 *Post Boy*, p. 1 reported that the garrison was granted the "Honour" of drums and colors for making "so brave a Defence," but were dispossessed of their flags and muskets upon exiting the town.

allowed evacuation honors.[100] One in Marlborough's intimate circle allows us rare insight into the proceedings:

> I can at last congratulate you upon the happy conclusion of our siege, sooner than I expected, and with more honour than I could have hoped for. The enemy offered to capitulate on Saturday morning before a breach was made, but in revenge for their flying off from their capitulation before [i.e. Louis had rejected an earlier proposed capitulation], and knowing the want they were in, they were told they must expect no other terms but prisoners of war. Since that the matter has again been in agitation, and was last night as it were concluded, that *they shall not be treated as prisoners, but accounted for as such*, and not serve again until exchanged. They are very unwilling to the last article of leaving behind them their arms and colours &c. but will I believe be forced to yield to it. His Grace is just going to Tournay to conclude this affair [...]. [P.S.] Tournay, 3 afternoon. The articles are signed by which the garrison are to leave their arms, but in compliment to Monsieur Surville they will, I believe, be given them.[101]

Thus, Surville could argue that he forced recognition by receiving all the marks of honor for his defense. The besieger's honor was just as implicated in the end result. Marlborough defended his last-minute leniency by reiterating that despite being immediately exchanged, the garrison was still "*in effect* [...] prisoners of war."[102] Such graces were within the rights of the magnanimous conqueror, though much of the ensuing discussion over Tournai would revolve around how to characterize the garrison's honor.

So, what did others make of giving "all the honors of war" to prisoners? As we should expect, it depended on their point of view. Of the dozens of descriptions of the citadel's surrender, most Allied accounts failed to note controversy of any kind, happy that they had taken its garrison prisoner and acquired one of the strongest citadels in Flanders a month earlier than expected. Their numerous descriptions disagreed over whether the garrison had acquired "honor," and if so, how that honor was marked. Some accounts merely described the "complaisances and honnêtetés" given to the French generals without

100 1709.09.05 SHD, GR A[1] 2160 #84bis.

101 Historical Manuscripts Commission, ed., *The Manuscripts of the Earl of Buckinghamshire* [...] (London, 1895), p. 228. My emphasis.

102 Murray, ed., *Letters and Dispatches*, 4:588. My emphasis. A camp journal noted that the garrison surrendered as prisoners "but desired some marks of honor." BL, Add. MSS 61404, f. 118, 1709.09.03.

mentioning the capitulation's confusing negotiations, while none expressed interest in evacuation honors.[103] Despite the honorable judgment cited in the *Post Boy* (footnote 99), within a few weeks London papers spoke only of an evacuation with swords and baggage (no mention of arms). *The Tatler* also ignored the evacuation marks, noting that the particular honor that the garrison had gained was their immediate exchange, "a seeming condition granted to the besieged above that of being prisoners of war."[104] Narratives of the campaign published in England the next year also failed to mention any controversy over the "usual" evacuation of swords and baggage "according to the capitulation."[105] This collective apathy over the evacuation honors due the garrison was fed by contradictory reports of the capitulation's contents: some circulating versions of the surrender included a postscript noting the granting of arms, while others did not. Yet despite such broad disagreements, no one bothered to clarify the confusion, and the marks of honor faded from English memory.

No surprise that French observers were more conflicted about the 'premature' surrender of one of their strongest fortresses. Once again, the official French version was indifferent to the capitulation process, grossly misrepresenting these unusual proceedings as a surrender in the "ordinary style," with the capitulation containing the "accustomed articles of all capitulations."[106] One Versailles journalist praised Surville for negotiating "one of the most singular" capitulations, acquiring "all the Honors of war" and reconciled this with their fate by describing them as "not quite prisoners."[107] Other French observers, however, once again focused on the garrison's ultimate fate, and the implications for personal, professional, corporate, and national honor if such a mediocre defense were deemed honorable. Saint-Hilaire referred to the capitulation as the strangest he had ever seen, where prisoners were treated like a free garrison.[108] The relief commander Villars, who lost face when the town fell 'under his nose,' was livid at the "weak and imbecilic" defense, and complained

103 Deane, *Journal*, p. 88; Millner, *Compendious Journal*, p. 270. Quote in the later compilation of Lamberty, *Mémoires*, 5:359.

104 1709.09.01 *Tatler* #62.

105 *Briefwisseling Heinsius*, 9:234 #508; David Jones, *A Compleat History of Europe, 1709* (London, 1710), pp. 235–236; Abel Boyer, *History of the Reign of Queen Anne Year Eight* (London, 1710), p. 46. Later mid-century English histories were just as likely to report an evacuation with only swords and baggage, e.g. John Bancks, *History of John, Duke of Marlborough* (London, 1741), p. 303.

106 1709.09 *Mercure galant*, p. 224. One narrative described the capitulation as "honorable." Saint-Simon, *Mémoires*, 18:485.

107 Sourches, *Mémoires*, 12:54.

108 Saint-Hilaire, *Mémoires*, 5:180. He primarily focused on the details of the exchange process, referring in passing to their being allowed swords and baggage.

of Surville's Court allies rationalizing his incompetence.[109] Saint-Simon again noted the defeated commanders dining with their conquerors, but focused most on the dispute between Villars and Surville.[110] One cantankerous French veteran was alone in explicitly calling out the manipulations of evacuation marks witnessed at Ghent and Tournai, where poor conduct was papered over by an appeal to honorifics: "I will never approve of the conduct of governors who so manage a capitulation as to acquire what are falsely called marks of honor.… I hold these marks of honor to be, in truth, marks of shame."[111] In his view, the marks Surville acquired from the enemy were trumped by more meaningful measures of conduct, but many others used different markers to make the same point. All these conflicting judgments highlight how confusing Tournai's status was: many ignored the evacuation marks, others misreported them, a few praised them, a few excoriated them, and some quickly forgot the entire controversy while others debated Surville's conduct for decades.[112] The confusing case of Tournai reinforces the lessons drawn thus far: first, that there was an utter lack of consensus over which criteria to judge a garrison by, with personal honor and factional disputes shaping many responses; second, that a large number of contemporaries didn't care about evacuation details; and, third, that most observers ultimately judged a garrison commander's honor based on his fate and conduct.

The next five French garrisons were all granted honorable surrenders, requesting and receiving the same marks (and fate) that Lille had received – the confusion of Tournai likely encouraged future garrisons to be consistent in their demands for evacuation honors. Mons (1709) as well as Douai, Aire and Béthune (1710) all put up competent defenses, though none resisted even nine weeks.[113] The weak place of Saint-Venant gained its favorable terms since the

109 The debate raged in SHD, GR A[1] 2152 #126–#155. Despite such partisan support, Surville's military service and Court presence ended soon after Tournai's surrender.

110 Boislisle, ed., *Mémoires de Saint-Simon*, 18:146–149.

111 Feuquières, *Mémoires*, p. 444. Recall Feuquières' disinterest in capitulations elsewhere: creating a capitulation required "an infinity of considerations too long to enumerate" (406).

112 With the publication of Feuquières' memoirs in 1735, Surville's conduct would continue to be debated, though evacuation marks played almost no role. Pinard, *Chronologie historique-militaire*, 8 vols (Paris, 1761), 4:496 referenced Henri Griffet's defense of Surville against Feuquières' 'baseless' accusations in Père Daniel, *Histoire de France*, nouvelle ed. (Amsterdam, 1758), 22:150–164. Griffet focused on Surville's conduct, mentioning briefly that the garrison was allowed its baggage and swords, yet was disarmed of its muskets and colors outside the town (158). The publication of the *Vie du maréchal duc de Villars* in 1784 renewed the controversy (2:76–81).

113 French accounts claimed that Mons' defenders received their requested cannon (article 5), but only Allied accounts reported that this request was denied in the capitulation. See Vault and Pelet, eds., *Mémoires militaires*, 9:395.

Allies chose to besiege both it and Aire at the same time. Villars praised Lieutenant-General Albergotti's "heroic valor" defending Douai beyond all expectations, contrasting it with Surville's underwhelming defense.[114] Newspapers continued to disagree, but not dispute, whether these garrisons had won "the usual" or "all" the marks of honor. Indicative of the insignificance of the evacuation performance to broader audiences, accounts equated "usual" with "all", as an English military dictionary illustrates: "The *ordinary and most honourable* Conditions are to march out at the Breach, with Arms and Baggage, Drums beating, Colours flying, Match lighted at both ends, Ball in Mouth."[115] But such marks were only a sop to a defeated foe. Marlborough reportedly consoled Goesbriand by telling him that "such a brave officer deserved a better fate [than losing Aire]. He added that he didn't doubt that his Master the King would agree, and that his good defense would gain him esteem."[116]

Regardless of evacuation performance and recorded compliments, outsiders contested garrison honor claims by pointing to more reliable markers of shame. Goesbriand's honor, and that of his compatriots, was besmirched by our English sergeant who described all of these towns as being "terrified" into surrendering.[117] The *Present State of Europe* belittled French boasts of the "long Defence of their Places," since they held out "only owing to their Situation in Bogs and Morasses, which must be drain'd before the Allies can come at them. The French used formerly to defend their Places at another Rate."[118] In addition to emphasizing French failure to live up to past glories, English papers also pointed out that terror led both Douai and Aire to surrender not just their beleaguered towns, but to also give up neighboring forts the Allies had not even attacked yet (Douai's Fort de l'Escarpe and Aire's Fort Saint-François). An over-eagerness to surrender reinforced accusations of cowardice and reversed the message sent via honorable evacuation.

A garrison commander's claims to honorable surrender might even be undermined by subordinates. Inside Béthune, a French *maréchal-de-camp* on the walls considered his superior's decision to surrender as dishonorably premature, and actually refused to beat the chamade at his post.[119] This tension

114 SHD, GR A[1] 2215 #222, #233.

115 *A Dictionary in 3 Parts* (London, 1705), "Capitulation" entry. My emphasis. This is another instance of sources describing contemporary practice impressionistically, rather than repeat the granted evacuation marks.

116 Testimony repeated in Lamberty, *Mémoires*, 6:135.

117 Millner, *Compendious Journal*, pp. 298, 303, 308, 312.

118 *Present State of Europe*, 1710.08, p. 311.

119 SHD, AG Article 15, Section 2 #4bis, Des Forges, *Journal du siège de Béthune*, f. 66. Treatises warned governors to gain the signatures of all the general officers when discussing the decision to capitulate; it was more honorable if the governor/commander was portrayed as being convinced to surrender against his wishes.

between individuals' honor then propagated beyond the French chain-of-command: in response, the Allied approach commander outside the town became incensed at the 'insult' that *his* approach was not included in the surrender process, and redoubled his fire against the town until his superiors included him in the negotiation. Here, a subordinate officer's sense of personal honor trumped his duty to follow his superior's orders – clearly they differed in their interpretation of when one should surrender. In turn, the approach commander's honor was brought into question by an enemy's act of insubordination, requiring his own vigorous response. Honor was invested in innumerable aspects of the siege; evacuation marks played a minimal role in these negotiations over honor.

By 1711, the faith between belligerents completely collapsed, and evacuation marks became moot as successful besiegers divorced a garrison's fate from its conduct once again. One of the largest siege-related controversies of the entire war erupted over the Bouchain garrison's status. After 23 days of trenchwork, the French defenders delivered up one of their gates, claiming that the attack commander had verbally promised a free evacuation. Once Allied troops had secured the gate, however, Marlborough declared the French were prisoners and immediately marched them to Holland. The two sides exchanged a flurry of correspondence and written testimony back and forth, with each side providing testimony (sworn on their word of honor) attesting contradictory recollections of events.[120] Faced with a *fait accompli*, Bouchain's defenders were forced to accept "the law of the victor," though the relief commander Villars considered its defense the most vigorous he had seen.[121] Once more, Louis complained of an enemy refusing to acknowledge his garrison's "honorable" defense with a free evacuation. English sources were more interested in promoting Marlborough's masterful conduct in overcoming the marshy terrain and turning back Villars' numerically-superior relief army.[122] Yet again, a garrison's fate failed to reflect its conduct.[123]

The next year, the Allies, with the English unilaterally negotiating an end to the war, refused Le Quesnoy's garrison the same terms as Tournai's citadel, and made them prisoners via a single-column capitulation. The French commander, upon his parole, was imprisoned in the Bastille for his disappointing

120 Widely reproduced in the press, e.g. *Present State of Europe*, 1711.11, pp. 420–428.

121 Quote in Veenendaal, Jr., ed., *Briefwisseling Heinsius*, 9:215 #443.

122 1711.09.17 *British Mercury*; 1711.09.19 HMC, *Buckinghamshire MSS*, p. 233.

123 In 1703, Louis had complained of the same injustice (imprisonment) imposed on French defenders whose conduct merited a better fate. The Bourbons did similarly in Spain, while the Allies complained of the Bourbons denying their Italian garrisons the free evacuations they had "earned" in 1705–1706.

defense.[124] Three weeks after Le Quesnoy fell, Villars surprised a Dutch entrenched camp at Denain; the French finally returned to the offensive. Villars captured four fortresses in quick succession and reciprocated. All of these Allied garrisons were taken prisoner with dictated terms, though Louis regretted that reciprocity mandated that enemy officers "who have done their duty must be forced to surrender as prisoners."[125] The year 1712 would end the fighting in Flanders, and provide several more examples where reciprocity and military expediency divorced garrisons' fates from their conduct. This disconnect between fate and conduct was far more troubling to contemporaries, and far more often remarked upon, than whether a garrison was granted specific evacuation marks it had not "earned."

The above cases illustrate that the surrender ceremony might have been meaningful in the moment, but the collective memory of a garrison's defense quickly forgot such evacuation marks and moved on to more meaningful measures of its honor and conduct. Observers assessed the defender's conduct for themselves by choosing from a dozen different indicators, rather than simply assume that requested (or granted) 'marks of honor' would objectively mirror its fate (much less its conduct). Nor did the praise of besiegers willing to bribe a garrison into surrendering end the debate. Few could agree on how much merit any given garrison deserved. Contemporaries praised well-conducted defenses regardless of the marks or fate given the garrison, and mocked and excoriated those who claimed (and were given) "honors" their conduct did not merit. Such judgments were inevitably made by parties seeking to defend their own status and reputation: whether or not an observer accepted a garrison's claims to honor depended on their own interests. Public accounts defended the conduct and honors of their own side and denigrated that of the enemy; privately, some held reservations about the comparative conduct of their own defenders as well.

3 Conclusion

The abstract concept of honor, the respect of one's peers and superiors, was highly prized, competitive, and difficult to define. While Louisquatorzian France may have developed a sophisticated, yet contested, system of courtesies

124 Joseph Sevin, chevalier de Quincy, *Mémoires du chevalier de Quincy*, ed. Léon Lecestre, 3 vols (Paris, 1898), 3:127–128. The commander's subordinates, themselves tarnished with shame, bewailed their unfair fate and defended their superior's conduct.

125 SHD, GR A[1] 2383 #195.

and *honnêtetés* to regulate social interactions within France, the Sun King could not enforce these practices on outsiders. After all, his enemies had been viciously mocking his royal person for decades.[126] Surrendering garrisons faced particular challenges making honor claims. They were unlikely to gain true glory in defeat – the loss of the town was the one indisputable 'fact' of a siege, and, as Vauban declared, *gloire* had to be based on "real and solid actions."[127] Garrisons could not deny their defeat, but they could at least claim some honor for themselves, assuming the honor won in a siege was not zero-sum. A minority of garrisons could point to a lesser but still significant indication of honor: a free surrender that allowed them to resume military service and evacuate with the marks of honor of so much interest to scholars. But historians, by privileging the garrison's capitulation requests and by focusing almost exclusively on their evacuation performance, have let the losers write the history. This chapter reads against the grain of these textual honor claims, emphasizing the conflicts between the multiple audiences assessing the garrison's honor. The ceremony spectators, the townspeople and the besieging troops were only the first of many groups that garrisons would have to convince of their honorable conduct. This broader perspective returns honor back to where most contemporaries believed it lay, with the conquerors, and with those defenders who could prove that they had truly *conducted* an honorable defense, not just claim one through capitulation language and ceremonial performance.

Evacuation marks were one very small piece of the contemporary discourse on honor, one minor example of potential garrison honor. Such ceremonial details were theoretically derived from the garrison's status, but since the reality of evacuation honors rarely matched the theory (nor did theorists agree with each other), such evacuation marks could only be coarse, rather than fine, distinctions. Contemporaries theoretically wanted a garrison's conduct to be indicated by its fate, and its fate to then be mirrored in the evacuation terms; they expressed frustration the many times when their own garrisons were denied a fate they had 'earned,' and Louis XIV even fretted when his own commanders denied a worthy enemy honorable surrender. As we have seen, however, this was almost always a question of its fate, whether a well-conducted defense acquired a garrison free or POW status. Most contemporaries could not be bothered to report the evacuation marks with any precision, perhaps because they were quite aware that these performative marks were easily counterfeited by self-interested defenders, and by opportunistic attackers will-

126 Tony Claydon, Charles-Edouard Levillain, eds., *Louis XIV Outside In* (London, New York, 2015).

127 E. A. A. de Rochas d'Aiglun, ed., *Vauban: Sa famille et ses écrits*, 2 vols (Paris, 1910), 1:627.

ing both to abbreviate sieges with a few token compliments and to represent themselves as magnanimous conquerors of a worthy foe.

Because of these interpretive problems, contemporaries, especially those beyond the siege, turned to testimonies and comparisons to assess whether the garrison had done its duty or not, the minimum threshold for honor in defeat. The most common markers of honorable conduct came from independent testimony that the defenders had conducted a technically sound defense, had bravely defended themselves until the last extremity (a practicable breach), had chosen to surrender in a state of firm resolve rather than panic, had forced the enemy to accept their terms, had fulfilled the expectations based off their (often desperate) situation, had marched out in a condition which indicated continued defiance but not desperation, and that they were treated honorably by their conqueror. The degree of honor won was based on comparison with other defenses – recent, historical and counterfactual – particularly regarding the expected duration of resistance and the damage inflicted on the besieging force. Nonetheless, such arguments were unresolvable, not only because of self-interested witnesses and the inability to agree on a single evidentiary standard, but also because these debates took place within a charged wartime environment. After deciding the strategic and operational benefits of the capture, multiple audiences then judged the merits of a capitulating garrison's conduct, arguing about who won the greater glory. In the ensuing debates, each interpreted the surrender in the context of their own honor.

Bibliography

Archival Sources

British Library, London, Additional Manuscripts (BL, Add. MSS)

Volume 4742

Volume 61309

Volume 61314

Volume 61397

Service historique de la Défense (SHD), Paris

GR A[1] 1938; 1939; 1988; 2154; 2152; 2160; 2215; 2215; 2383

GR A[4] carton 8, Flandres 1712 folder

AG Article 15, Section 2

Newspapers

The British Mercury

The Daily Courant
The English Post
Lettres historiques contenant ce qui se passe de plus important en Europe
Le Mercure Galant
The Observator
The Post Boy
The Post Man
The Present State of Europe
The Supplement
The Tatler

The Montaigne Project
Si le Chef d'une Place Assiegée Doit Sortir pour Parlementer, <http://artflsrv02.uchicago.edu/cgi-bin/philologic/getobject.pl?c.0:2:5.montaigne.58263> (accessed 23 October 2018)

University of Michigan Libraries, Special Collections
D281.A2 A74

Printed Sources

An Epistle from the Duke of Burgundy to the French King (London, 1709).
Bancks, John, *History of John, Duke of Marlborough* (London, 1741).
Bland, Humphrey, *A Treatise of Military Discipline* (2nd ed., London, 1727).
Bouchet, Louis François du, marquis de Sourches, *Mémoires du marquis de Sourches sur le règne de Louis XIV*, ed. G. J. Cosnac, É. Pontal, 13 vols (Paris, 1882–1893).
Boyer, Abel, *History of the Reign of Queen Anne Year Eight* (London, 1710).
Brewster, Samuel, *Jus Feciale Anglicum* (London, 1725).
Briquet, Pierre de, *Code militaire ou Compilation des ordonnances des roys de France concernant les gens de guerre*, 2 vols (Paris,1708).
Chandler, David, ed., *Military Memoirs of Marlborough's Campaigns, 1702–1712* (London, 1968; repr. London, 1998).
Chandler, David, ed., *Military Miscellany II: Manuscripts from Marlborough's Wars, the American War of Independence and the Boer War* (Gloucestershire, 2005).
Churchill, John, Duke of Marlborough, *The Letters and Dispatches of John Churchill, First Duke of Marlborough from 1702 to 1712*, ed. G. Murray, 5 vols (London, 1845).
Churchill, John, Duke of Marlborough; Godolphin, Sidney, *The Marlborough-Godolphin Correspondence*, ed. H. Snyder, 3 vols (Oxford, 1975).
Colonie, Jean Martin de la, *Mémoires de Monsieur de La Colonie: maréchal de camp des armées de l'Electeur de Bavière*, ed. Anne-Marie Cocula (Paris, 1992).

Cuper, Gisbert, *Het Dagboek van Gisbert Cuper, gedeputeerde te velde, gehouden in de Zuidlijke Nederlanden in 1706*, ed. A. J. Veenendaal, Sr. (The Hague, 1950).

Courcillon, Philippe de, marquis de Dangeau, *Journal du marquis de Dangeau*, 19 vols (Paris, 1854–1860).

Daniel, Gabriel, *Histoire de France*, nouvelle ed. (Amsterdam, 1758).

Gaya, Louis de, *La Science militaire* (The Hague, 1689).

Mormes de Saint-Hilaire, Armand de, *Mémoires de Saint-Hilaire*, ed. L. Lecestre, 6 vols (Paris, 1903)

Godefroy, Théodore, *Le cérémonial François*, 2 vols (Paris, 1649).

Guignard, Pierre-Claude, *L'école de Mars*, 2 vols (Paris, 1725).

Hare, Francis, *The Conduct of the Duke of Marlborough during the Present War* (London, 1712).

Historical Manuscripts Commission, ed., *The manuscripts of the Earl of Buckinghamshire, the Earl of Lindsey, the Earl of Onslow, Lord Emly, Theodore J. Hare, esq., and James Round, esq., M.P.* (London, 1895).

Jones, David, *A Compleat History of Europe, 1709* (London, 1710).

Lamberty, Guillaume de, *Mémoires pour servir à l'histoire du XVIII siècle*, 14 vols (The Hague, 1724).

Manesson Mallet, Alain, *Les Travaux de Mars*, 3 vols (Paris, 1672).

Matthieu, Ernest, ed., *Relation du siège de la ville d'Ath en 1697* (Mons, 1910).

Millner, J. A., *A Compendious Journal of All the Marches, Famous Battles, Sieges* [...] (London, 1733; repr. East Sussex, 2005).

Nève, Joseph E., *Gand sous l'occupation de Louis XIV, 1678–1679, 1701–1706, 1708* (Ghent, 1929).

Osorio, Alvaro Navia, marqués de Santa Cruz de Marcenado, *Réflexions militaires et politiques*, 12 vols (Paris, 1738).

Pas, Antoine de, marquis de Feuquières, *Memoirs Historical and Military*, 2 vols (London, 1735).

Pinard, [François-Joseph-Guillaume], *Chronologie historique-militaire*, 8 vols (Paris, 1761).

Rechberger, Josef Ritter von Rechcron, ed., *Spanischer Successions-Krieg. Feldzug 1709* (Vienna, 1886).

Rochas d'Aiglun, E. A. A. de, ed., *Vauban: Sa famille et ses écrits*, 2 vols (Paris, 1910).

Saint-Simon, Louis de Rouvroy, duc de, *Mémoires de Saint-Simon*, ed. A. de Boislisle, 41 vols (Paris, 1879).

Sautai, Maurice, *Le siège de la ville et de la citadelle de Lille* (Lille, 1899).

Sevin, Charles, marquis de Quincy, *L'art de la guerre ou Maximes et Instructions sur l'art Militaire*, 2 vols (The Hague, 1728).

Sevin, Joseph, chevalier de Quincy, *Mémoires du chevalier de Quincy*, ed. Léon Lecestre, 3 vols (Paris, 1898).

[Unknown], *A Dictionary in 3 Parts* (London, 1705).
[Unknown], *The History of the Campaign in Flanders, in the Year 1708* (London, 1709).
Vauban, Sébastien Le Prestre de, *Oeuvres militaires*, 3 vols (Paris, 1795).
Vault, Eugène de; Pelet, Jean-Jacques-Germain, eds., *Mémoires militaires relatifs à la succession d'Espagne sous Louis XIV*, 11 vols (Paris, 1835).
Veenendaal, A. J. Jr., ed., *De briefwisseling van Anthonie Heinsius 1702–1720*, 15 vols (The Hague, 1976).
Ville, Antoine de, *De la Charge des gouverneurs des places* (Paris, 1639).

Secondary Sources

Chandler, David, *The Art of Warfare in the Age of Marlborough* (London, 1976; repr. New York, 1995).
Charles, J.-L., "Le sac des villes dans les Pays-Bas au XVIe siècle: Étude critique des règles de la guerre," *Revue internationale d'histoire militaire* 24 (1965), 288–301.
Childs, John, "Surrender and the Laws of War in Western Europe, c. 1650–1783," in *How Fighting Ends: A History of Surrender*, ed. Holger Afflerbach, Hew Strachan (New York, 2012), pp. 153–168.
Childs, John, "The Laws of War in Seventeenth-Century Europe and Their Application during the Jacobite War in Ireland, 1688–1691," in *Age of Atrocity: Violence and Political Conflict in Early Modern Ireland*, ed. David Edwards et al. (Dublin, 2007), pp. 283–300.
Claydon, Tony; Levillain, Charles-Edouard, eds., *Louis XIV Outside In* (London, New York, 2015).
Cornette, Joël, *Le roi de guerre: Essai sur la souveraineté dans la France du Grand Siècle* (Paris, 1993).
Donagan, Barbara, "The Web of Honour: Soldiers, Christians, and Gentlemen in the English Civil War," *The Historical Journal* 44, no. 2 (2001), 365–389.
Drévillon, Hervé, *L'impôt du sang: Le métier des armes sous Louis XIV* (Paris, 2005).
Drévillon, Hervé; Venturino, Diego, eds., *Penser et vivre l'honneur à l'époque moderne* (Rennes, 2011).
Duffy, Christopher, *Fire and Stone: The Science of Fortress Warfare 1660–1860* (Newton Abbott, 1976; repr. Edison, NJ, 1975).
Dabhoiwala, Faramerz, "The Construction of Honour, Reputation and Status in Late Seventeenth- and Early Eighteenth-Century England," *Transactions of the Royal Historical Society* 6 (1996), 201–213.
Hugill, J. A. C., *No Peace Without Spain* (Oxford, 1991).
Keegan, John, *The Face of Battle* (New York, 1976).
Ladurie, Emmanuel Le Roy; Fitou, Jean-François, *Saint-Simon and the Court of Louis XIV*, trans. A. Goldhammer (University of Chicago Press, 2001).

Lesaffer, Randall, “Siege Warfare and the Early Modern Law of War,” SSRN Scholarly Paper (Rochester, 2006), <http://papers.ssrn.com/abstract=926312> (accessed 23 October 2018).

Lynn, John, *Giant of the Grand Siècle* (Cambridge, 1997).

Lynn, John A., “Introduction: Honourable Surrender in Early Modern European History, 1500–1789,” in *How Fighting Ends: A History of Surrender*, ed. Holger Afflerbach, Hew Strachan (New York, 2012), pp. 99–112

Madunić, Domagoj, “Taming Mars: Customs, Rituals and Ceremonies in the Siege Operations in Dalmatia during the War for Crete (1645–69),” *Hungarian Historical Review* 4, no. 2 (2016), 445–470.

Manning, Roger B., *Swordsmen: The Martial Ethos in the Three Kingdoms* (Oxford, 2004).

Muir, Edward, *Ritual in Early Modern Europe* (New York, 1997).

Nimwegen, Olaf van, *The Dutch Army and the Military Revolutions, 1588–1688* (Rochester, 2010).

Ostwald, Jamel, *Vauban under Siege: Engineering Efficiency and Martial Vigor in the War of the Spanish Succession* (Leiden, 2007).

Pepper, Simon, “Siege Law, Siege Ritual, and the Symbolism of City Walls in Renaissance Europe,” in *City Walls: The Urban Enceinte in Global Perspective*, ed. James D. Tracy (Cambridge, 2000), pp. 573–604.

Sandberg, Brian, *Warrior Pursuits: Noble Culture and Civil Conflict in Early Modern France* (Baltimore, 2010).

Sternberg, Giora, *Status Interaction during the Reign of Louis XIV* (New York, 2014).

Swart, Erik, “Defeat, Honour and the News: The Case of the Fall of Breda (1625) and the Dutch Republic,” *European History Quarterly* 46, no. 1 (2016), 6–26.

Vo-Ha, Paul, *Rendre les armes – Le sort des vaincus XVI–XVIIe siècles* (Champ Vallon, 2017).

Wright, John, “Sieges and Customs of War at the Opening of the Eighteenth Century,” *American Historical Review* 39 (1934), 629–644.

PART 2

Imperial Boundaries

CHAPTER 5

Straddling Empires: Revolt and Religion in Early Modern Dalmatia

Eric R. Dursteler

The notion of a siege presupposes the existence of boundaries – an inside and an outside.* Contemporary representations of sieges, which were widely popular in the early modern era, visually emphasize this fundamental characteristic for both practical and aesthetic purposes. Before the start of hostilities, siege plans were often prepared on the fly to serve a strategic purpose, namely the "prime requirement" of achieving "deadly accuracy." After a siege was lifted, paintings and prints with "spectacular visual effect[s]" were produced to celebrate and memorialize the event.[1] These siege views, particularly those from an elevated perspective, clearly delineate the assembled combatants: The besiegers encircle the city, castle or fortress, whose walls cocoon and mark off the defenders inside.

These orderly, often aesthetically satisfying depictions of what were in the end highly violent and messy affairs, give the impression of unambiguously clearcut battle-lines. However, abandoning the bird's-eye view for the worm's perspective of the social and cultural context on the ground reveals a much more equivocal reality. This is evident in the noted capture and subsequent siege of the mighty Ottoman fortress of Clissa (Klis, Croatia) in 1596, an episode that raises a variety of questions about the tension between the representations and realities of conflict, the complexity and imprecision of boundaries, and their relationship to the malleability of political and religious identity along the frontiers of the early modern Mediterranean. It also suggests the ways in which the portrayal of such conflicts by contemporaries, as well as by subsequent generations of scholars, can distort and decontextualize these events in the service of conflicting narratives and imperatives.

1 The Capture and Siege of the Fortress of Clissa

The fortress of Clissa is located at the intersection of the political boundaries that divided the Ottoman, Venetian and Habsburg empires, in the frontier

* Unless otherwise noted, all archival sources are located in the Archivio di stato di Venezia.

1 Martha Pollak, *Cities at War in Early Modern Europe* (Cambridge, UK, 2010), pp. 109–54.

 | DOI:10.1163/9789004395695_006

region between northern Croatia, Dalmatia, and Bosnia.[2] It is situated atop a 360 meter outcropping of rock that looms over the mouth of the best pass through the Dinaric Alps into the Croatian hinterland. The great Ottoman traveler Evliya Çelebi described it evocatively as "like a cone of pilaf in a bowl."[3] The fortress was located "a vigorous stone's throw" (about 10 km) from the coast and the chief town of the region, Spalato (Split), which during this time was developing into an important Venetian commercial entrepôt for the overland caravan trade with Istanbul.[4] The stronghold's founding stretched back to Roman times, and it had been a lynchpin in the control of the region since the Middle Ages, indeed it was the last stronghold to fall to the Ottomans in 1537, following repeated failed attempts.[5] For the next century, Clissa played an essential role in the projection of Ottoman power in Dalmatia, as it was considered the strongest fortress of the empire's "entire western frontier."[6] It was garrisoned by a *disdar* (castellan), several cannon, and as many as 250 men, who defended it and kept watch over the two hundred houses and resident Ottoman *kadi*, or judge, in the town at the foot of the bluff.[7]

In terms of its defenses, the fortress of Clissa was decidedly medieval in its vertical design.[8] With its bare curtain walls, Clissa contrasted with the *trace italienne* bastions then beginning to appear in Venetian and Habsburg territories, which Geoffrey Parker has called "the heartland" of the military revolution.[9] While the analytical utility of the notion of the military revolution has of

---

2 Wendy Bracewell, "The Historiography of the Triplex Confinium: Conflict and Community on a Triple Frontier, 16th–18th Centuries," in *Frontiers and the Writing of History, 1500–1850*, ed. Steven G. Ellis, Raingard Esser (Hannover, 2006), p. 211.

3 Cemal Kafadar, "Evliya Çelebi in Dalmatia: An Ottoman Gentleman's Encounter with the Arts of the Franks," in *Dalmatia and the Mediterranean: Portable Archeology and the Poetics of Influence*, ed. Alina Payne (Leiden, 2014), p. 73.

4 British Library (BL), *MS 8605*, c. 7v, "Relazione di Nicolo Donado."

5 Catherine Wendy Bracewell, *The Uskoks of Senj: Piracy, Banditry and Holy War in the Sixteenth-Century Adriatic* (Ithaca, 1992), pp. 20, 48–49; Angelo de Benvenuti, "Fortezze e castelli di Dalmazia: la fortezza di Clissa," *La rivista dalmatica* 16, no. 4 (1935), pp. 28, 37–39; 17, no. 1 (1936), p. 16; Fred Singleton, *A Short History of the Yugoslav Peoples* (Cambridge, UK, 1985), pp. 5, 60–61; Biblioteca del Museo civico Correr di Venezia (BMCV), *Cicogna* 3098, "Compendio di varie revolutioni della famosa fortezza di Clissa."

6 Rita Tolomeo, "La fortezza di Clissa," *Atti e memorie della Società dalmata di storia patria* 34 (2012), 47.

7 "Relazione di Nicolo Correr," in Grga Novak, ed., *Commissiones et Relationes Venetae*, tome 4 (Monumenta spectantia historiam slavorum meridionalium), vol. 47 (Zagreb, 1964) 4:338; "Revisti Dalmati confini del dragoman Salvago," in Grga Novak, ed., *Commissiones et Relationes Venetae*, tome 7 (Monumenta spectantia historiam slavorum meridionalium), vol. 47 (Zagreb, 1972) 7:32; "Relazione di Pietro Basadonna," in Simeon Ljubić, ed., *Commissiones et Relationes Venetae*, tome 2 (Monumenta spectantia historiam slavorum meridionalium), vol. 8 (Zagreb, 1877), 2:224–225.

8 Tolomeo, "La fortezza di Clissa," p. 47.

9 Geoffrey Parker, *The Military Revolution: Military Innovation and the Rise of the West, 1500–1800* (2nd ed., Cambridge, UK, 1996), p. 24; Benjamin Arbel, "Venice's Maritime Empire in

course been much debated,[10] Venice's northeastern Adriatic territories were unquestionably the site of important innovations in military architecture, most evident in the new fortress at Palmanova, begun in 1594, and just up the coast in Zara (Zadar).[11] Similar developments were also gradually taking place among some of the defenses on the Habsburg side of the Hungarian-Croatian frontier.[12] Bastions were more effective against artillery barrages, but they were also expensive, and Clissa's elevation and natural defenses made bombards fairly ineffectual against it.[13] The fortress was, as one contemporary observer noted, "surrounded by deep valleys, and enclosed by inaccessible, alpine mountains," and located on "a precipitous outcropping of stone between mountains, but so distant that it cannot be defeated, and only with difficulty assaulted."[14] Contemporaries were unanimous in describing the fortress as "impregnable," though the several times that it changed hands from 1537 to 1648, including twice in 1596, suggest that it was not an impossible nut to crack.[15]

The strategic importance of Clissa is illustrated in the hybrid map and view entitled "Clissa. Chief Fortress of the Turk in Dalmatia, and Key to the Kingdom of Bosnia. 5 miles Distant from Spalato," now located in the Newberry Library [Figure 5.1]. Christofaro Tarnowskij created, or at least signed, the image in the summer of 1605, near the end of the grueling and ruinous Long War (1593–1606) between the Ottomans and the Austrian Habsburgs. Little is known about Tarnowskij, except that he was a knight of Malta, and may have been from Bosnia. The map of Clissa, along with two others depicting Castelnovo (Hercegnovi, Montenegro) and Scutari (Shkodër, Albania), was part of a set of maps by Tarnowskij that seem to have been produced as part of a broader effort to reinvigorate support for military action against the Ottomans in the region. It was

the Early Modern Period," in *A Companion to Venetian History, 1400–1797*, ed. Eric Dursteler (Leiden, 2013), p. 207.

10 On this see Clifford J. Rogers, ed., *The Military Revolution Debate: Readings on the Military Transformation of Early Modern Europe* (Boulder 1995); and more recently, Carol B. Stevens, "Warfare on Land," in *The Oxford Handbook of Early Modern European History, 1350–1750*, vol. 2: *Cultures and Power*, ed. Hamish Scott (Oxford, 2015), pp. 566–568.

11 Simon Pepper, "Defending the Frontiers of Venice: Fortification and Defensive Strategy in the Friuli before Palmanova," in *L'architettura militare di Venezia in terraferma e in Adriatico fra XVI e XVII secolo: Atti del Convegno internazionale di studi (Palmanova, 8–10 novembre 2013)*, ed. Francesco Paolo Fiore (Florence, 2014), pp. 3–20.

12 Gábor Ágoston, "Empires and Warfare in East-Central Europe, 1550–1750: The Ottoman-Habsburg Rivalry and Military Transformation," in *European Warfare, 1350–1750*, ed. Frank Tallett, D. J. B. Trim (Cambridge, UK, 2010), pp. 110–134.

13 Parker, *The Military Revolution*, pp. 7, 12.

14 Tolomeo, "La fortezza di Clissa," p. 53.

15 Biblioteca Marciana (BM), *It VI* 105 (5728), "Viaggio da Venezia a Costantinopoli," c. 5r; "Itinerario di Giovanni Battista Giustinian," in *Commissiones et Relationes Venetae*, 2:211; Evlijā Čelebī, *Putopis: Odlomci o júgoslavenskim zemljama*, ed. Hazim Šabanović (Sarajevo, 1967), pp. 152–153.

FIGURE 5.1 Clissa. Chief Fortress of the Turk in Dalmatia, and Key to the Kingdom of Bosnia. 5 miles Distant from Spalato
NEWBERRY LIBRARY, NOVACCO 2F 205, SHEET 3 OF 3

hoped that seizing these three key defensive points along the Veneto-Ottoman frontier might trigger a popular uprising. The maps were intended to provide a reasonably accurate, and at the same time aesthetic, depiction of the frontier. They include a "detailed network of roads and tracks," and indications of defended river crossings. No political boundaries are sketched out, rather religious signs – "crosses for Christians … stars and half moons for Muslims" – are intended to delineate political allegiance, but do not correspond to communal religious identity. The fortress of Clissa is depicted in exaggerated scale in comparison to Spalato, which is actually the larger of the two towns, most likely to emphasize to potential imperial and papal supporters the threat it presented to Christians in the region.[16]

The military and symbolic significance of Clissa were key factors in the celebrated capture of the fortress in 1596. Because of the notoriety of the event

16 James P. Krokar, "New Means to an Old End: Early Modern Maps in the Service of an Anti-Ottoman Crusade," *Imago Mundi* 60 (2008), 23–25.

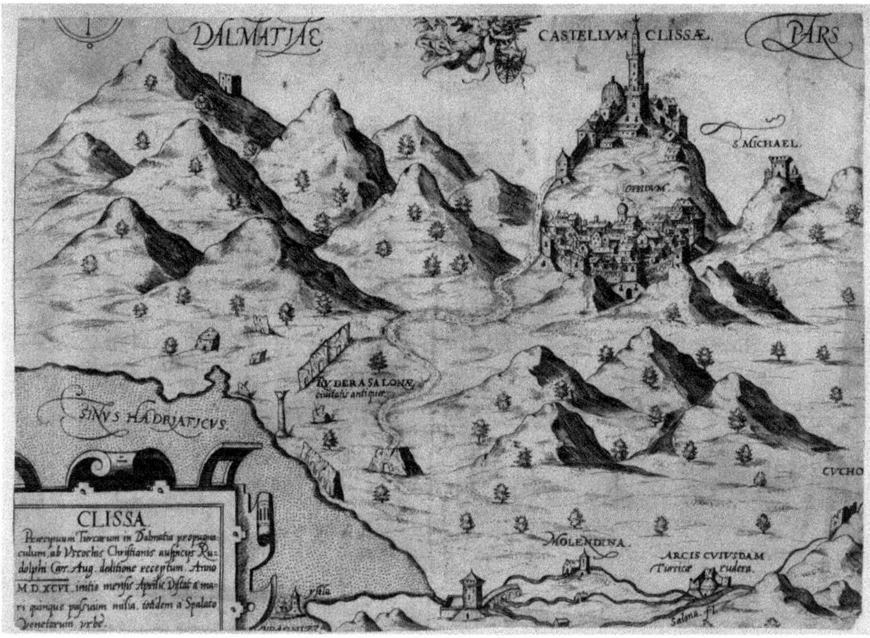

FIGURE 5.2 Clissa, Praecipuum Turcarum in Dalmatia popugnaculum, ab Uscochis Christianis auspicys Rudolphi Caes. Aug Deditione recptum Anno M.D.XCVI initio mensis April 15. Distat a mari quinque passuum misia totidem a Spalato Venetorum urbe, [n. p., c. 1596]
COURTESY OF BARRY LAWRENCE RUDERMAN ANTIQUE MAPS

among contemporaries, a number of accounts have survived,[17] as well as at least one contemporary map [Figure 5.2]. One of these narratives, the "Relatione del'infelice avvenimento dell'impresa di Clissa fatta l'anno della nostra salute M.D.XC.VI.," was composed in the early summer of that year by Hieronimo Pavon, a canon in Venetian Spalato, and a "virtuous and most honorable gentleman."[18] We know very little about Pavon or his report, but it is clear that he was a supporter and intimate observer of the conspiracy, knew its leaders, and served as an intermediary between them and individuals in Venice and Rome.[19] The report was obtained in August 1596 by Constantino Prosperi in

17 G. E. Rothenberg, "Christian Insurrections in Turkish Dalmatia 1580–96," *Slavonic and East European Review* 40 (1961), 144 n. 52. For one of these accounts, see Vladimir Lamansky, ed., *Secrets d'etat de Venise*, vol. 1 (St. Petersburg, 1884; reprint New York, 1968), pp. 502–515.

18 "Relatione del'infelice avvenimento dell'impresa di Clissa fatta l'anno della nostra salute M.D.XC.VI.," in *Spomenici hrvatske Krajine*, ed. R. Lopasić (Monumenta spectantia historiam Slavorum meridionalium), vol. 15 (Zagreb, 1884), 1:236–255.

19 Carolus Horvat, ed., *Monumenta historiam Uscocchorum illustrantia*, 1 (Monumenta spectantia historiam Slavorum meridonalium), vol. 32 (Zagreb, 1910), pp. 100–103.

Ferrara, and given to Ippolito Bentivoglio, a leading commander of the forces of Alfonso II d'Este, Duke of Ferrara. Pavon's "Relatione," supplemented by numerous Venetian archival sources, provides a clear sense of the broad course of events that led to the capture of the great citadel of Clissa.

According to Pavon, during the night of April 6–7, a group of 40–50 men "holding onto each other with great difficulty, ascended the mount by way of [a] steep cliff face. At two hours before daybreak they were let into the most important upper part of the [fortress] through an opening on the south, from which the garbage of the city was dumped." On the night in question, "the fortress was almost empty of soldiers": Some "had gone to a fair in the interior of the land, others [were] with the kadi in Spalato, others [were] in Almissa and Suzuraz, a castle in the jurisdiction of Spalato." The few guards who remained in the upper fortress "were found slumbering," and thus the infiltrators were able to slip in without detection.[20]

They lay in wait quietly until just before daybreak, and then "after first invoking the help of Jesus three times," they burst out of their hiding places and their leader "planted their standard with a crucifix in the middle of the fortress." They then "began to run through the upper city with great ardor, killing everyone they came in contact with, including those who were still in their houses half-asleep, and others who were preparing to flee. And they pushed forward so forcefully that very quickly they seized control of the most important part of the fortress."[21]

Encouraged by their initial success, the invaders broke into the house of the disdar, "whose head they cut off, and they took his keys along with his head," and threw open the doors to the lower fortress and turned their fury on the small town outside the walls, "cutting to pieces everyone that they found in the houses." And so, Pavon notes without irony, "on the 7th of April, Palm Sunday, this fortress came into the hands of the Christians almost without any spilling of blood."[22]

The day following the capture, a group of Uskok reinforcements from Habsburg Segna (Senj) joined the fray, and a small force of Ottomans made several vain attempts to regain the fortress, with significant loss of life and captives. When word of this second victory spread, 200 Ottoman Christian subjects and "almost all the priests" from Polizza (Poljica) in Ottoman territory, as well as "many" Venetian subjects "from the territory of Spalato and also Trau

20 "Relatione del'infelice avvenimento," p. 242.

21 "Relatione del'infelice avvenimento," p. 242.

22 "Relatione del'infelice avvenimento," p. 242; Letter from Hieronimo Pavon to Nicola Alberti, 20 April 1596, in Horvat, ed., *Monumenta historiam Uscocchorum*, 1:101.

[Trogir]" flocked to Clissa. In preparation for the inevitable Ottoman response, Pavon reports that the new proprietors of the citadel "flattened" the small town outside the fortress and "knocked down all the houses up to the walls," to deny eventual besiegers any "accommodation." They carved out passageways along the interior of the fortifications so that they could move about "comfortably" during a siege, gathered food and weapons, and filled every possible container with water. Finally, "in a sign of victory and to strike fear in their foes," the new lords of Clissa "adorned" the fortress walls with numerous decapitated heads.[23]

Over the next weeks, Ottoman forces gathered outside the bulwarks and steadily increased in number. On April 12, the *sanjakbeg* of Clissa arrived with 600 men. The pasha of the *eyalet* of Bosnia, the chief Ottoman official in the northern Balkans, fearing that the uprising would spread more widely, sent 150 men who arrived on April 15. By the 17th the besiegers numbered 2000, divided equally between "Turks" and "Christian Morlachs," or *vlachs*, as the amorphous, largely Orthodox, transhumant cattle herder subjects of the sultan from inland Dalmatia were called.[24] By April 25, 8000, and by May 12, between 12,000 and 15,000 men encircled Clissa, heavily outnumbering the several hundred defenders within the fortress walls.[25]

Initially, the objective of this "inexperienced," "badly armed," and "poor[ly] organiz[ed]" Ottoman force was simply "to frighten" the defenders, not to regain the citadel. As their numbers grew, however, the Ottomans became more bellicose.[26] On April 17, 150 janissaries approached the walls and attracted a "tempest" of gunfire and rocks. Forty men charged out of the fortress and wounded a number of janissaries, killed and decapitated four, and forced the others to retreat. As Pavon notes, "from that point forward they did not come so close."[27]

On April 22 two small cannon hauled from neighboring areas bombarded Clissa, though to minimal effect. Several days later, three larger cannon caused

23 "Relatione del'infelice avvenimento," pp. 242–243.

24 Larry Wolff, *Venice and the Slavs: The Discovery of Dalmatia in the Age of Enlightenment* (Stanford, 2001), p. 11; Cristian Luca, "The Vlachs/Morlaks in the Hinterlands of Traù (Trogir) and Sebenico (Šibenik), Towns of the Venetian Dalmatia, during the 16th Century," in *Miscellanea Historica et Archaelogica in Honorem Professoris Ionel Câdea*, ed. Valeriu Sîrbu, Cristian Luca (Brăila, 2009), pp. 315–321. See also the discussion in Bracewell, "The Historiography of the Triplex Confinium," pp. 219–225; and James D. Tracy, *Balkan Wars: Habsburg Croatia, Ottoman Bosnia, and Venetian Dalmatia, 1499–1617* (Lanham, 2016), p. 159.

25 PTM, b. 417, 23 Apr 1596; PTM, b. 417, 26 Apr 1596; PTM, b. 417, 12 May 1596; "Relatione del'infelice avvenimento," p. 243.

26 PTM, b. 417, 23 Apr 1596.

27 "Relatione del'infelice avvenimento," p. 244.

some damage, though because of the defenders' effective system of trenches at the outer walls of the fortress, the Ottomans could not take advantage, and the breaches and burned-out doors of the fortress were successfully blocked with debris. A week later, fifty janissaries made "a great push to climb up to the fortress," however, they too were dispersed with "arquebus rounds."[28]

By the first days of May, the pasha of Bosnia's chief lieutenant arrived with additional men, and on the 5th at sunrise, the Ottomans made their most concerted attempt yet to storm the fortress. Pavon reports, "having made terrible shouts and barbaric cries that made it seem that the sky might fall to the earth, they approached the fortress from all sides with many ladders … it seemed like everything was on fire because there was a large cloud of smoke everywhere, which made a dark fog, … [from the] continual arquebus and musket fire, which lasted until midday." Three hundred assailants were killed, mostly "Turks," who were targeted specifically by the defenders, whereas the vlachs among the attackers were encouraged to "keep their distance," and indeed to abandon the siege entirely, which would have significantly weakened the Ottoman forces. The failure of this assault led to a change in tactics: realizing they could not storm the fortress, the Ottomans decided to settle in and drive the defenders out "with hunger and thirst."[29]

This proved more effective as the situation inside Clissa's walls was already dire. The Ottomans had blocked all water sources going into the fortress with dirt and stones, and thirst drove some defenders to drink "their own urine." Circumstances became so desperate that some of the besieged men began plotting to kill their leaders and abandon Clissa. On May 7, the feast of Saint Domnius, patron saint of Spalato, "God miraculously sent them a good rainstorm," and two relatively small relief convoys also managed to slip past the Ottomans and delivered food to the fortress, which provided temporary relief. The additional mouths to feed ultimately worsened the situation, however, as they and the starving defenders quickly consumed the additional provisions. They were soon reduced to eating "wild plants," horses, donkeys, cats, dogs, boiled cowhides, and even "the stinking hooves of [dead] horses." Soon, almost every day "someone was dying of hunger."[30]

The confrontation came to a head on May 26–27 when a relief force of perhaps 1000 men led by Georg Lenkovich, the Habsburg general of

28 "Relatione del'infelice avvenimento," p. 244; PTM, b. 417, 26 Apr 1596. PTM, b. 417, 30 Apr 1596; PTM, b. 417, 2 May 1596.

29 "Relatione del'infelice avvenimento," pp. 244–245; PTM, b. 417, 6 May 1596.

30 "Relatione del'infelice avvenimento," pp. 245–247; PTM, b. 417, 30 Apr 1596; PTM, b. 417, 12 May 1596; PTM, b. 417, 18 May 1596; Vjekoslav Klaić, *Povijest Hrvatâ od najstarijih vremena de svršetka XIX stoljeća* (Zagreb, 1981), 5:524.

Croatia, landed on the coast, boldly marched past Venice's amassed forces, and attacked the Ottoman camp at daybreak in a brazen attempt to break the siege. The surprise initially favored his forces, who overran one of the besiegers' main camps. The Ottomans feigned retreat, moving to higher ground where their artillery was located, and a significant portion of Lenkovich's forces, believing their adversaries had retreated "out of fear" and were defeated, began plundering their tents. Seeing this and their enemies' small numbers, the Ottomans counterattacked and "cut to pieces and made prisoners the greater part" of the looters. Lenkovich countered by charging the hill where the cannon were located, but the Ottomans turned these on him, and their light cavalry drove his men from the battlefield with heavy losses. The survivors fought their way back to the coast, and fled by boat.[31]

With all hope of relief lost, the remaining defenders sought terms for the return of Clissa. The Ottomans promised them safe passage with their arms and possessions, horses to carry the wounded, and the release of all slaves taken during the siege, terms guaranteed with two sanjakbegs as hostages. On May 31, the siege was lifted, and Ottoman flags once more flew over the walls of Clissa.[32]

2 The Clissa Affair Reconsidered

On its surface, the celebrated capture and siege of Clissa seems yet another example of the sort of religious rivalry and violence that was endemic to the triplex confinium of the Veneto-Ottoman-Habsburg frontier in Dalmatia.[33] Contemporaries often depicted the region as the tectonic fault line between Islam and Christendom, the eastern front in the so-called "clash of civilizations." Dalmatia also occupied a disproportionate place in larger political rivalries of the day, particularly between the Ottomans and their papal and Habsburg antagonists. Pope Clement VIII was full of crusading zeal and entertained hopes that the Balkans might be the tinder box where Christian passions might be enflamed to mobilize collective action against the sultans.[34] For his part, Holy Roman Emperor Rudolph II entertained a sense of mission vis-à-vis the Ottomans, and hoped to weaken their position on his southern

31 "Relatione del'infelice avvenimento," pp. 250–254; *PTM*, b. 417, 27 May 1596.

32 "Relatione del'infelice avvenimento," pp. 253–255; *PTM*, b. 417, 31 May 1596.

33 Bracewell, "The Historiography of the Triplex Confinium," pp. 211–227. Also, Egidio Ivetic, Drago Roksandić, eds., *Tolerance and Intolerance on the Triplex Confinium: Approaching the "Other" on the Borderlands Eastern Adriatic and Beyond, 1500–1800* (Padua, 2007).

34 See *Senato dispacci Roma*, b. 37, passim.

borders, but realized that he was not powerful enough to accomplish this on his own.[35]

When we shift from the halls of geo-political power, however, and examine actual conditions on the ground, we discover a much more equivocal situation that requires a more fine-grained analysis. To begin, Dalmatia's political borders were disputed throughout the sixteenth and seventeenth centuries, despite recurring joint Veneto-Ottoman efforts to demarcate them.[36] This lack of a clearcut division produced a situation in which Venetian, Ottoman and Habsburg subjects mingled freely.[37] Venetians worked Ottoman fields and ground their wheat in Ottoman mills, while Ottomans worked and even rented Venetian fields. Venetian soldiers lived in Ottoman territory, and the transhumant vlachs moved freely throughout the region and proffered their military services to both Istanbul and Venice.[38] Families straddled political and religious frontiers,[39] and cross-border friendship and marriages were ubiquitous.[40] Migration and resettlement between Venetian, Ottoman and Habsburg lands and beyond were also common among desperate and ambitious people throughout the region.[41]

35 Zdenko Zlatar, *Between the Double Eagle and the Crescent: The Republic of Dubrovnik and the Origins of the Eastern Question* (Boulder, 1992), pp. 81–89; R. J. W. Evans, *Rudolf II and His World: A Study in Intellectual History 1576–1612* (London, 1997), pp. 24, 75–78; Gunther Erich Rothenberg, *The Austrian Military Border in Croatia, 1522–1747* (Urbana, 1960), pp. 58–59; Jan Paul Niederkorn, *Die europäischen Mächte und der 'Lange Türkenkrieg' Kaiser Rudolfs II. (1593–1606)* (Vienna, 1993), pp. 85–86.

36 V. Solitro, *Documenti sull'Istria e la Dalmazia* (Venice, 1844), pp. 257–268; "Relatione di Nicolo Priuli," in *Commissiones et Relationes Venetae*, 4:429. Also Maria Pia Pedani, "Beyond the Frontier: The Ottoman-Venetian Border in the Adriatic Context from the Sixteenth to the Eighteenth Centuries," in *Zones of Fracture in Modern Europe: The Baltic Countries, the Balkans and Northern Italy*, ed. Almut Bues (Wiesbaden, 2005), p. 174; Seid M. Traljić, "Tursko-mletačke granice u Dalmaciji u XVI. i XVII. Stoljeću," *Radovi* 20 (1973), 447–458.

37 Maria Pia Pedani, *Dalla frontiera al confine* (Rome, 2002), pp. 39–55.

38 Krešimir Kužić, "Prilog biografiji nekih Kačićevih vitezova te podrijetlu stanovništva njihova kraja," *Radovi Zavoda za povijesne znanosti HAZU u Zadru* 47 (2005), 223–224; BM, *It VII* 1217 (9448), #18, c. 249r, "Relatione di Dalmatia;" British Library (BL), *MS 8613*, cc. 81r–88r, "Relatione sopra i Morlacchi."

39 Ivan Vitezić, *La prima visita apostolica postridentina in Dalmazia (nell'anno 1579)* (Rome, 1957), pp. 19, 31–34.

40 "Relazione di Marco Barbarigo," in *Commissiones et Relationes Venetae*, 4:342; BL, *MS 8605*, cc. 9r–v, "Relazione di Nicolo Donado;" "Relazione di Andrea Rhenier," in Grga Novak, ed., *Commissiones et Relationes Venetae*, tome 6 (Monumenta spectantia historiam slavorum meridionalium) vol. 49 (Zagreb 1970), 6:67; "Relazione di Marino Mudazzo," in *Commissiones et Relationes Venetae*, 6:202–203; *Collegio – Risposte di fuori*, b. 352, 12 Jul 1599; Bracewell, *The Uskoks of Senj*, pp. 35–36; Pedani, "Beyond the Frontier," p. 46.

41 Alberto Marani, ed., *Atti pastorali di Minuccio Minucci arcivescovo di Zara (1596–1604)* (Rome, 1970), pp. 50, 52–53; Karlo Horvat, "Glagolaši u Dalmaciji početkom 17. vijeka t.

There was a significant cross-border trade in Dalmatia's urban centers and its regional fairs, which approached 400,000 ducats annually by the mid-sixteenth century.[42] Fish, oil, spices, and salt from the Venetian coastal enclaves were exchanged for Ottoman cattle, cheese, hides, and especially grain, which was essential to the survival of land-poor Venetian Dalmatia.[43] Venetian and Ottoman officials also collaborated in making Spalato a lucrative pivot in the overland caravan trade to Istanbul from 1590 onward.[44]

Because of the porosity of the frontier, a shared culture that transcended the region's gossamer political and religious borders endured.[45] In terms of religion, Christians (largely Roman Catholic) were the majority, with small communities of Jews, and a few concentrations of Muslims in Ottoman territory. Shared popular beliefs and practices blurred these religious lines, manifested in many common holy sites and popular religious practices. Religious nomadism, or "strategic syncretism," in the form of conversion was also common.[46] On his visit, Evliya Çelebi was "astonished and bemused" by the common frontier custom of "swapping religions": "If a Christian soldier were to befriend a Muslim captive, or vice versa, one would promise to save the other from captivity, with one so saved pledging to return the favor one day if necessary; in a private ceremony, each swore to 'take the religion of the other' and both would take oaths by blood, thus becoming 'brothers-in-religion.'"[47] The shared frontier culture and identity trumped official efforts to delineate clear borders, and the persistent tension between "necessity and ideology" resulted at times in what we might cautiously describe as a sort of Dalmatian *convivencia*.[48]

This latter point should not be exaggerated, as the Dalmatian frontier was also the site of considerable friction: in addition to the occasional mur-

---

j. godine 1602.–1603." *Starine* 33 (1911), 557–559.

42 Pedani, "Beyond the Frontier," p. 46; Seid M. Traljić, "Zadar i turska pozadina od XV do potkraj XIX stoljeća," *Radovi* 11–12 (1965), 226; Marani, *Atti pastorali di Minuccio Minucci*, pp. 60–61.

43 Renzo Paci, *La scala di Spalato e il commercio veneziano nei Balcani fra Cinque e Seicento* (Venice, 1971), pp. 14, 101; BL, *MS 8610*, c. 14r, "Relazione di Giovanni Tiepolo," 14r; "Relazione di Jacopo Foscarini," in *Commissiones et Relationes Venetae*, 4:35–36; Grga Novak, "Poljoprivreda na dalmatinskom primorju i otocima u XVIII stoljeću," *Starine* 51 (1962), 75; *PTM*, b. 422, 18 Oct 1628.

44 Maren Frejdenberg, "Venetian Jews and Ottoman Authorities on the Balkans (16th Century)," *Études balkaniques* 30, no. 3 (1994), 56; Paci, *La scala di Spalato*, pp. 17, 71, 92–93; Tracy, *Balkan* Wars, pp. 259–260.

45 Fernand Braudel, *The Mediterranean and the Mediterranean World in the Age of Philip II* (New York, 1972), 2:759, 889; Linda Colley, *Captives: Britain, Empire, and the World, 1600-1850* (New York, 2002), p. 121.

46 Palmira Brummett, *Mapping the Ottomans: Sovereignty, Territory, and Identity in the Early Modern Mediterranean* (Cambridge, UK, 2015), p. 161.

47 Kafadar, "Evliya Çelebi in Dalmatia," p. 66.

48 Bracewell, *The Uskoks of Senj*, p. 174.

der, animals were rustled, contraband goods transferred surreptitiously, and individuals kidnapped for ransom.[49] Indeed, the intersection of three major polities and the ill-defined borders facilitated illegal activities, and were manipulated by bandits, common folk, and occasionally officials on all sides.[50] The reality on the Dalmatian frontier was that, despite attempts to assert centralized authority, the metropoles often had relatively little sway over conditions on the periphery. Local authorities often ignored diplomatically negotiated "periods of peace", and had limited control over their men, who acted more like "frontier freebooters" and raided across the borders with impunity. Their objective was booty and prisoners for ransom or sale, rather than the destruction of their adversaries, who quite often came from the same lineages, faiths, and villages.[51]

Conflict did not, however, define Veneto-Ottoman relations. Both Ottoman and Venetian officials were regularly directed to preserve the "*bona pace* (good peace)" on the frontier.[52] In the Ottoman case, maintaining "tranquility" was "advantageous for both sides," and to this end Istanbul attempted to keep "a tight rein" on local officials in Dalmatia "to prevent them from violating treaties" with Venice, though this was not always possible.[53] Local officials acknowledged exaggerating the problems on the frontier for diplomatic effect, and reported that their subjects along the border generally lived "in peace and quiet and with excellent understanding."[54] Ottoman and Venetian officials regularly collaborated in dealing with border issues, in particular the disruptions of the piratical Uskoks who swarmed the Adriatic coast, as well as the actions of their own restive subjects.[55]

When considered in the light of Dalmatia's muddled boundaries, the events set in motion on Palm Sunday, 1596, take on a decidedly different hue. As was common in these frontier confrontations, the conspirators' supporters marshaled religious rhetoric and imagery to represent and defend the actions of

49 *Commemoriali*, r. 25; *Lettere e scritture turchesche*, b. 5; *Senato deliberazioni Costantinopoli*, b. 5, 24 Feb 1581 (MV).

50 Florence Sabine Fabijanec, "Le développement commercial de Zadar et de Split aux XVe–XVIe siècles," Ph.D. Dissertation (Université Paris I, 2002), p. 77; Paci, *La scala di Spalato*, p. 45.

51 David Nicolle, *Ottoman Fortifications 1300–1710* (Oxford, 2010), pp. 32, 40.

52 *Senato deliberazioni Costantinopoli*, b. 9, 8 Aug 1598; *Bailo a Costantinopoli*, b. 252, cc. 7v–8r, 12 Apr 1593.

53 Brummett, *Mapping the Ottomans*, p. 161.

54 "Relazione di Antonio Pasqualigo," in Simeon Ljubić, ed., *Commissiones et Relationes Venetae*, tome 3 (Monumenta spectantia historiam slavorum meridionalium), vol. 11 (Zagreb 1880), 3:187; *Archivio proprio Costantinopoli*, b. 20/II, c. 107v, 3 Feb 1626 (MV); *Collegio – Relazioni*, b. 72, c. 4v, "Relazione di Giulio Contarini."

55 Angelo de Benvenuti, *Storia di Zara dal 1409 al 1797* (Milan, 1944), p. 131; Rothenberg, "Christian Insurrections in Turkish Dalmatia," pp. 136–147.

the motley crew who captured Clissa – from the symbolism of the date of the takeover, to the flag with a crucifix, from the prayers offered by the conspirators, to a request for the papal grant of a plenary indulgence for "those who defend the faith of Christ."[56] Claims such as these, purporting to act "in the service of a war for the faith" were commonly articulated on the frontier to justify and glorify or to mobilize support for similar actions. As with so much, however, the reality was much more ambiguous than the words and symbols suggest.[57]

Although he attempts to drape the affair in the cloak of purely religious conflict, Pavon's account makes clear that the seizure of the seemingly impregnable fortress of Clissa was carried out by a diverse group. The plan was the brainchild of Giovanni Alberti, a Spalatine noble, a Christian, and a Venetian subject, whose brother, Nicola, was archdeacon of Spalato. Another key figure in the plot, Francesco Antonio Bertucci, was a nobleman from the island of Lesina (Hvar), and like Alberti, a Venetian subject, as well as a Knight of Malta, and a papal and imperial agent.[58] Knowing full well that "his natural prince [Venice] was at peace with the Great Turk and that no matter how great the enterprise, would never break its treaty," in the months leading up to the incident, Alberti and his co-conspirators enlisted promises of support from other quarters, including Rudolph II and Clement VIII.[59]

While Alberti and his men articulated their aim in religious terms, as "a feat that would be pleasing to the Divine Majesty and to the Christian republic" and to throw off the "arduous tyranny" and the "bitter yoke" of the Ottomans, many subjects of the sultan also played a decisive role in the conspiracy's success. For instance, Alberti bought the favor of several Ottoman Christians who lived in Clissa, and the support of Paolo Pavich, the Grand Prince of Polizza, and an Ottoman Christian subject. The conspiracy was not a strictly Christian affair either. According to Pavon, four Muslim brothers, and a deputy of the disdar of Clissa, played a decisive role in the capture of the fortress.[60] Two women enslaved in the initial attack reported that these men became involved, in part at least, because of strained relations with local Ottoman officials.[61] As

56 Letter from Giovanni Alberti to Clement VIII, 10 April 1596, in Horvat, ed., *Monumenta historiam Uscocchorum*, 1:95.

57 Bracewell, "The Historiography of the Triplex Confinium," p. 218.

58 Krokar, "New Means to an Old End," p. 30; Noel Malcolm, *Agents of Empire: Knights, Corsairs, Jesuits and Spies in the Sixteenth-Century Mediterranean World* (Oxford, 2015), pp. 410–411.

59 "Relatione del'infelice avvenimento," p. 239; Zlatar, *Between the Double Eagle and the Crescent*, pp. 91–92.

60 Hieronimo Pavon to Nicola Alberti, 20 April 1596, in Horvat, ed., *Monumenta historiam Uscocchorum*, 1:100.

61 *Cariche da mar – Processi*, b. 8, 13–17 Apr 1596.

this motley crew of Venetian, Ottoman, Habsburg, Muslim, and Christian conspirators suggests, religion was not the sole motivation for the taking of Clissa. Even more important, I would argue, the capture of Clissa was a product of economic despair and political frustration with Venetian and Ottoman rule, as well as an expression of the ongoing tensions between center and periphery in this borderland.

Dalmatia held a central place both militarily and commercially in Venice's eastern Mediterranean *stato da mar*, or maritime empire. Throughout much of the sixteenth century, however, its subjects "simmered with revolt at [Venice's] indifference to their welfare."[62] The peace of 1573, following the second war of the Holy League and the great victory at Lepanto, only exacerbated already deep-rooted problems: Venice's key Dalmatian coastal cities, known as the *Sottovento*, were reduced to isolated outposts with almost no hinterland.[63] As a result, urban population plummeted, many villages were abandoned, infrastructure collapsed, trade was diminished, and these enclaves struggled to provision themselves.[64] Spalato in this time was characterized as full of "unhappiness and misery."[65] Poverty was widespread, beggars filled the streets, Venice had to send food and clothing to the city, and civic and ecclesiastical officials "distribute[d] bread and broth, vegetables and rice to the poor, who come to beg." A Venetian official reported that Spalato's inhabitants "cry out that they are abandoned," and as a result "have various seditious ballads always in their mouth ... which they sing at night, under our very palace windows."[66]

Venice's Dalmatian subjects at the end of the sixteenth century were also deeply exasperated by "Venice's careful relations with the Ottomans," and with what was perceived as its failure to defend and preserve the well-being of the region. This was exacerbated by perceptions of Venetian mismanagement and administrative abuses, which many held responsible for the region's decline and the collective despair.[67] The Dalmatian nobility, particularly in Spalato, were sympathetic toward the Austrian Habsburgs, because they "had been divested of their former political power and economic fortunes," and saw their op-

62 Rothenberg, "Christian Insurrections in Turkish Dalmatia," p. 138.

63 Arbel, "Venice's Maritime Empire in the Early Modern Period," p. 226.

64 Vera Graovac, "Populacijski razvoj Zadra," *Geoadria* 9 (2004), 54–55; *Collegio – Relazioni*, b. 72, "Relatione de Lunardo Zulian;" also BMCV, *Cicogna*, b. 2854, cc. 107r–v, 123r–125r, "La discretion delle anime nella Città de Zara et suo Territorio, fatta li 24 Dec 1579;" de Benvenuti, *Storia di Zara*, pp. 128–129.

65 *Collegio – Risposte di dentro*, b. 10, #131, 11 July 1597; Frejdenberg, "Venetian Jews and Ottoman Authorities on the Balkans (16th Century)," p. 56.

66 John Gardner Wilkinson, *Dalmatia and Montenegro* (London, 1848), 2:335–337.

67 Bracewell, *The Uskoks of Senj*, pp. 26–29, 77.

portunities blocked by Venice.[68] In short, to many in Dalmatia, Venice seemed more interested in keeping the peace with the Ottomans and profiting from commerce, rather than addressing its own subjects' trying circumstances.[69]

On the Ottoman side there were parallel frustrations. The barren hinterland beyond the coastal regions was "incapable of supporting its population," and various "economic pressures" gave rise to widespread and generalized banditry.[70] After 1573, the district was divided administratively into two *sanjaks*: The result, as Ottoman officials themselves complained, was that it "began to decline, and to deteriorate," which created "a thousand bad effects, where before there was no problem whatsoever."[71] Unable to support themselves on the income from the split region, the local officials became more aggressive in their extractions. This led to a chain reaction: "Unable to tolerate or put up with the tyranny of these sanjakbegs, the inhabitants either rebelled or fled the sultan's lands," and taxes plummeted. The officials imposed heavier burdens on those who remained, going "from villa to villa" seizing large quantities of wheat, animals, butter and honey with no payment, which they in turn sold, and holding peasants prisoner to extort money from them. The situation reached such a low point that several minor Ottoman officials wrote to the Venetian Senate requesting its intervention on their behalf, otherwise "this land will descend into complete misfortune [...] and all the inhabitants of the [sultan] will abandon the region, and will move elsewhere."[72] This local discontent fit into a wider pattern throughout the empire toward what were perceived as increasingly distant sovereigns, and was evidenced in repeated uprisings in the Balkans,[73] and most notably the Celali revolts that raged in the Anatolian heartland in the late sixteenth and early seventeenth centuries.[74]

Against this troubled backdrop, Dalmatia's inhabitants, regardless of religion or ruler, articulated their dissatisfaction through various "language[s] of the unheard,"[75] from open revolt to attempts to elicit outside intervention. In

68 Zlatar, *Between the Double Eagle and the Crescent*, pp. 91–92; Michael Knapton, "Tra Dominante e dominio (1517–1630)," in *La Repubblica di Venezia nell'età moderna*, vol. 2: *Dal 1517 alla fine della Repubblica*, ed. Gaetano Cozzi, Michael Knapton (Turin, 1992), p. 382.

69 Tolomeo, "La fortezza di Clissa," p. 46.

70 Rothenberg, "Christian Insurrections in Turkish Dalmatia," p. 138.

71 *PTM*, b. 922, 4 Jan 1598 MV.

72 *PTM*, b. 922, 4 Jan 1598 MV.

73 Dorothy M. Vaughan, *Europe and the Turk: A Pattern of Alliances, 1350–1700* (Liverpool, 1954; reprint, New York, 1976), p. 218.

74 Kemal Karpat, "The Stages of Ottoman History: A Structural Comparative Approach," in *The Ottoman State and its Place in World History*, ed. Kemal Karpat (Leiden, 1974), pp. 97–98.

75 Martin Luther King, *Where Do We Go From Here?* (Boston, 2010), 119.

1544, for example, a Franciscan from Spalato claiming connections with Ottoman guards garrisoned in Clissa, contacted the Habsburg ambassador in Venice, but before anything developed, the soldiers were removed.[76] In 1580, and again in 1586, a Franciscan from Trau, "claiming to represent a dissident group in Dalmatia," sought assistance from the emperor in return for delivering Clissa to imperial control. In 1583 Clissa was attacked by Venetian subjects, which generated "much anxiety" among Venice's ruling elite, who sent "reinforcements to Dalmatia and decreed heavy penalties against Venetian subjects implicated in the raid."[77] Much more frequently, however, misery, desperation, and rage were manifested through acts of theft, violence and flight, forms of what James Scott has described as "weapons of the weak," which were endemic on the frontier.[78]

The most notable expression of this was the Uskok phenomenon. Deriving from the Slavic verb *uskòčiti*, "to jump over … [or] to cross," Uskoks were refugees or exiles who traversed the region's boundaries.[79] Rather than a distinct ethnicity, they were a disparate and fluid community of people who came from across the Venetian, Ottoman and Habsburg Balkans.[80] Their base was in Segna, in Habsburg territory, and while nominally in the service of the empire, they primarily served their own ends. For several decades, the Uskoks seriously and quite indiscriminately disrupted trade and shipping in the Adriatic, and were heavily involved in the slave trade. Ottomans (including Christians) were their primary target, but Venetians and Ragusans were also considered fair game because of their close relations with the Ottomans. The Uskoks "enjoyed great popularity" and were "often supported" by the Christian population throughout the region.[81] For Venice, the Uskoks were "thieves by nature," to Ottomans they represented the most "detested of the 'infidel bandits.'"[82]

---

76 Emrah Safa Gürkan, "The Efficacy of Ottoman Counter-Intelligence in the 16th Century," *Acta Orientalia Academiae Scientiarum Hungaricae* 65 (2012), 11.

77 Rothenberg, "Christian Insurrections in Turkish Dalmatia," pp. 139–141.

78 James C. Scott, *Weapons of the Weak: Everyday Forms of Peasant Resistance* (New Haven, 1985).

79 Maria Pia Pedani, *In nome del Gran Signore: inviati ottomani a Venezia dalla caduta di Costantinopoli alla guerra di Candia* (Venice, 1994), p. 179; Gunther E. Rothenberg, "Venice and the Uskoks of Senj: 1537–1618," *Journal of Modern History* 33 (1961), 148; Bracewell, *The Uskoks of Senj*, pp. 51–88.

80 John V. A. Fine Jr., *When Ethnicity Did Not Matter in the Balkans: A Study of Identity in Pre-Nationalist Croatia, Dalmatia, and Slavonia in the Medieval and Early-Modern Periods* (Ann Arbor, 2006), pp. 216–219.

81 Tea Perinčić, "The Slave Trade on the Adriatic in the 17th Century," *Miscellanea Hadriatica et Mediterranea* 1 (2013), 112.

82 PTM, b. 417, 27 May 1596; Kafadar, "Evliya Çelebi in Dalmatia," p. 65.

Parallel to the rise of the Uskoks, many individuals in the region expressed their dissatisfaction with their feet by migrating to Ottoman lands and Ottoman service. Some converted to Islam, others did not, but many found alluring the increased opportunities, particularly in Istanbul. A variation on this is the many vlachs who migrated to Bosnia from the area of Clissa around the start of the seventeenth century, and ended up defending the Ottoman frontier, while other vlachs served the Habsburgs in the same capacity on the other side of the border.[83] In a similar vein, Muslim subjects of the sultans migrated northward from Bosnia to Habsburg lands.[84]

3 The Siege and Reconquest of Clissa Revisited

Given the persistently volatile situation in Dalmatia, it is no surprise that the Venetian *Signoria* viewed the taking of Clissa as a revolt against itself as much as an attack on the Ottomans. Venice was well aware of both its own subjects' discontent, and the interest of the papacy and the Habsburgs in fomenting dissent and disruption in the hope they would lead to a general Christian revolt in the Balkans.[85] Venetian officials thus worked actively to neutralize the imperial and papal machinations, to assuage the angry Ottomans, and to reassert authority among their restive subjects.

This is evident in the Venetian response to the capture of Clissa: It only took a few days for the news to reach Venice, where the Senate and *Signoria* responded immediately with a flurry of correspondence to key officials. The newly elected *Provveditore Generale in Dalmazia ed Albania*, Benetto Moro, was ordered to depart "immediately" for Dalmatia to supervise the Venetian response and to ensure "the security of public affairs in that province."[86] The Provveditore Generale was the senior official and authority in Venice's territories in the northern and central Adriatic, and though newly elected, Moro was a trusted figure who had held many important positions in the Venetian hierarchy.[87]

When he arrived in Zara on April 23, Moro was warned that his presence in Spalato might exacerbate the volatile situation there, so he delayed his transfer for several days. When he finally did take up residence, he and other top Vene-

83 Tracy, *Balkan* Wars, p. 350.

84 Marani, "*Atti pastorali di Minuccio Minucci*," pp. 50, 52–53; Horvat, "Glagolaši u Dalmaciji početkom 17. vijeka," pp. 557–559.

85 Rothenberg, "Christian Insurrections in Turkish Dalmatia," p. 136.

86 *Senato deliberazioni Costantinopoli*, b. 9, 13 Apr 1596.

87 Arbel, "Venice's Maritime Empire in the Early Modern Period," p. 152.

tian officials were deeply uncertain about the course of action to pursue, particularly in relation to Venice's own subjects, whom they feared might revolt, flee, or join the Habsburg forces gathering to relieve Clissa. Moro reported that the region was "greatly distanced from the faith and love that used to exist toward" Venice, and that everyone favored Vienna. Many subjects had fled to the mountains upon his arrival, and he was convinced that "the whole province would rise up" if he were too aggressive or heavy-handed. Anonymous placards posted in Zara declaring "Viva, Viva, Viva Clissa and those who hold it" underlined the challenge Moro faced.[88]

As rumors of an imminent imperial intervention grew, Venetian subjects continued to lend their support to the besieged defenders of Clissa. A cook in a Trau monastery removed his tunic and accompanied several Venetian sailors and a soldier headed to join the defenders, while the "greater part" of the relief force that broke through to the fortress with provisions consisted of Venetian subjects led by a captain formerly in Venetian service.[89] According to Moro, "everyone" in the region felt "enormous happiness" at the success of this mission: the churches were open all night, "crowded with all the *popolo*, who pray with bare knees on the ground … that the Clissa affair turns out favorably."[90] A priest went from house to house carrying women's spinning implements, shaming men who had not joined the cause: "Here you go, coward, this is your weapon, leave the guns and sabers. Put on an apron and take a staff and head off. Don't you see that your comrades have left for a holy battle of faith, and you have stayed home alone to be lazy?"[91] The subversive role of many clerics in the region, who went about "fomenting, inflaming, and exciting the laypeople to riots and uprisings of all sorts" was of such concern that Venice's ambassador in Rome complained of the republic's "enormous displeasure" directly to the pope.[92]

Paralleling suspicions about their subjects, Venetian officials also harbored doubts about the loyalty of their own military forces, which included numerous Albanians, Croats, and Italians. A majority were from Polizza, and were therefore Ottoman subjects, though in Venetian service. To further complicate

88 *PTM*, b. 417, 27 Apr 1596; *PTM*, b. 417, 28 Apr 1596; *PTM*, b. 417, 24 Apr 1596; *PTM*, b. 417, 27 May 1596.

89 *PTM*, b. 417, 26 May 1596; *PTM*, b. 417, 12 May 1596.

90 *PTM*, b. 417, 12 May 1596.

91 Klaić, *Povijest Hrvatâ*, 5:523.

92 *Senato dispacci Roma*, b. 37, 1 June 1596, c. 255r. On the growing leadership role of clerics in early modern revolts, see Samuel K. Cohn, Jr., "Authority and Popular Resistance," in *The Oxford Handbook of Early Modern History, 1350–1750*, vol. 2: *Cultures and Power*, ed. Hamish Scott (Oxford, 2015), pp. 427–428.

matters, many men from the same region of Polizza played an important, early role in defending Clissa against their Ottoman lords, following the initial seizure of the fortress in April. As a result, Moro preferred not to entrust them with "very important" duties. The many Venetian "Croat" subjects serving in the fleet were similarly suspect, "because of the relationship that they have with the Uskoks." Moro was convinced that he could not "trust them, or utilize them."[93] Thus, Venice's military strength was deeply compromised because of the fluid situation of the frontier.

Part of the fear of a popular uprising was due to the fact that from the outset of the crisis Venetian officials worked closely with the Ottomans to resolve the situation quickly and cleanly. This fit squarely with the longstanding Veneto-Ottoman cooperation to preserve the "bona pace" in Dalmatia. So for example, the Venetians provided critical logistical assistance in a variety of ways. When the Ottomans initially besieged Clissa, they suffered from a severe shortage of provisions, and so, in order "to be good neighbors with the Turks, and so that they know that in all things [Venice] had no knowledge of the taking of Clissa," Moro permitted them to buy "wine, figs, carob, and rice" in Spalato. He also provided an armed escort to accompany them back to their camps.[94] It was whispered that Venice also supplied the Ottomans "with a great quantity of barrels of powder."[95] When the affair was resolved, the chief Ottoman military men acknowledged that Venice had done everything necessary to "faithfully preserve the good peace" with the sultan.[96]

Veneto-Ottoman collaboration also involved information sharing. When rumors of a conspiracy against Clissa began circulating in 1595, Venetian officials in Dalmatia warned their Ottoman counterparts.[97] During the siege, when Moro was informed that several Uskok ships "loaded with provisions to succor Clissa" had landed near Spalato, he "immediately sen[t] a person directly to Salona to warn the Turks to be vigilant."[98] The collaboration carried over to Istanbul, where the Venetian representative, Bailo Marco Venier, worked closely with Grand Vizier Damat Ibrahim Pasha to resolve the situation without harming Veneto-Ottoman relations, and to preserve the commercial "fruits[...] that the subjects of both rulers receive on that border." To accomplish this, the two consulted an Ottoman "design of the gulf badly painted by hand" to get a sense of the situation in the region. And, while Ibrahim pushed for Venice to

---

93 PTM, b. 417, 2 May 1596; Tolomeo, "La fortezza di Clissa," p. 48.
94 PTM, b. 417, 2 May 1596; PTM, b. 417, 23 Apr 1596.
95 *Senato dispacci Roma*, b. 37, 8 June 1596, cc. 216r–218v.
96 PTM, b. 417, 6 May 1596.
97 *Cariche da mar – Processi*, b. 8, 13–17 Apr 1596.
98 PTM, b. 417, 26 Apr 1596.

resolve the Uskok situation permanently by attacking Segna, he also assured Venier that the Ottoman fleet would not "enter into your house," that is the Adriatic, without permission, which was a principal Venetian concern.[99]

The most consequential Venetian effort was establishing a blockade to prevent the reinforcing and re-provisioning of Clissa. As soon as word of its fall reached Rome, Clement ordered provisions and munitions sent from Ancona, and Giovanni Alberti requested papal intervention so that Venice would not "impede the assistance and provisions" coming by sea.[100] Instead, the Venetian Senate "explicitly ordered" that no aid be given to the defenders, and directed Moro to blockade the coastal areas around Spalato "to impede the navigation of the Uskoks" and prevent anyone coming by sea from joining or relieving the defenders of Clissa.[101] Some in "influential quarters" in Venice even advocated for direct military intervention to restore the fortress to Ottoman rule, leading to familiar, and not entirely inaccurate, accusations that Venice was guilty "of outright collaboration with the common enemy of Christendom."[102] And indeed early in the siege the grand vizier made the quite astonishing offer to put the Ottoman fleet under Venetian control in order to expand the blockade and strengthen Venice's ability to control the Adriatic.[103]

While not watertight, the blockade ultimately proved effective. When the Viceroy of Naples, for example, dispatched a "royal frigate" with letters regarding provisions he intended to send to Clissa, it "was taken by a galley of the Republic" and all its men "put to the oar."[104] When the imperial commander Lenkovich set out to relieve Clissa on May 19, he was delayed almost an entire week by "disasters on the sea" and the Venetian blockade.[105] In the aftermath of the recapture of the fortress, Lenkovich claimed that his relief effort was doomed when the Venetian *Capitano del Golfo* warned the Ottomans of his coming with three cannon shots.[106] While this was publicly denied by the Venetians, in private, the Ottoman commander acknowledged that the "recov-

---

99 *Senato dispacci Costantinopoli*, b. 43, 13 June 1596, cc. 271r–272v; *Senato dispacci Costantinopoli*, b. 43, 19 May 1596, cc. 204r–v; *Senato dispacci Roma*, b. 37, 27 Apr 1596, cc. 143v–144v.

100 *Senato deliberazioni Roma ordinaria*, r. 11, 20 Apr 1596, c. 16r; Letter from Giovanni Alberti to Clement VIII, 10 April 1596, in Horvat, ed., *Monumenta historiam Uscocchorum*, 1:94.

101 PTM, b. 417, 6 May 1596; *Senato deliberazioni Roma ordinaria*, r. 11, 10 May 1596, cc. 22r–v.

102 Bracewell, *The Uskoks of Senj*, pp. 28, 48, 107; Rothenberg, "Christian Insurrections in Turkish Dalmatia," pp. 144–146; Hieronimo Pavon to Nicola Alberti, 20 April 1596, in Horvat, ed., *Monumenta historiam Uscocchorum*, 1:102.

103 *Senato dispacci Costantinopoli*, b. 43, 13 June 1596, cc. 271r–272v.

104 *Senato dispacci Roma*, b. 37, 8 June 1596, cc. 216r–218v.

105 Klaić, *Povijest Hrvatâ*, 5:525.

106 *Senato dispacci Roma*, b. 37, 8 June 1596, cc. 216r–218v.

ery" of the fortress had been made possible, at least in part, by unspecified Venetian "actions."[107] In his own analysis, the Venetian commander Moro contended that the blockade both hindered relief from reaching Clissa, and prevented the Dalmatian revolt against Venice from spreading. Indeed, he believed that "without a doubt" the people of Spalato would have rebelled and "plant[ed] the imperial standards on the city's towers," were it not for Venice's decisive actions.[108]

The popular uprising manqué was central to the ultimate failure of the conspirators to hold Clissa. They assumed that the capture of the most important Ottoman fortress on the frontier would encourage oppressed Dalmatians to revolt and drive off their Ottoman and Venetian overlords, and that this would quickly spread throughout the Balkans, and perhaps into the heartland of the empire itself.[109] While many in the region supported the capture of Clissa, and even joined the fray, others "vehemently criticized" the conspirators' actions as being disruptive to the social and political order, or to the fragile economy of the region, while many others simply lay low waiting to see how the situation would play out. The end result was that the hoped-for general uprising never materialized.[110]

This was crucial to Lenkovich's failed attempted to relieve the fortress on May 27. He had been convinced that the Ottoman Christian vlachs would turn on their Muslim masters, and thus made the fatal decision to immediately attack the besieging Ottoman forces head on, despite a 15 to 1 disparity in numbers. For a time the vlachs did in fact hold back, waiting to see "which direction the victory swung." But seeing the tide turning against the small imperial force, and unwilling to compromise their position with the Ottomans, the vlachs began rolling stones onto the imperial force, and with arquebuses made "a horrible massacre," and this proved pivotal in the Ottoman rout.[111]

In the final stages and aftermath of the siege, the leader of the conspiracy, Giovanni Alberti, was killed in battle trying to rally his hunger-weakened troops.[112] The Venetians punished some of their subjects and the clergy who had participated in the affair, largely as a show to mollify the Ottomans when they discovered how important the involvement of Venetian subjects had been

107 *Senato dispacci Costantinopoli*, b. 43, 21 June 1596, c. 310r.

108 PTM, b. 417, 3 June 1596.

109 Zlatar, *Between the Double Eagle and the Crescent*, p. 92.

110 Knapton, "Tra Dominante e dominio," p. 382; Rothenberg, *The Austrian Military Border in Croatia*, pp. 60–61.

111 PTM, b. 417, 27 May 1596; Lazzaro Soranzo, *L'Ottomanno* (4th ed., Naples, 1600), pp. 108–181; Klaić, *Povijest Hrvatâ*, p. 5.

112 "Relatione del'infelice avvenimento," 1:251–252; Minuccio Minucci, *Historia degli Uscochi scritta da Minvcio Minvci arcivescovo di Zara* (Venice, 1683), p. 60.

to the whole enterprise.[113] In general, however, Moro thought it wise "to correct [the people] as a pious father, rather than castigate them as a severe judge," given the still volatile situation in Dalmatia.[114] For the Ottomans, the successful recapture of Clissa unsurprisingly was not celebrated publicly in any fashion.[115] They too settled some scores: the sanjakbeg of Clissa was replaced, and the Polizza region was raided to punish the families of the men who had participated in the conspiracy.[116] These Venetian and Ottoman actions, combined with the unresolved issues that had inspired the uprising in the first place, served to radicalize many more people in the region, who joined the Uskoks in Segna intent on seeking revenge. This ensured that many of the same issues that produced the Clissa affair would continue to plague the region.[117] Indeed, less than a year later, another plan to recapture Clissa materialized, but was cut short when a local Venetian official warned an Ottoman counterpart.[118]

4 Conclusion

The famed capture, siege and ultimate loss of the fortress of Clissa in the spring of 1596 is usually portrayed as a minor, but symbolically potent moment in the age-old "clash of civilizations," with proto-nationalist undertones.[119] There is ample documentary evidence of ways in which contemporaries availed themselves of religious rhetoric and imagery in representing the recurring conflicts along the Dalmatian frontier, the triplex confinium where the political boundaries of the three major eastern Mediterranean empires intersected. The actual situation on the ground, however, was much more subtly shaded than allowed for by the stark chiaroscuro of this traditional representation, advanced by contemporaries and embraced by generations of scholars. The actual borders were ill-defined and widely ignored by the region's inhabitants, who shared a fluid cross-border culture that transcended all attempts to sketch out clearcut political or religious lines of demarcation. This is evident in the mix-and-match composition of Muslim, Christian, Ottoman, Habsburg and Venetian conspirators who initially captured Clissa, as well as the equally grab-bag character of the several military forces involved in the siege, from the nominally Habsburg

113 *Senato deliberazioni Costantinopoli*, b. 9, 6 July 1596.
114 PTM, b. 417, 27 June 1596; PTM, b. 417, 30 May 1596.
115 *Senato dispacci Costantinopoli*, b. 43, 15 June 1596, cc. 302v–303r.
116 PTM, b. 417, 4 July 1596; *Senato Mar*, b. 56, 19 July 1596, cc. 67v–68r.
117 Klaić, *Povijest Hrvatâ*, 5:526; Tracy, *Balkan* Wars, p. 340.
118 Malcolm, *Agents of Empire*, p. 411.
119 See for example, Tracy, *Balkan Wars*, pp. 2–3; Klaić, *Povijest Hrvatâ*, 5:523.

Uskoks to the Ottoman vlachs to the Venetian subjects of coastal Dalmatia. It is similarly manifest in the collaboration between Venetian and Ottoman officials in Dalmatia, Venice and Istanbul, to return the frontier to the status quo. The roots of the Clissa affair are better understood not as primarily religious in nature, the opening salvo in a new crusade as the conspirators and their allies attempted to depict it, but rather as another in a long series of revolts and frontier disruptions growing out of the deeply-rooted economic and political tensions between imperial centers and the distant peripheries of their composite states, linked to the significant changes affecting the Mediterranean and Adriatic during the early modern era.

Bibliography

Archival Sources

Archivio di stato di Venezia
Archivio proprio Costantinopoli, b. 20/II
Bailo a Costantinopoli, b. 365 I
Cariche da mar – Processi, b. 8
Collegio – Relazioni, b. 72
Collegio – Risposte di dentro, b. 10
Collegio – Risposte di fuori, b. 352
Commemoriali, r. 25
Lettere e scritture turchesche, b. 5
Provveditori di terra e mar (PTM), b. 417, 422, 922
Senato deliberazioni Costantinopoli, b. 5
Senato deliberazioni Roma ordinaria, r. 11
Senato dispacci Costantinopoli, b. 43
Senato dispacci Roma, b. 37
Biblioteca Marciana (BM)
It VI 105 (5728)
It VII 1217 (9448)
Biblioteca del Museo civico Correr di Venezia (BMCV)
Cicogna, b. 2854
Cicogna, b. 3098
British Library (BL)
MS 8605
MS 8610
MS 8613

Printed Sources

Čelebī, Evlijā, *Putopis: Odlomci o júgoslavenskim zemljama*, ed. Hazim Šabanović (Sarajevo, 1967).
Horvat, Carolus, ed., *Monumenta historiam Uscocchorum illustrantia* (Monumenta spectantia historiam Slavorum meridonalium), vol. 32 (Zagreb, 1910).
Lamansky, Vladimir, ed., *Secrets d'état de Venise*, vol. 1 (St. Petersburg, 1884; reprint New York, 1968).
Lopasić, R., ed., *Spomenici hrvatske Krajine* (Monumenta spectantia historiam Slavorum meridonalium), vol. 15 (Zagreb, 1884).
Ljubić, Simeon, ed., *Commissiones et Relationes Venetae*, tome 2 (Monumenta spectantia historiam slavorum meridionalium), vol. 8 (Zagreb, 1877).
Ljubić, Simeon, ed., *Commissiones et Relationes Venetae*, tome 3 (Monumenta spectantia historiam slavorum meridionalium), vol. 11 (Zagreb, 1880).
Marani, Alberto, ed., *Atti pastorali di Minuccio Minucci arcivescovo di Zara (1596–1604)* (Rome, 1970).
Minucci, Minuccio, *Historia degli Uscochi scritta da Minvcio Minvci arcivescovo di Zara* (Venice, 1683).
Novak, Grga, ed., *Commissiones et Relationes Venetae*, tome 4 (Monumenta spectantia historiam slavorum meridionalium), vol. 47 (Zagreb, 1964).
Novak, Grga, ed., *Commissiones et Relationes Venetae*, tome 6 (Monumenta spectantia historiam slavorum meridionalium), vol. 49 (Zagreb, 1970).
Novak, Grga, ed., *Commissiones et Relationes Venetae*, tome 7 (Monumenta spectantia historiam slavorum meridionalium), vol. 50 (Zagreb, 1972).
Solitro, V., *Documenti sull'Istria e la Dalmazia* (Venice, 1844).
Soranzo, Lazzaro, *L'Ottomanno* (4th ed., Naples, 1600).

Secondary Sources

Ágoston, Gábor, "Empires and Warfare in East-Central Europe, 1550–1750: The Ottoman-Habsburg Rivalry, Military Transformation," in *European Warfare, 1350–1750*, ed. Frank Tallett, D. J. B. Trim (Cambridge, UK, 2010), pp. 110–134.
Arbel, Benjamin, "Venice's Maritime Empire in the Early Modern Period," in *A Companion to Venetian History, 1400–1797*, ed. Eric Dursteler (Leiden, 2013), pp. 125–254.
Benvenuti, Angelo de, "Fortezze e castelli di Dalmazia: la fortezza di Clissa," *La rivista dalmatica* 16, no. 4 (1935), 28–40; 17, no. 1 (1936), 16–27.
Benvenuti, Angelo de, *Storia di Zara dal 1409 al 1797* (Milan, 1944).
Bracewell, Wendy, "The Historiography of the Triplex Confinium: Conflict and Community on a Triple Frontier, 16th–18th Centuries," in *Frontiers and the Writing of History, 1500–1850*, ed. Steven G. Ellis, Raingard Esser (Hannover, 2006), pp. 211–227.

Bracewell, Catherine Wendy, *The Uskoks of Senj: Piracy, Banditry and Holy War in the Sixteenth-Century Adriatic* (Ithaca, 1992).
Braudel, Fernand, *The Mediterranean and the Mediterranean World in the Age of Philip II*, 2 vols (New York, 1972).
Brummett, Palmira, *Mapping the Ottomans: Sovereignty, Territory, and Identity in the Early Modern Mediterranean* (Cambridge, UK, 2015).
Cohn, Jr., Samuel K., "Authority and Popular Resistance," in *The Oxford Handbook of Early Modern History, 1350–1750*, vol. 2: *Cultures and Power*, ed. Hamish Scott. (Oxford, 2015), pp. 418–439.
Colley, Linda, *Captives: Britain, Empire, and the World, 1600-1850* (New York, 2002).
Dursteler, Eric R., *Venetians in Constantinople: Nation, Identity, and Coexistence in the Early Modern Mediterranean* (Baltimore, 2006).
Evans, R. J. W., *Rudolf II and His World: A Study in Intellectual History 1576–1612* (London, 1997).
Fabijanec, Florence Sabine, "Le développement commercial de Zadar et de Split aux XVe–XVIe siècles," Ph.D. Dissertation (Université Paris I, 2002).
Fine Jr., John V. A., *When Ethnicity Did Not Matter in the Balkans: A Study of Identity in Pre-Nationalist Croatia, Dalmatia, and Slavonia in the Medieval and Early-Modern Periods* (Ann Arbor, 2006).
Frejdenberg, Maren, "Venetian Jews and Ottoman Authorities on the Balkans (16th Century)," *Etudes Balkaniques* 30 (1994), 56–66.
Graovac, Vera, "Populacijski razvoj Zadra," *Geoadria* 9 (2004), 51–72.
Gürkan, Emrah Safa, "The Efficacy of Ottoman Counter-Intelligence in the 16th Century," *Acta Orientalia Academiae Scientiarum Hungaricae* 65 (2012), 1–38.
Horvat, Karlo, "Glagolaši u Dalmaciji početkom 17. vijeka t. j. godine 1602.–1603," *Starine* 33 (1911), 537–564.
Ivetic, Egidio; Roksandić, Drago, eds., *Tolerance and Intolerance on the Triplex Confinium: Approaching the "Other" on the Borderlands Eastern Adriatic and Beyond, 1500–1800* (Padua, 2007).
Kafadar, Cemal, "Evliya Çelebi in Dalmatia: An Ottoman Gentleman's Encounter with the Arts of the Franks," in *Dalmatia and the Mediterranean: Portable Archeology and the Poetics of Influence*, ed. Alina Payne (Leiden, 2014), pp. 59–78.
Karpat, Kemal, "The Stages of Ottoman History: A Structural Comparative Approach," in *The Ottoman State and its Place in World History*, ed. Kemal Karpat (Leiden, 1974), pp. 79–98.
King, Martin Luther, *Where Do We Go From Here?* (Boston, 2010).
Klaić, Vjekoslav, *Povijest Hrvatâ od najstarijih vremena de svršetka XIX stoljeća* (Zagreb, 1981).

Knapton, Michael, "Tra Dominante e dominio (1517–1630)," in *La Repubblica di Venezia nell'età moderna*, vol. 2: *Dal 1517 alla fine della Repubblica*, ed. Gaetano Cozzi, Michael Knapton (Turin, 1992), pp. 201–549.

Krokar, James P., "New Means to an Old End: Early Modern Maps in the Service of an Anti-Ottoman Crusade," *Imago Mundi* 60 (2008), 23–38.

Kužić, Krešimir, "Prilog biografiji nekih Kačićevih vitezova te podrijetlu stanovništva njihova kraja," *Radovi Zavoda za povijesne znanosti HAZU u Zadru* 47 (2005), 191–224.

Luca, Cristian, "The Vlachs/Morlaks in the Hinterlands of Traù (Trogir) and Sebenico (Šibenik), Towns of the Venetian Dalmatia, during the 16th Century," in *Miscellanea Historica et Archaelogica in Honorem Professoris Ionel Câdea*, ed. Valeriu Sîrbu, Cristian Luca (Brăila, 2009), pp. 311–322.

Malcolm, Noel, *Agents of Empire: Knights, Corsairs, Jesuits and Spies in the Sixteenth-Century Mediterranean World* (Oxford, 2015).

Nicolle, David, *Ottoman Fortifications 1300–1710* (Oxford, 2010).

Niederkorn, Jan Paul, *Die europäischen Mächte und der 'Lange Türkenkrieg' Kaiser Rudolfs II. (1593–1606)* (Vienna, 1993).

Novak, Grga, "Poljoprivreda na dalmatinskom primorju i otocima u XVIII stoljeću," *Starine* 51 (1962), 61–111.

Paci, Renzo, *La scala di Spalato e il commercio veneziano nei Balcani fra Cinque e Seicento* (Venice, 1971).

Parker, Geoffrey, *The Military Revolution: Military Innovation and the Rise of the West, 1500–1800* (2nd ed., Cambridge, UK, 1996).

Pedani, Maria Pia, "Beyond the Frontier: The Ottoman-Venetian Border in the Adriatic Context from the Sixteenth to the Eighteenth Centuries," in *Zones of Fracture in Modern Europe: The Baltic Countries, the Balkans and Northern Italy*, ed. Almut Bues (Wiesbaden, 2005), pp. 45–60.

Pedani, Maria Pia, *Dalla frontiera al confine* (Rome, 2002).

Pedani, Maria Pia, *In nome del Gran Signore: inviati ottomani a Venezia dalla caduta di Costantinopoli alla guerra di Candia* (Venice, 1994).

Pepper, Simon, "Defending the Frontiers of Venice: Fortification and Defensive Strategy in the Friuli before Palmanova," in *L'architettura militare di Venezia in terraferma e in Adriatico fra XVI e XVII secolo: Atti del Convegno internazionale di studi (Palmanova, 8–10 novembre 2013)*, ed. Francesco Paolo Fiore (Florence, 2014), pp. 3–20.

Perinčić, Tea, "The Slave Trade on the Adriatic in the 17th Century," *Miscellanea Hadriatica et Mediterranea* 1 (2013), 109–120.

Pollak, Martha, *Cities at War in Early Modern Europe* (Cambridge, UK, 2010).

Praga, Giuseppe, *History of Dalmatia* (Pisa, 1993).

Rogers, Clifford J., ed., *The Military Revolution Debate: Readings on the Military Transformation of Early Modern Europe* (Boulder 1995).

Rothenberg, G. E., "Christian Insurrections in Turkish Dalmatia 1580–96," *Slavonic and East European Review* 40 (1961), 136–147.

Rothenberg, Gunther Erich, *The Austrian Military Border in Croatia, 1522–1747* (Urbana, 1960).

Rothenberg, Gunther E., "Venice and the Uskoks of Senj: 1537–1618," *Journal of Modern History* 33 (1961), 148–156.

Scott, James C., *Weapons of the Weak: Everyday Forms of Peasant Resistance* (New Haven, 1985).

Singleton, Fred, *A Short History of the Yugoslav Peoples* (Cambridge, UK, 1985).

Stevens, Carol B., "Warfare on Land," in *The Oxford Handbook of Early Modern European History, 1350–1750*, vol. 2: *Cultures and Power*, ed. Hamish Scott (Oxford, 2015), pp. 561–590.

Tolomeo, Rita, "La fortezza di Clissa," *Atti e memorie della Società dalmata di storia patria* 34 (2012), 31–72.

Tracy, James D., *Balkan Wars: Habsburg Croatia, Ottoman Bosnia, and Venetian Dalmatia, 1499–1617* (Lanham, 2016).

Traljić, Seid M., "Tursko-mletačke granice u Dalmaciji u XVI. i XVII. Stoljeću," *Radovi* 20 (1973), 447–458.

Traljić, Seid M., "Zadar i turska pozadina od XV do potkraj XIX stoljeća," *Radovi* 11–12 (1965), 203–227.

Vaughan, Dorothy M., *Europe and the Turk: A Pattern of Alliances, 1350–1700* (Liverpool, 1954; reprint, New York, 1976).

Vitezić, Ivan, *La prima visita apostolica postridentina in Dalmazia (nell'anno 1579)* (Rome, 1957).

Wilkinson, J. Gardner, *Dalmatia and Montenegro*, vol. 1 (London, 1848).

Wolff, Larry, *Venice and the Slavs: The Discovery of Dalmatia in the Age of Enlightenment* (Stanford, 2001).

Zlatar, Zdenko, *Between the Double Eagle and the Crescent: The Republic of Dubrovnik and the Origins of the Eastern Question* (Boulder, 1992).

CHAPTER 6

The 1689 Siege of Bombay in Global Historical Perspective

Philip J. Stern

On February 15, 1689, the East India Company's fort, settlement, and town at Bombay (today, Mumbai) received an unwelcome surprise. Along with a large retinue of sailors and soldiers, Sidi Qasim Yaqut Khan, a tributary naval commander for the Mughal Empire and a power in his own right in the western Indian littoral, landed at the village of Sewri to the north of Bombay fort, and proceeded to occupy and force evacuation of the outlying fortlets around the archipelago – at Sewri, as well as Mahim, Sion, and Worli. Ultimately making his way to the southeastern hill forts at Mazagaon and Dongri, for the next sixteen months, his forces laid siege to the fort and town of Bombay, which only two years earlier the East India Company had effectively declared the capital of its operations throughout Asia.[1]

The siege of Bombay has barely registered as an event in the history of the "world of the siege" and has been consigned to an historiographical footnote to the East India Company's early history in India, as a symbol of the "total fiasco" of the short-term, temporary and ill-advised policy of late seventeenth-century Company leadership to turn away from irenic commerce and instead seek war with the Mughal Empire.[2] In this sense, the siege was an emergency and an exception – but one that supposedly proved the Company could *not* be a sovereign, fundamentally different from Giorgio Agamben's notion of the siege, often fictive, as revealing the capacity to produce a state of emergency that was the very mark of sovereignty.[3]

1 For elaboration on the course and consequences of the siege, including a critical edition of the only comprehensive contemporary account of the siege, see Margaret R. Hunt, Philip J. Stern, eds., *The English East India Company at the Height of Mughal Expansion: A Soldier's Diary of the 1689 Siege of Bombay* (Boston and New York, 2016). I have also discussed the Sidi's invasion in Philip J. Stern, *The Company-State: Corporate Sovereignty and the Early Modern Foundations of the British Empire in India* (New York, 2011).

2 Bruce Lenman, *England's Colonial Wars 1550–1688: Conflicts, Empire and National Identity* (London, 2001; reprint: Abingdon, 2013), p. 210. See, also, for example, Bruce Lenman, "The East India Company and the Emperor Aurangzeb," *History Today* 37, no. 2 (1987), 23–29; Philip Lawson, *The East India Company: A History* (Abingdon, 1993), pp. 49–50.

3 Giorgio Agamben, *State of Exception* (Chicago, 2005), pp. 17, 28.

 | DOI:10.1163/9789004395695_007

For the East India Company, this chapter suggests, the siege was exceptional in neither sense. Though treated as catastrophic by the Company's critics and many historians since, much about the siege was indeed unremarkable – emblematic of siege warfare in both Europe and Asia, and reflective of the nature of early modern Eurasian claims to authority, jurisdiction, and sovereignty. The siege resulted from sustained attempts by the East India Company to insert itself as both a commercial and political power in the western Indian littoral, exemplifying the fuzzy boundaries between "inside" and "outside" that shaped political and religious allegiance and obligation in the early modern world. In this sense, the siege was a normal articulation of the ways in which law and politics were forged in the Indian Ocean world: a balance of armed conflict, diplomacy, and fluid sovereign forms. As such, the Bombay government did not use the siege so much to override or suspend the usual procedures of law but rather saw their challenge as how to maintain the rule of law amidst the pragmatic and emotional stresses of mortality, desertion, disruption, and the mind-numbing set of quotidian routines that marked daily life in an early colonial settlement, but which were no doubt amplified by the sustained siege of Bombay town and fort. As such, the siege was no interruption in the East India Company's history, but rather a continuous part of the making of sovereignty and jurisdiction in what one might call British "India."

If the Company capitalized upon the siege as a state of emergency, it was in its aftermath. It was the history of the siege as crisis, rather than the siege itself, that propelled the development and buildup of Bombay's defenses and infrastructure into the eighteenth century. In the decades following the siege, its memory and history became a constant reminder of the need for more aggressive defense and expansion of the Company's sovereign claims, and almost became embedded in the political culture and anxieties of the eighteenth-century colonial state, at Bombay and across maritime British India more generally.[4] It was only after the territorial expansion of the East India Company in the late eighteenth century and the firm establishment of the so-called British Raj in the nineteenth that the siege could be safely historiographically abandoned as a curiosity and an artefact of a distant and antiquarian past.

Both the immediate importance and the historiographical forgetting of the Bombay siege thus form a case study in the intersections of historiography, culture, and siege warfare reflected in this volume more generally. It also

4 For a parallel case of the ways in which military defeat leads to reassessment and transformation of colonial fortification, see Pedro Luengo, "Military Engineering in Eighteenth-Century Havana and Manila: The Experience of the Seven Years War," *War in History* 24, no. 1 (2017), 4–27.

represents a key example of the importance of recovering the histories of "failure" that, as Alison Games has argued, are so critical to the story of empire: not a series of progressive victories punctuated by the exceptions, outliers, or pre-histories of defeat but rather a "complex process, one riddled with trial and error, success and failure, triumph and despair."[5] Furthermore, this conflict opens up a range of questions as to how to define and understand "victory" and "defeat" in the context of siege warfare – and its memory – in the first place, not to mention military and political conflict itself in the religiously, ethnically, and culturally porous world of seventeenth-century Eurasian imperial expansion. The siege was not simply a bipolar clash between English and Sidi forces but rather resulted from a complex of interactions among a range of political actors, all jostling for position and power in an Indian Ocean world defined neither by partnership nor outright states of war but rather "bounded" and "contained conflict."[6] The siege itself was a political act, continuous with rather than a result of the failure of diplomacy. In turn, while the engagement itself certainly prompted any number of instances of discordant goals and misunderstandings across cultures and camps, ultimately the experience revealed far more often efforts at translation, communication, as well as diplomacy and exchange that were endemic to order in the early modern world, a form of what Lauren Benton and Adam Clulow have called the "interpolity law" centered on "protocol, jurisdiction and protection."[7]

Finally, the Bombay siege becomes a prime focal point for a much wider and more global story about the role of violence and diplomacy in shaping the interaction of early modern Eurasian empires. It reveals the ways in which that violence was shaped by a range of cultural and religious border-crossings that cause us to question our received and static notions of what constituted ethnic, national, political, or religious identities. Revisiting the history of the Bombay siege offers an opportunity to explore the methodological problem of scale in world history, an exercise in what has been called, paradoxically but productively, "global microhistory."[8] This siege, like all sieges, was an intensely local

5 Alison Games, *The Web of Empire: English Cosmopolitans in an Age of Expansion, 1560–1660* (New York, 2008), p. 14.

6 As suggested by Sanjay Subrahmanyam, *The Political Economy of Commerce: Southern India 1500–1650* (Cambridge, UK, 1990), p. 254, and taken up by many since.

7 Lauren Benton, Adam Clulow, "Legal encounters and the origins of global law," in *The Cambridge World History*, vol. 6: *The Construction of a Global World 1400–1800 CE*, part 2: *Patterns of Change*, ed. Jerry Bentley, Sanjay Subrahmanyam, Merry E. Wiesner-Hanks (Cambridge, UK, 2015), p. 82.

8 See, e.g., Carlo Ginzburg, "Microhistory and World History," in *The Cambridge World History*, vol. 6: *The Construction of a Global World 1400–1800 CE*, part 2: *Patterns of Change*, ed. Jerry Bentley, Sanjay Subrahmanyam, Merry E. Wiesner-Hanks (Cambridge, UK, 2015),

affair that illuminated the complex cultural, social, and ethnic fabric of a fledgling European colony on the outskirts of a major Asian empire. Yet its causes and effects were deeply embedded in regional histories in which such colonies were always situated, which in turn were produced by a variety of global forces, from centuries of European expansion in Asia to the highly competitive market for soldiers and settlers that stretched from Persia to China. This story is, in this sense, an example of the European blending of "global consciousness" with the historian's "concrete story," which Natalie Zemon Davis has argued is perhaps best exemplified in "cases of cultural crossing."[9] Thus, this moment in Bombay's history suggests one way in which the siege can have overlapping but also competing and evolving histories that are at the same time local and global in nature.

1 Making "English" Bombay in the Early Modern Indian Ocean World

The long-term roots of Sidi Yaqut's invasion can be argued to have extended back decades, to the "transfer" of Bombay from the Portuguese to the English Crown in 1661, if not even earlier. The treaty itself, in which the English Crown also took possession of the North African town of Tangier, originated in circumstances and concerns far removed from the politics of the western Indian littoral. At its root a dynastic alliance forged by the marriage of the English King Charles II and the Portuguese princess Catherine of Braganza, the Luso-English strategic arrangement certainly had potential consequences for Europeans in India, especially in theoretically aligning English interests with the century-long Portuguese conflict with the Dutch throughout South and Southeast Asia. Yet, in the end, the arrangement had far greater implications for European dynastic politics and Atlantic commerce. In the end, it was Spain – which had at best a limited presence in the East Indies – not the Netherlands that seemed most immediately troubled about the treaty and militated against its promulgation.[10]

---

pp. 446–473; Filippo de Vivo, "Prospect or Refuge? Microhistory, History on the Large Scale," *Cultural and Social History: The Journal of the Social History Society* 7, no. 3 (2010), 387–397; Tonio Andrade, "A Chinese Farmer, Two Black Boys, and a Warlord: Towards a Global Microhistory," *Journal of World History* 21, no. 4 (2011), 573–591; Lara Putnam, "To Study the Fragments/Whole: Microhistory and the Atlantic World," *Journal of Social History* 39, no. 3 (2006), 615–630.

9 Natalie Zemon Davis, "Decentering History: Local Stories and Cultural Crossings in a Global World," *History and Theory* 50, no. 2 (2011), p. 197.

10 Gerald L. Belcher, "Spain and the Anglo-Portuguese Alliance of 1661: A Reassessment of Charles II's Foreign Policy at the Restoration," *Journal of British Studies* 15, no. 1 (1975),

Bombay was somewhat more of an afterthought in this exchange. The English state's ambitions were far more focused on Tangier, and the Crown seemed either uninterested or unable to govern effectively in India, very soon granting Bombay to the English East India Company, for a nominal rent of £10 per year, as a sort of proprietary colony in 1668. Whether under Crown or Company rule, rendering Bombay into an English colony was not an easy task. The sparsely inhabited island was still home primarily to Portuguese-speaking Catholics, not to mention polyglot communities of Hindus, Muslims, Zoroastrians, and others. In the late seventeenth century, much of the land, labor, and littoral was still controlled by the dozen or so powerful Luso-Indian *fidalgos*, as well as Bombay's Jesuit mission, which resided on the northwest portion of Bombay as well as on the island of Salsette to the north, regions disputed by local Portuguese officials as not having been included in the original grant of jurisdiction. English governors left in place various aspects of Portuguese law and systems of adjudication, including the continued use of *vereadores*, local grandees who served as a sort of regional Justice of the Peace (the term usually translates as "aldermen"). Most notably, of course, English officials, bound by the terms of the 1660 Anglo-Portuguese treaty and the Company's proprietary charter as well as their own principles of political economy, continued to allow relatively free exercise of Catholicism as well as continued functioning of the main Catholic church just outside the town walls, known to contemporaries, tellingly, as the "Portuguese church." The Portuguese language, or some version of it, was also ubiquitous – as it was in many places around the commercial south Asian littoral. The continued presence of Portuguese language, people, and Catholics on the island was by its very nature never simply an inward-facing issue. It also formed the foundation for continued attempts by officials in the *Estado da India*'s nearby regional stronghold of Bassein, its capital in Goa, and even as far away as Lisbon to continue to dispute the terms of the original transfer of the island. Both sides disagreed vehemently over the degree of sovereignty the English had over Bombay's inhabitants, especially its Catholic subjects. Moreover, Portuguese officials continued to exact levies on maritime traffic to the mainland, as well as to support inhabitants of nearby Salsette, Diu, and Bassein in their continued claims to fishing rights in the vicinity. This in turn emboldened a number of Bombay's larger and more well-connected landowners to seek out justice beyond the East India Company's authority, appealing to

67–88; Clyde L. Grose, "The Anglo-Portuguese Marriage of 1662," *The Hispanic American Historical Review* 10, no. 3 (1930), 313–352; Rodrigo Bentes Monteiro, "Overseas Alliances: The English Marriage and the Peace with Holland in Bahia (1661–1725)," in *Polycentric Monarchies: How Did Early Modern Spain and Portugal Achieve and Maintain a Global Hegemony*, ed. Pedro Cardim et al. (Eastbourne, 2012), pp. 54–69.

authorities in Goa, Lisbon, and London, a right the government at Bombay resolutely denied they had.

At the same time, the assumption of Bombay's government and fort by English officials also seemed to open an opportunity for other residents to adopt and absorb languages of English subjecthood in order to challenge the existing and entrenched power of Portuguese landholders. As early as 1663, over two-hundred Catholic, Muslim, and Hindu inhabitants of the island had petitioned Charles II to dispossess revenue farmers in various "districts" of Bombay of their power, they being "men powerfull, arrogant, and Exorbitant violators, Ecclesiastiques as well as Civil."[11] In July 1674, the Company's government came to an agreement with appointed representatives of the Portuguese community on the island about the terms on which land titles and revenues could be secured. The opening lines of the ensuing document declared its origins not in an affirmative action of government but in the "disquiet" amongst landowners about the unsettled state of affairs. Even if only a conceit, the agreement insisted that it arose when the "people thought fit, of their own free motion, by mutual assent, in a public declaration and manifesto to propose [the settlement] to the Governor in Council." Company Governor Gerald Aungier and his council called "A general assembly of the chief representatives of the said people," 120 of the "eminents of the Povo [or people] in behalf of the whole Povo of the Island," which included a number of Luso-Indian inhabitants, including the attorney-general's assistant, the Jesuits' agent, and other major landholders. From this followed twelve articles, in which these large landholders essentially agreed to an annual quit-rent in exchange for confirmation of their rights in land. Following this, another "General assembly" was convened "whereunto all the people in general interested in the affair were invited to appear," and even a second almost town-hall-type meeting to hear supposed objections to the agreement, "whereupon the Povo in general said they [...] were thoroughly satisfied therewith and of the justice thereof."[12]

Such sentiments sounded good in principle but masked the continuing tensions over authority and jurisdiction, both with many of the Portuguese landholders at Bombay and with the *Estado da India*. It was this, rather than conflicts with Mughal officials, that as much as anything drove the early Company government at Bombay to regard establishing and defending Bombay's territorial and maritime jurisdiction as a high priority. Moreover, the ambiguity

11 S. A. Khan, *Anglo-Portuguese Negotiations Relating to Bombay, 1660–1677* (London, 1922), p. 451.

12 *Materials Towards a Statistical Account of the Town and Island of Bombay*, vol. III: *Administration* (Bombay, 1894), p. 262.

of the Company's claim to territory was compounded by the mobility of its inhabitants. Whether Catholic landholders, Hindu *bhandaris,* Muslim laborers, or others, Bombay's residents maintained enduring connections to the "Main" that rendered their ties and allegiances to English Bombay somewhat tenuous and uncertain. This was particularly evident in times of emergency, real or imagined. For example, in 1673, (unfounded) fears of a Dutch invasion led many among the Portuguese population, including several large landowners and the captain of the militia in the outlying village of Mazagoan, to flee to safety in other jurisdictions, particularly those under Portuguese control. Sovereignty in this sense was a constantly moving target: the challenges of producing "English" subjects locally depended upon establishing "English" authority regionally, and vice versa.

Attempts to stabilize English rule at Bombay, both as a territorial jurisdiction and as a set of claims over the allegiance of its inhabitants, began to draw the East India Company into an even more extensive conflict raging in western India: a decades-long struggle between the expanding Mughal Empire under the Emperor Aurangzeb, which was relentlessly extending its control into Southern India, and the increasingly powerful Maratha confederation. Though often seen as a battle over territory – which in part there is no doubt it was – the so-called Deccan Wars had a significant maritime dimension, which was worked out in the western Indian littoral by a complex arrangement of feudatory coastal powers. The primary clash was between Maratha tributaries – whom the English referred to as "Sevajees," after the Maratha Empire's patriarch Sivaji – and coastal naval power under the command of the Sidis of Janjira. The Sidis, who traced their lineage to east African slaves and servants brought to western India centuries earlier, were both allied with Mughal forces as well as incorporated into their structures of power, holding titles and responsibilities, such as those of military commanders (*faujdars*), among others.[13]

From a territorial perspective, Bombay was on the extreme margins of this conflict but when one resituates the Mughal-Maratha wars in maritime perspective, the island's geographical and political position renders it far less insignificant. Maratha and Sidi forces had been jostling for positions in the Gujarati littoral for decades, and had occupied the adjoining rocky outposts of Underi and Khanderi in Bombay harbor by the early 1680s. The Company

13 Rahul C. Oka, Chapurukha M. Kusimba, "Siddi as Mercenary or as African Success Story on the West Coast of India," in *India in Africa, Africa in India: Indian Ocean Cosmopolitanisms*, ed. John C. Hawley (Bloomington and Indianapolis, 2008), pp. 203–229; Kenneth Robbins, John McLeod, eds., *African Elites in India: Habshi Amarat* (Ahmedabad, 2006); Pedro Machado, *Oceans of Trade: South Asian Merchants, Africa and the Indian Ocean, c. 1750–1850* (Cambridge, UK, 2014), esp. chap 5.

claimed these islands as part of Bombay's jurisdiction, but in practice they were powerless to enforce such a claim against either the Maratha or Sidi fleets. Moreover, Sidi Yaqut Khan also saw Bombay's harbor as an ideal spot for wintering his fleet. Under some duress, the English allowed Sidi Yaqut periodic access to its port, bazaar, and other markets, while also attempting to strike up an alliance with Sivaji and his son (and ultimate successor) Sambhaji. Tensions with Sidi Yaqut's presence at Bombay reached a fever pitch in 1683, when a fight that broke out in the bazaar between Englishmen and some of the Sidi's sailors resulted in the death of one of the English. Sidi Yaqut's refusal to turn the murderer over to the Company led to a foolish attempt by some of the English to board his ship; the ineffectiveness of the Bombay government's attempt to redress the situation formed one of the grievances that led, later that year, to a mutiny and rebellion among the English soldiery that took control of the island for upwards of a year.

Thus, by the mid-1680s, East India Company Bombay sat at the intersection of a number of ongoing conflicts, some of recent origin and others that stretched back decades if not more. In 1685, the Bombay government's superiors in London took bold, if perhaps overly optimistic, action. Concerned to establish their position with respect to the Mughal Empire, other Europeans, and its own supposed English subjects, the Company's court of committees (or directors) declared war on the Mughal Empire in both eastern and western India. This was part of a larger strategy that involved a declaration of hostilities with the Kingdom of Siam as well. Yet, conflict with the Company's Asian competitors was not the only objective. As the EIC's "secret committee," established in London to handle the affairs of this war, made clear in private instructions to Surat and Bombay, any war with Mughal forces in western India would be an opportunity to pull the Portuguese into a conflict, and finally solidify Company control over "those adjacent Islands to Bombay which belong to Bombay & were formerly dependences upon Bombay (especially Salsett) and ought to have been delivered to his Majesty when Bombay was surrendered."[14]

By 1686, news of London's decision had reached Company stations in India. The Bombay government, led by its Governor John Child – who had also been recently appointed to a new position of General of the Company's forces in Asia – imagined this was an ideal time to apply pressure to the Mughal government at the Western Indian port town of Surat. The Company blockaded Surat, ultimately seizing twenty-one ships including probably one provision ship of Sidi Yaqut Khan. Such actions appeared to Mughal officials, merchants, and

14 Secret Committee to Surat and Bombay, 31 March 1686, British Library, Asia, Pacific, and Africa Collections, India Office Records [hereinafter IOR], E/3/91 f. 49.

religious leaders in Surat no different from the sorts of "piracy" the English were constantly complaining about; Surat's elite demanded reprisals.[15]

2 Violence, Identity, and Sovereignty in the Siege of Bombay

Thus, the historical conjuncture that led Sidi Yaqut's forces to besiege Bombay town and fort ranged from the immediate and personal to the larger forces of warfare and alliances both in the region and across the globe. In this sense, the origins of the conflict, though hardly world-historical in a colloquial sense, are situated within a global context on a number of levels. Yet, perhaps even more compelling than the overdetermined story of the siege's origins for a global microhistory is the story of the siege-as-experience: that is, the ways in which the day-to-day histories of those involved, as best as they can be recreated, also reveal the wide range of intersecting scales of history embedded in such an event – even challenging the conception of what bounds and defines the siege as an "event" in the first place. Most importantly, the siege proved to be an exercise not in the establishment of firm boundaries between English and Sidi forces but rather engendered the frequent border-crossings and flexible alliances and identities that one might argue were typical of the encounter of Eurasian empires.

Despite the surprise and frenzy of activity in its first days, by a few weeks into the siege, matters had settled into a sort of routine. The Sidi forces began to dig trenches and tunnels, and build batteries and earthworks, inching ever closer to Bombay Fort. The East India Company forces set up defensive bulwarks, built largely of palm boughs and dirt. There were also skirmishes at sea – seizures of ships, exchanges of artillery fire, and minor engagements at sea from both sides, usually involving small, armed single- or double-masted boats called *gallivats* as well as European long-boats.

Not unlike other sieges in Europe or Asia, it would appear that far more inhabitants of Bombay suffered or died from the effects of the siege than the battles themselves. Disease was of course an endemic problem for Europeans in early modern Asia, but the conditions of the siege amplified the problem; among other concerns, many at Bombay died from what may have been an outbreak of bubonic plague. Still even more fled. Just two days after the invasion, according to James Hilton, the adjutant to General John Child, "most or all" of the island's militia, including its officers, had abandoned their posts.[16]

15 See Stern, *Company-State*, pp. 122–24.

16 Hunt and Stern, eds., *Soldier's Diary*, p. 32.

Many inhabitants left their land and sought protection either on the mainland or in Portuguese territory. Worse than this, the Company's garrison soldiery – a motley crew of English, Maratha, *bhandari*, and Muslim soldiers – bled from desertions. While some did return seeking pardon, a year after Sidi Yaqut's departure from Bombay, according to one report, there were still sixteen or seventeen Englishmen with him, although not necessarily happy about it.[17]

Of course, desertion was a common problem in sieges in Europe as well, but issues of race, religion, and allegiance here complicated matters significantly. For English or other European soldiers, to abandon Bombay for the Sidi's forces often meant not simply abandoning their countrymen, but Christianity. One Isaac Scott, for example, not only apparently willingly joined the Sidi's army very early on in the conflict but actively took the lead in firing against his countrymen; wounded in battle, he was discovered by *bhandari* and Maratha soldiers loyal to the East India Company; the Company made a show of having him court-martialed before he was hanged for treason, though admittedly this was all done rather quickly, by nine o'clock the morning he was found.[18] One Evert Evertson, a Dutch surgeon who had absconded from the Mughal army on the mainland to East India Company service in July 1689 only to run away once more, offered his services to Sidi Yaqut in September, allegedly converting to Islam in the process.[19] John Stevens – or, as he was later known, Abd al-Allah – reported on his experience as a deserter in a deposition taken after he attempted to return to the Company's garrison. Stevens insisted he was taken as a prisoner of war and joined under duress, though he admitted ultimately to receiving a salary, fighting against the English, and converting to Islam, including submitting to circumcision. In fact, according to him, there were over sixty Europeans, mostly English, fighting for Sidi Yaqut, many of them also converts.[20] Only two Englishmen evidently took the Company up on its blanket offer to return with full pardons after hostilities had ended, and as Sidi Yaqut's ships were leaving the island in May 1690, seven more soldiers deserted to him, according to one English observer, "impudently" taunting Company soldiers in the fort from Sidi's ships as they sailed away.[21]

The fluidity with which these men could move among different military services – and, in the case of English converts, even religious affiliations – was hardly unique to this siege, seen both within Europe as well as in conflicts around the Mediterranean world. Among other examples, Geoffrey Parker has

17 George Weldon to Surat, 10 August 1692, IOR G/36/110/3 f. 113.
18 Hunt and Stern, eds., *Soldier's Diary*, p. 48.
19 Hunt and Stern, eds., *Soldier's Diary*, pp. 53, 64, 152.
20 Hunt and Stern, eds., *Soldier's Diary*, pp. 149–150.
21 Hunt and Stern, eds., *Soldier's Diary*, p. 99.

noted that between 1608 and 1619, more than four thousand Spanish soldiers garrisoned at Oran placed themselves into a voluntary captivity with their Muslim Algerian besiegers.[22] At Bombay, however, there was no infrastructure, as there was in the Mediterranean, for ransoming their return, nor was there the same sort of slave trade in Christians that underlay that infrastructure; moreover, rather than enter as prisoners of war, these English and other European soldiers seemed to have become at least ostensibly willing combatants. The ability of the European soldiers to pass back and forth between Company and Sidi forces, and the information they carried with them when they did, is a reminder of the fact that "identity" did not determine the composition or allegiance to military forces in this period, in Europe or Asia. Muslims no more served only Muslim polities and commanders than Christians only served Christian ones. Both military leadership and rank-and-file soldiery on both sides were products of a market for itinerant military labor for hire, while the ability of people more generally – from landowners to fishermen – to migrate in search of security and prosperity was indicative of a mobile and circulatory labor system more generally in early modern South Asia.[23]

The Company understood these circumstances quite well. After all, its garrison consisted of a motley crew of Protestants, Catholics, Hindus, Zoroastrians, Muslims and others; it also relied heavily during the siege on reinforcements by land and sea of Maratha auxiliaries. However, in a world in which "national" identity did not necessarily determine service, subjecthood and religion did nonetheless affect expectations of obedience. Questions of authority and obligation were crucial to any polity, let alone a colony as new and fragile as Bombay. As such, both desertion and conversion struck particular chords with Company leadership – not least because it reinforced the trenchant critiques from Company opponents at home that it failed to properly defend English and Christians abroad. To "turn Moor" was to express that rejection of English authority in the most uncertain terms; if it was surprisingly easy for some Christian soldiers, for Company officials such a decision was a loud expression

22 Geoffrey Parker, *The Military Revolution: Military Innovation and the Rise of the West, 1500–1800* (2nd ed., Cambridge, UK, 1996), p. 57.

23 Dirk H. A. Kolff, *Naukar, Rajput & Sepoy: The Ethnohistory of the Military Labour Market in Hindustan, 1450–1850* (Cambridge, UK, 1990); Ian Kerr, "On the Move: Circulating Labour in Pre-Colonial, Colonial, and Post-Colonial India," in Rana Behal, Marcel van der Linden, eds., *International Review of Social History, Supplement 14: Coolies, Capital and Colonialism: Studies in Indian Labour History* 51 (2006), 85–109; Claude Markovits, Jacques Pouchepadass, Sanjay Subrahmanyam, eds., *Society and Circulation: Mobile People and Itinerant Cultures in South Asia 1750–1950* (London, 2006).

of that subject's "wicked Soul" – less because of his apostasy and more because of his disobedience to Company authority.[24]

3 The Consequences of the Siege

The siege of Bombay was typical of European experiences in Asia as well as of a siege itself: the experience of entrenchment in fact revealed the great permeability of boundaries, both physical and cultural, that defined this early modern world. It also reflected the wider world of "contained conflict" in which diplomacy, commerce, and military force were part of an extended negotiation over the place of a commercial European empire on the edges of expanding and ambitious Mughal and Maratha power. As the peace negotiations that ensued seem to imply, the proximate cause, in some sense, of the siege itself was not necessarily Company aggression but its threatened withdrawal of its trading headquarters from Mughal-governed Surat to English-governed Bombay. In any event, in September 1689, the Bombay government sent three officials to meet with emissaries from Surat and then on to the Mughal court to settle on terms for peace. News of an agreement reached Bombay the following February and March, but the Sidi did not receive orders to leave the island until April 10.

Though the siege never breached Bombay fort, it had taken its toll on the island. Buildings had been razed. The outforts were in shambles. The population had been decimated by departures, disease, and death. What was more, the peace settlement that arrived at Surat – in the form of an imperial *farman*, command or order, from the Mughal Court – was not what the Company envoys thought had been promised at Court. It referred to the "shamefull manner" in which the Company had behaved, spoke of a pardon and forgiveness for their "faults," and demanded that John Child, the Governor of Bombay and General over the Company's affairs in Asia, "who did the disgrace be turned out and expelled."[25] (The last order was impossible to comply with, as Child had died in February.) Of course, the *farman* did not encapsulate the full extent of the negotiations, which also called for reparations paid to Surat merchants, and for the Company to resume its trading in Surat. For his part, Sidi Yaqut was simply redeployed to fight elsewhere along the coast against Maratha forces.[26]

---

24 See, e.g., Surat to London, 28 Nov 1691, IOR G/36/93 f. 71.

25 IOR E/3/48 f. 156.

26 IOR E/3/48 f. 152.

So, what, in the end was the consequence of this siege? Who "won"? There is no clear answer. Certainly back in England, the terms – and even more, the language – of the *farman* helped the Company's critics and political enemies to frame the invasion of Bombay island itself as an ignominious defeat.[27] Company officials in London countered that the very fact that Bombay Fort and Town still existed was in itself an unprecedented victory "wonder'd at by foreign Europeans that know the Power and greatness of the Mogull," and that the language of the *farman* was just that: formulaic "vanity [...] that are in all their Phirmaunds and writeings."[28] So, in some sense, the question of victory was in the somewhat subjective definition of what precisely had transpired: had the Sidi's forces pulled off a successful occupation of Bombay island or an ultimately unsuccessful siege of Bombay Town and Fort? Certainly the conflict had led the English to come to terms and resume its trade with the Mughal Empire, but over the next few decades, the Company nonetheless moved the center of its political and commercial operations to its sovereign island. Moreover, Bombay never made full restitution to the Surat merchants of the seizures it had made. Indeed, failure to do this led to repeated accusations from Surat commercial and political leaders through the 1690s that conflated the Company with the pirates from the Atlantic world – William Kidd and Henry Every most famous among them – who had begun flooding into the Indian Ocean and whom the Company was itself struggling to suppress. And while Bombay suffered from difficulties in restoring population, defenses, and infrastructure, ultimately the experience of the siege provided an excuse and argument for pouring more resources into the political and military establishment, rather than retreating into a purely commercial enterprise as Company critics demanded.[29]

In a way, the devastation of the siege did far more to allow the East India Company to remake Bombay as an English colony than any of its previous proactive policies had accomplished. Rebuilding presented an opportunity for treating Bombay's legal infrastructure as a blank slate. As had happened in the 1670s, when rumors of a possible Dutch invasion led many large landowners to flee, Bombay's government in the 1690s set upon a large-scale confiscation of private property, with the stated intention of reinstating those who "had bin faithful" and for those who had been "bad" to serve as an "Example to be degraded & loose theirs."[30] Yet, this was not so much of a suspension of the law in

27 See, for example, Alexander Hamilton, *A New Account of the East Indies* (Edinburgh, 1727) in Hunt and Stern, eds., *Soldier's Diary*, pp. 142–146.

28 Quoted in Stern, *Company-State*, pp. 125, 126.

29 These issues are explored in more detail in Stern, *Company-State*, chapter 6.

30 Bombay to London, 15 Jan 1690/1, IOR E/3/48 f. 232.

a state of emergency but rather an enforcement of what Bombay officials argued was simply the violation of the terms on which people held land in the first place. The failure of these inhabitants to perform their duty as a militia, "as all good Subjects are engaged to defend their Countrey and Estates as well as the Garrison Souldiers," was a "great Crime" and justification for condemning land in forfeit.[31] Yet, there was of course a politics to this as well. The Company particularly targeted, again, large and powerful landowners and especially the Jesuits, who Bombay leaders believed had materially aided Sidi Yaqut's forces.[32]

In practice, the Company was not able to confiscate and redistribute as much land as its local governors intended; not only did they lack a proper judge to oversee the process, but the time and disruptions of doing so would have devastated the already fragile production (and thus tax revenues and farms) of the property itself. However, the confiscations and reallocations of land revealed how the Company tried to capitalize upon the siege in its aftermath to secure loyalty of particularly powerful figures in its jurisdiction. Rama Comotin, a *saraf* and former tax farmer who had sustained a leg injury by friendly fire from a company mortar during the siege,[33] ultimately received title to sell, lease, or farm rice fields on the southern tip of the island, which had formerly belonged to Muslims accused of aiding Sidi Yaqut.[34] Inhabitants of Mahim directly challenged their alleged disloyalty, insisting they had remained "humble vassals Subjects" to the East India Company and John Child specifically.[35]

The "time of the Sidi" also became an argument for expansion and expenditure. A decade after the siege, the Bombay government wrote to its superiors in London, citing the experience as a reason to invest more, not less, in the island's defenses and infrastructure:

> Notwithstanding the small strength that has been added to this Island of late it will require a great number of people to defend it against the Invasion of the Mogull; for that they may land in the faire time of the year at any place round it; of the truth of this assertion your Honors had a

31 Bombay Consultations, 21 Aug 1694, IOR G/3/4 f. 23.

32 Minutes of the Court of Committees, 18 Mar 1691/2, IOR B/40 f. 103–104; Memorial of Vicomte de Fonte Arcada, 15 May 1695, IOR H/36 f. 99; Duke of Shrewsbury to John Fleet, 27 Aug 1695, IOR H/36 f. 99; John Fleet to Duke of Shrewsbury, 10 Sept 1695, IOR H/36 f. 97; Minutes of the Court of Committees, 27 Aug 1695, IOR B/40 f. 19; Earl of Dartmouth to EIC, 1 November 1711, E/1/3 f. 388; Portuguese Envoy Memorial, 23 September/4 October 1711, IOR E/1/3 f. 390–91.

33 Hunt and Stern, eds., *Soldier's Diary*, p. 85.

34 Bombay Dairy, 23 June and 21 Dec 1694, IOR G/3/10a, n.p.

35 Inhabitants of Mahim to Captain Michaell George, 7 Sep 1694, Bombay Consultations, 28 Aug 1694, IOR G/3/4 f. 25.

> Sufficient instance in February 1688/9 when the Garrison was so strong with English & the Island with Inhabitants & a small Navy of Ships in the road, notwithstanding all which the Siddye made himself Master of all except the fort the first day after his landing.[36]

Over the next several decades the fort was indeed rebuilt and strengthened with some sense of urgency and other forts built or refurbished across the archipelago. This was accompanied by a renewal of the Company's longstanding attempts to expand jurisdiction locally at the expense of the Portuguese, one major pretext for which was the fact that some Portuguese had offered material support to Sidi Yaqut. In the long run, the impulse to greater control over both the land and the surrounding waterways led to the drainage of the tidewater causeways between the islands of the Bombay Archipelago and their amalgamation into what became Greater Bombay. Through the early eighteenth century, the Bombay council instituted policies designed to attract more settlers, rebuild both population and infrastructure, and engage in shipbuilding and other efforts that would ultimately lead to the creation of the so-called Bombay Marine, a crucial aspect of eighteenth-century British dominance of the Indian Ocean.[37]

Indeed, naval power became, if anything, even more crucial in the aftermath of the siege. The decades between about 1690 and 1730 witnessed a dramatic rise in European maritime predation in the Indian Ocean, particularly (but not exclusively) from "pirates" coming from England and the English Americas. At the same time, new maritime powers arose in the western Indian Ocean, such as Qawasim on the northern coast of modern-day Oman, known to the Company as the "Muscat pirates," and revived Maratha power centers, particularly those associated with the person of Kanhoji Angre, variously described as a "pirate" by the British and a "naval admiral" by the Marathas. Numerous factors led to this rise in decentralized violence on the sea, yet in retrospect, it is clear that the need to combat both Euroamerican and Asian "piracy" became one of the key justifications for the British to systematically

36 Bombay to London, 22 May 1698, IOR E/3/54 f. 51.

37 See C. R. Low, *History of the Indian Navy: 1613–1863*, 2 vols (London, 1877); Anirudh Deshpande, "The Politics and Culture of Early Modern Warfare on the Konkan Coast of India During the Seventeenth and Eighteenth Centuries," in *Coastal Histories: Society and Ecology in Pre-Modern India*, ed. Yogesh Sharma (Delhi, 2010), pp. 43–74; Anirudh Deshpande, "The Bombay Marine: Aspects of Maritime History 1650–1850," *Studies in History* 11 (1995), 281–301; Andrew Lambert, "Strategy, Policy and Shipbuilding: The Bombay Dockyard, the Indian Navy and Imperial Security in the Eastern Seas, 1784–1869," in *The Worlds of the East India Company*, ed. H. V. Bowen, Margarette Lincoln, Nigel Rigby (Woodbridge, 2002), pp. 137–151.

and dramatically extend the use of their navy and their criminal justice system into the Indian Ocean world.[38] East India Company dominance of the maritime littoral of western India would be solidified by the 1750s – around the same time it achieved territorial dominance in Bengal after the Battle of Plassey – when it managed both to subdue the "piratical" threat of the Angres and to take official command of Mughal Surat, through a Mughal appointment as *qilidar* of the Surat fort – goals both ironically accomplished with the critical help of a new ally: the Sidis of Janjira.

Thus, from a long-term perspective, it could be argued, ever so cautiously, that the siege of Bombay was one of the defining moments in the making of the British Empire in India. Certainly, the Company had endured and perhaps even repelled a much larger and well-organized army, but the experience of the siege itself was by no means a decisive victory. Instead, its significance arose not from any great military triumph, about which the stories of empires are usually told, but rather from the ways in which the siege itself served as an historical representation of failure, a lesson and a political tool that helped mobilize the Company into an even more aggressive pursuit of its political and military goals in the Western Indian Ocean. Far from the evidence, as its critics at the time and since have suggested, that the occupation of Bombay, the terms of the peace, and the devastation that followed were evidence that the Company had overstepped its legitimate role as a commercial enterprise, the siege revealed the Company's nature – and perhaps the nature of early modern commerce itself – as inherently political and martial in nature. Moreover, it suggests a form of history of the modern British Empire in India that sees its origins as arising slowly and not necessarily simply in the progression of armies or sovereign territorial power, but in the fits, starts, and struggles of the intertwined worlds of diplomacy, violence, and sovereignty in the early modern world.

---

38 See Stern, *Company-State*, pp. 186–191; on the late seventeenth-century rise of piracy in the British Empire and Indian Ocean, see also, Lauren Benton, *A Search for Sovereignty: Law and Geography in European Empires, 1400–1900* (Cambridge, UK, 2009); Mark Hanna, *Pirate Nests and the Rise of the British Empire, 1570–1740* (Chapel Hill, 2015); Kevin P. MacDonald, *Pirates, Merchants, Settlers and Slaves: Colonial America and the Indo-Atlantic World* (Berkeley, 2015); Robert Ritchie, *Captain Kidd And the War Against the Pirates* (Harvard, 1986).

Bibliography

Archival Sources

British Library, Asia, Pacific, and Africa Collections, India Office Records (IOR)
B/40
E/3/48
E/3/54
E/3/91
G/3/4
G/3/10a
G/36/93
G/36/110/3
H/36

Secondary Sources

Agamben, Giorgio, *State of Exception* (Chicago, 2005).

Andrade, Tonio, "A Chinese Farmer, Two Black Boys, and a Warlord: Towards a Global Microhistory," *Journal of World History* 21, no. 4 (2011), 573–591.

Belcher, Gerald L., "Spain and the Anglo-Portuguese Alliance of 1661: A Reassessment of Charles II's Foreign Policy at the Restoration," *Journal of British Studies* 15, no. 1 (1975), 67–88.

Benton, Lauren, *A Search for Sovereignty: Law and Geography in European Empires, 1400–1900* (Cambridge, UK, 2009).

Benton, Lauren; Clulow, Adam, "Legal encounters and the origins of global law," in *The Cambridge World History*, vol. 6: *The Construction of a Global World 1400–1800 CE*, part 2: *Patterns of Change*, ed. Jerry Bentley, Sanjay Subrahmanyam, Merry E. Wiesner-Hanks (Cambridge, UK, 2015), pp. 80–100.

Davis, Natalie Zemon, "Decentering History: Local Stories and Cultural Crossings in a Global World," *History and Theory* 50, no. 2 (2011), 188–202.

Deshpande, Anirudh, "The Bombay Marine: Aspects of Maritime History 1650–1850," *Studies in History* 11 (1995), 281–301.

Deshpande, Anirudh, "The Politics and Culture of Early Modern Warfare on the Konkan Coast of India During the Seventeenth and Eighteenth Centuries," in *Coastal Histories: Society and Ecology in Pre-Modern India*, ed. Yogesh Sharma (Delhi, 2010), pp. 43–74.

Games, Alison, *The Web of Empire: English Cosmopolitans in an Age of Expansion, 1560–1660* (New York, 2008).

Ginzburg, Carlo, "Microhistory and World History," in *The Cambridge World History*, vol. 6: *The Construction of a Global World 1400–1800 CE*, part 2: *Patterns of Change*, ed.

Jerry Bentley, Sanjay Subrahmanyam, Merry E. Wiesner-Hanks (Cambridge, UK, 2015), pp. 446–473.

Grose, Clyde L., "The Anglo-Portuguese Marriage of 1662," *The Hispanic American Historical Review* 10, no. 3 (1930), 313–352.

Hanna, Mark, *Pirate Nests and the Rise of the British Empire, 1570–1740* (Chapel Hill, 2015).

Hunt, Margaret; Stern, Philip J., eds., *The English East India Company at the Height of Mughal Expansion: A Soldier's Diary of the 1689 Siege of Bombay* (Boston and New York, 2016).

Kerr, Ian, "On the Move: Circulating Labour in Pre-Colonial, Colonial, and Post-Colonial India," in Rana Behal, Marcel van der Linden, eds., *International Review of Social History, Supplement 14: Coolies, Capital and Colonialism: Studies in Indian Labour History* 51 (2006), 85–109.

Khan, S. A., *Anglo-Portuguese Negotiations Relating to Bombay, 1660–1677* (London, 1922).

Kolff, Dirk H. A., *Naukar, Rajput & Sepoy: The Ethnohistory of the Military Labour Market in Hindustan, 1450–1850* (Cambridge, UK, 1990).

Lambert, Andrew, "Strategy, Policy and Shipbuilding: The Bombay Dockyard, the Indian Navy and Imperial Security in the Eastern Seas, 1784–1869," in *The Worlds of the East India Company*, ed. H. V. Bowen, Margarette Lincoln, Nigel Rigby (Woodbridge, 2002), pp. 137–151.

Lawson, Philip, *The East India Company: A History* (Abingdon, 1993).

Lenman, Bruce, *England's Colonial Wars 1550–1688: Conflicts, Empire and National Identity* (London, 2001; reprint: Abingdon, 2013).

Lenman, Bruce, "The East India Company and the Emperor Aurangzeb," *History Today* 37, no. 2 (1987), 23–29.

Low, C. R., *History of the Indian Navy: 1613–1863*, 2 vols (London, 1877).

Luengo, Pedro, "Military Engineering in Eighteenth-Century Havana and Manila: The Experience of the Seven Years War," *War in History* 24, no. 1 (2017), 4–27.

MacDonald, Kevin P., *Pirates, Merchants, Settlers and Slaves: Colonial America and the Indo-Atlantic World* (Berkeley, 2015).

Machado, Pedro, *Oceans of Trade: South Asian Merchants, Africa and the Indian Ocean, c. 1750–1850* (Cambridge, UK, 2014).

Markovits, Claude; Pouchepadass, Jacques; Subrahmanyam, Sanjay, eds., *Society and Circulation: Mobile People and Itinerant Cultures in South Asia 1750–1950* (London, 2006).

Monteiro, Rodrigo Bentes, "Overseas Alliances: The English Marriage and the Peace with Holland in Bahia (1661–1725)," in *Polycentric Monarchies: How Did Early Modern Spain and Portugal Achieve and Maintain a Global Hegemony*, ed. Pedro Cardim et al. (Eastbourne, 2012), pp. 54–69.

[N.N.], *Materials Towards a Statistical Account of the Town and Island of Bombay*, vol. III: *Administration* (Bombay, 1894).

Oka, Rahul C.; Kusimba, Chapurukha M., "Siddi as Mercenary or as African Success Story on the West Coast of India," in *India in Africa, Africa in India: Indian Ocean Cosmopolitanisms*, ed. John C. Hawley (Bloomington, Indianapolis, 2008), pp. 203–229.

Parker, Geoffrey, *The Military Revolution: Military Innovation and the Rise of the West, 1500–1800* (2nd ed., Cambridge, UK, 1996).

Putnam, Lara, "To Study the Fragments/Whole: Microhistory and the Atlantic World," *Journal of Social History* 39, no. 3 (2006), 615–630.

Stern, Philip J., *The Company-State: Corporate Sovereignty and the Early Modern Foundations of the British Empire in India* (New York, 2011).

Subrahmanyam, Sanjay, *The Political Economy of Commerce: Southern India 1500–1650* (Cambridge, UK, 1990).

Ritchie, Robert, *Captain Kidd And the War Against the Pirates* (Harvard, 1986).

Robbins, Kenneth; McLeod, John, eds., *African Elites in India: Habshi Amarat* (Ahmedabad, 2006).

Vivo, Filippo de, "Prospect or Refuge? Microhistory, History on the Large Scale," *Cultural and Social History: The Journal of the Social History Society* 7, no. 3 (2010), 387–397.

CHAPTER 7

Narratives of Akbar's Sieges and the Construction of Mughal Universal Sovereignty

Pratyay Nath

> The world conqueror, Akbar Ghazi, than whose sword,
> Of a certainty, there is no key for the fortresses of the world.
>
> NIZAMUDDIN AHMAD[1]

History of warfare is a growing field of research for South Asia. As far as the early modern period is concerned, much of the existing body of literature focuses on the Mughal Empire. Historians have concentrated primarily on three themes here – pitched battles, gunpowder technology, and military organisation.[2] It is only recently that scholars have started to go beyond these issues and explore other dimensions of the Mughal military experience.[3] Only a few

1 '*Kishwar kushāyī, Akbar g͟hāzī, ki bī suk͟hun/ jaz teg͟h-i u, qilāʿ-i jahān-rā kalīd nīst.*' Khwaja Nizamuddin Ahmad, *Ṯabaqāt-i Akbarī*, ed. Brajendranath De, 3 vols (Calcutta, 1931), 2:248; *The Ṯabaqāt-i Akbarī of K͟hwājah Niẓāmuddīn Aḥmad*, trans. Brajendranath De, 3 vols (Delhi, 1992), 2:387.

2 Works that study Mughal battles include Jadunath Sarkar, *Military History of India* (Delhi, 1970); Kaushik Roy, *India's Historic Battles: From Alexander the Great to Kargil* (New Delhi, 2004), pp. 54–79. Prominent studies of technology include Iqtidar Alam Khan, "Early Use of Cannon and Musket in India, AD 1442–1526," *Journal of Economic and Social History of the Orient* 24, no. 2 (1981), 146–164; Iqtidar Alam Khan, "Origin and Development of Gunpowder Technology in India, AD 1250–1500," *The Indian Historical Review* 4, no. 1 (1977), 20–29; Iqtidar Alam Khan, *Gunpowder and Firearms: Warfare in Medieval India* (New Delhi, 2004); Iqbal Ghani Khan, "Metallurgy in Medieval India – The Case of Iron Cannons," *Proceedings of the Indian History Congress 1984* 45 (1984), 464–471; Irfan Habib, "Akbar and Technology," *Social Scientist* 20, no. 9–10 (1992), 3–15; G. N. Pant, *Mughal Weapons in the Baburnama* (Delhi, 1989). Finally, works on army organisation include William Irvine, *The Army of the Indian Moghuls* (Delhi, 2004); Abdul Aziz, *The Mansabdari System and the Mughal Army* (Lahore, 1945).

3 Jos Gommans and Andrew de la Garza have thrown valuable light on several new areas of Mughal warfare, including the nature of imperial frontiers, military mobilisation, hunt-

 | DOI:10.1163/9789004395695_008

historians have explored sieges – the mode of warfare the present volume is interested in. This is in spite of the fact that siege warfare was very frequent in early modern South Asia and it played a key role in Mughal territorial expansion.[4] In several cases, scholars have referred to sieges as a part of their broader works on Mughal warfare.[5] The richness of these discussions notwithstanding, sieges have not received the kind of detailed attention they deserve.

Against this backdrop, the present chapter will look at Mughal sieges through the lens of literary representation.[6] The focus will remain on five imperial sieges – ones that the third Mughal emperor Jalaluddin Muhammad Akbar (r. 1556–1605) led in person in course of his long rule spanning almost half a century. These were the sieges of Mankot (1557), Chitor (1567–68), Ranthambhor (1569), Surat (1572–73), and Patna (1574). The goal will be to analyse the politics of the narrativisation of these sieges in three imperial court chronicles – Muhammad Arif Qandahari's *Taʾrīkh-i Akbarī*, Khwaja Nizamuddin Ahmad's

---

ing, tactics, logistics, recruitment, and training. See Jos Gommans, *Mughal Warfare: Indian Frontiers and High Roads to Empire, 1500–1700* (London, New York, 2002); Andrew de la Garza, *The Mughal Empire at War: Babur, Akbar and the Indian Military Revolution, 1500–1605* (London, New York, 2016).

4 Publications on Mughal forts and sieges include Ayesha Begum, "Mughal Fort Architecture in Bengal with an Introduction to Some Important River Forts," *Journal of Asiatic Society of Bangladesh* 47, no. 1 (2002), 1–24; Pratyay Nath, "Siege Warfare in Mughal India, 1519–1538," in *Warfare and Politics in South Asia from Ancient to Modern Times*, ed. Kaushik Roy (New Delhi, 2011), pp. 121–144; Pratyay Nath, "Through the Lens of War: Akbar's Sieges (1567–69) and Mughal Empire-Building in Early Modern North India," *South Asia: Journal of South Asian Studies* 41, no. 2 (2018), 245–258; Margaret R. Hunt, "The 1689 Mughal Siege of East India Company Bombay: Crisis and Historical Erasure," *History Workshop Journal* 84, no. 1 (2017), 149–169. For some of the important works on the history of fortress warfare in early modern South Asia in general, see Stuart Gordon, "Forts and Social Control in the Maratha State," *Modern Asian Studies* 13 (1979), 1–17; Jean Deloche, *Studies on Fortification in India* (Pondicherry, 2007); Richard M. Eaton, Philip B. Wagoner, *Power, Memory, Architecture: Contested Sites on India's Deccan Plateau, c. 1300–1600* (New Delhi, 2014).

5 The interventions of both Jos Gommans and Kaushik Roy fall under this category. Gommans, *Mughal Warfare*, pp. 136–141; Kaushik Roy, *Military Transition in Early Modern Asia, 1400–1750: Cavalry, Guns, Government and Ships* (London/New York, 2014), pp. 127–160. Douglas Streusand was the first historian to analyse Akbar's sieges in considerable detail. Douglas Streusand, *The Formation of the Mughal Empire* (New Delhi, 1990), pp. 57–64.

6 Very few scholars have worked on the representation and memorialisation of Mughal war and conquests. Important works include Sandhya Sharma, *Literature, Culture and History in Mughal North India, 1550–1800* (Delhi, 2011), pp. 158–203; Cynthia Talbot, "Justifying Defeat: A Rajput Perspective on the Age of Akbar," *Journal of the Economic and Social History of the Orient* 55, no. 2–3 (2012), 329–368; Audrey Truschke, "Setting the Record Wrong: A Sanskrit Vision of Mughal Conquests," *South Asian History and Culture* 3, no. 3 (2012), 373–396.

Ṭabaqāt-i Akbarī, and Abul Fazl's *Akbar-nāma*.[7] I will trace how exactly these chroniclers used their literary descriptions of these sieges as sites for the construction and memorialisation of the greatness of Akbar as a universal ruler. This claim to universal rule animated Mughal imperial ideology since the second half of the 16th century.[8] The present chapter begins by discussing the nature of this claim under Akbar and his successors. This will provide the political and cultural context for the subsequent analysis of the politics of representation of the imperial sieges by these chroniclers. Next, I will briefly discuss the three texts, their authors, and the five sieges under consideration. This will familiarise the reader will all the major players and figures involved in the history we are going to study. Next, the different narratives of the five imperial sieges provided by them will be discussed. In the process, the chapter will tease out the various patterns and strategies of representation deployed by the three authors and discuss how these narratives collectively fed the broader imperial project of the public construction of Mughal universal sovereignty.

1 Universal Sovereignty and Mughal Kingship

In order to comprehend the concerns and objectives of the narratives in question, it is imperative to understand the official ideological paradigm of the Mughal court in the late-16th century, when all of the three chronicles under consideration were composed. One of its key dimensions was a vigorous claim towards universal sovereignty. In Mughal imperial discourse – shaped substantially and articulated vociferously by Abul Fazl, Akbar was projected as *insān-i kāmil* – the perfect man, the lord of time and space, divinely mandated and destined to rule all of mankind. Drawing heavily upon the writings of the 13th century-Persian political philosopher Nasiruddin Tusi, the figure of the Mughal emperor was conceptualized as God's representative on earth, deputed with the express task of establishing and preserving social harmony and order, eradicating the evil of multiplicity of human intentions by teaching them the merit of single-minded devotion to the Mughal sovereign, and nurturing all of humanity with his wisdom, care, and justice. In such a moral universe, war did not remain a royal whim or even a matter of choice for the ruler. The declared commitment of the emperor towards establishing justice all over the world on

7 Muhammad Arif Qandahari, *Taʾrīkh-i Akbarī*, ed. Haji Syed Muinuddin Nadwi, Syed Azhar Ali, Imtiaz Ali Arshi (Rampur, 1962); *Taʾrīkh-i Akbarī*, trans. Tasneem Ahmad (Delhi, 1993); *Ṭabaqāt-i Akbarī*, ed. De; *Ṭabaqāt-i Akbarī*, trans. De; *Akbar-nāma*, ed. Rahim; *Akbarnama*, trans. Beveridge.

8 Harbans Mukhia, *The Mughals of India* (Malden, Oxford, Carlton, 2004), pp. 14–71.

his own terms made war an unavoidable reality. The commitment of the caring, if strict, figure of the universal sovereign transformed war into a necessary tool for eliminating whatever he considered to be evil and bringing all of humanity under his watchful and just rule.[9]

Based on recent scholarship, one can point out that such claims to universal sovereignty were repeatedly and spectacularly made by a large number of rulers across early modern Eurasia.[10] In the Mughal Empire, imperial biographies, autobiographies, and dynastic histories comprised one of the main vehicles for the articulation of these claims. By way of telling and retelling the official version of the history of the empire and the emperor, these texts presented the literary performance of Mughal sovereignty to its imperial elite as well as to the large secretarial and scribal class that served it. Notwithstanding substantial variations among these texts in terms of both form and content, these official histories shared the basic concern of justifying the Mughal emperors' claim to universal rule through a laudatory narration of the exploits and achievements of them and their armies, among other things. As Muzaffar Alam points out, Persian normative texts, like Nasiruddin Tusi's *A<u>kh</u>lāq-i Nāṣirī*, comprised a central part of the advanced school curriculum under Akbar. Later, Abul Fazl's *Akbar-nāma* and *inshā* – themselves deeply shaped by Tusi's writings on kingship – were also included in this curriculum. Educated in this system, many

9 The rich and growing body of literature on Mughal imperial ideology and the location of claims to universal sovereignty therein include Ebba Koch, "The Symbolic Possession of the World: European Cartography in Mughal Allegory and History Painting," *Journal of the Economic and Social History of the Orient* 55, no. 2–3, 547–580; Ebba Koch, "How the Mughal *Padshahs* Referenced Iran in Their Visual Construction of Universal Rule," in *Universal Empire: A Comparative Approach to Imperial Culture and Representation in Eurasian History*, ed. Peter Fibiger Bang, Dariusz Kołodziejczyk (Cambridge, UK, 2012), pp. 194–209; Mukhia, *The Mughals of India*, pp. 14–112; A. Azfar Moin, *The Millennial Sovereign: Sacred Kingship and Sainthood in Islam* (New York, 2012); Lisa Balabanlilar, "Lords of the Auspicious Conjunction: Turco-Mongol Imperial Identity on the Subcontinent," *Journal of World History* 18, no. 1 (2007), 2–39; Lisa Balabanlilar, *Imperial Memory and Dynastic Politics in Early Modern South and Central Asia* (New Delhi, 2013); Sumathi Ramaswamy, "Conceit of the Globe in Mughal Visual Practice," *Comparative Studies in Society and History* 49, no. 4 (2007), 751–782. For the influence of Nasirean political norms on Mughal imperial thought and culture, see Muzaffar Alam, "Akhlaqi Norms and Mughal Governance," in *The Making of Indo-Persian Culture: Indian and French Studies*, ed. Muzaffar Alam et al. (New Delhi, 2000), pp. 67–95; Muzaffar Alam, *The Languages of Political Islam in India, c. 1200–1800* (Delhi, 2004).

10 See, for instance, Bang and Kołodziejczyk, eds., *Universal Empire*; Anthony Pagden, *Lords of All the World: Ideologies of Empire in Spain, Britain and France c. 1500–c. 1800* (New Haven, London, 1995); Pamela Kyle Crossley, *A Translucent Mirror: History and Identity in Qing Imperial Identity* (Berkeley, Los Angeles, London, 1999).

students eventually found employment in the imperial secretarial ranks.[11] Rajeev Kinra's recent work on Chandar Bhan Brahman – a scribe who served the Mughal court from the 1630s through the 1660s – indicates that the reading and re-reading of these texts over decades within these circles led to a considerable dissemination of the ideals of Mughal universal sovereignty.[12] On the other hand, the writings of Mirza Nathan – a Mughal commander who served in Bengal and Assam mainly under Emperor Jahangir (r. 1605–1627) – bear evidence to how the same ideals had also percolated to the junior ranks of the military aristocracy by the early-seventeenth century.[13]

Within the narratives of military campaigns that these imperial biographies presented, sieges occupy a special position. In contrast to day-long battles and skirmishes, sieges would usually unfold over months. This much-longer duration meant that sieges could fit a much higher number of events worth describing in imperial histories in comparison to battles.[14] In turn, this implied that both the garrison and the besiegers had the chance to demonstrate the skill of their soldiery and the effectiveness of their weapons over a much longer period of time in sieges. This longer duration also offered the various players involved enough time to engage each other in complicated and codified rituals of power. All this would also give the Mughal biographers greater opportunity and more material to memorialize the actual military exploits of the army as well as construct the symbolic greatness of imperial rule and divine kingship. The fact that Mughal emperors like Akbar fought in a greater number of sieges than big battles also made it necessary for their biographers to focus on sieges while expounding the military glory of their patrons.[15] Let us now

11 This points to a considerable readership of both Persian normative texts and imperial biographies within – and even beyond – Mughal imperial and official circles. Muzaffar Alam, "The Pursuit of Persian: Language in Mughal Politics," *Modern Asian Studies* 32, no. 2 (1998), 317–349, see pp. 326–327.

12 Rajeev Kinra has studied the works of Chandar Bhan Brahman in great detail in his recent monograph. Rajeev Kinra, *Writing Self, Writing Empire: Chandar Bhan Brahman and the Cultural World of the Indo-Persian State Secretary* (Oakland, 2015). See especially pp. 95–158.

13 Mirza Nathan, *Bahāristān-i Ghā'ibī*, transcribed copy of the original Persian manuscript preserved in Bibliotheque Nationale, Paris, JS 60–62, Jadunath Sarkar Collection, National Library, Kolkata; *Bahāristān-i Ghaybī: A History of the Mughal Wars in Assam, Cooch Behar, Bengal, Bihar and Orissa during the Reigns of Jahangir and Shah Jahan,* 2 vols, English trans. M. I. Borah (Guwahati, 1992).

14 I am thankful for this point to Prof. Peter Wilson, who discussed this in course of his keynote speech in the "The World of the Siege" conference at Duke University, 2014.

15 It is Douglas Streusand who first noted that in Mughal warfare, sieges were much more common than pitched battles. For example, Akbar, the Mughal emperor under whom the

quickly take a look at the nature of the three texts in question and the identity of their authors.

2 The Three Chronicles

Arif Qandahari, the author of *Taʾrīkh-i Akbarī,* was a low-ranking Mughal bureaucrat. He initially joined Mughal service as the *Mīr-i sāmān* of Bairam Khan, the regent of Akbar. In the late-1570s Qandahari also briefly served as *Dīwān* in Punjab. Commissioned by a Mughal commander called Muzaffar Khan and completed around 1580, Qandahari's text was the first biography of Akbar. It narrates events of the first two and a half decades of his rule. It is the most flamboyant and the least comprehensive of the three texts under focus. Quoting Quranic verses amply, Qandahari uses a distinct religious idiom to narrate Akbar's rule. He repeatedly calls the emperor *Bādshāh-i Islām* (the Emperor of Islam) and the protector of *shariʿā*. He was also a firm believer in the divinity of kingship and this idea finds repeated expression in his text.[16]

Like Qandahari, Nizamuddin Ahmad was also a member of the imperial officialdom under Akbar. As a high-ranking commander, he took a leading part in Mughal campaigns in Western India. He served as *Bakhshī* in Gujarat throughout the 1570s and 1580s.[17] His biography of Akbar comprises only a part of a much larger work entitled *Ṭabaqāt-i Akbarī.* Through a narration of political histories of Islamic rule in nine different regions of South Asia, this text aims at chronicling how these regions eventually came to be conquered by his patron. In contrast to the other two authors under consideration, Nizamuddin scarcely gave in to literary flamboyance or explicit value judgments. His narrative has a pronounced matter-of-fact tone to it and sticks to the barebones of the political events. His personal involvement in many of the campaigns he describes means that these are eye-witness accounts.[18]

---

empire saw the most dramatic territorial expansion, fought only one big pitched battle – the Second Battle of Panipat (1556). In comparison, he led five sieges from the front.

16 *Taʾrīkh-i Akbarī*, trans. Ahmad, 1–2; Khaliq Ahmad Nizami, *On History and Historians of Medieval India* (New Delhi, 1983), pp. 230–232. Syed Athar Abbas Rizvi, *Religious and Intellectual History of the Muslims in Akbar's Reign, with Special Reference to Abul Fazl* (New Delhi, 1975), pp. 250–253.

17 M. Athar Ali, *The Apparatus of Empire: Awards of Ranks, Offices and Titles to the Mughal Nobility 1574–1658* (New Delhi, 1985), pp. 3, 9, 11, 14, 19.

18 Harbans Mukhia, *Historians and Historiography during the Reign of Akbar* (New Delhi, 1976), pp. 132–153; Nizami, *On History and Historians*, pp. 239–240; Rizvi, *Religious and Intellectual History*, pp. 277–278.

Finally, Abul Fazl was a confidant, aide, and close friend of Akbar's. His *magnum opus Akbar-nāma* is the most comprehensive account of Akbar's rule as well as the most vivid embodiment of the imperial philosophy of rule as it stood in the 1590s. The author's enduring personal association with the emperor lent a personal flavour to the whole account. Among the three chroniclers, Abul Fazl's account is the most complex. It presents a most sophisticated blend of secular information and imperial rhetoric. Written in the late-16th century, it articulates a mature form of Akbar's political ideology as it had evolved over the past four decades of his rule.[19]

Next, let us proceed to look at the five imperial sieges in question. The following section will locate their significance within the broader picture of Mughal siege warfare and trace their importance for the personal career of emperor Akbar. It will highlight why these imperial sieges represent key moments of Mughal political history and crucial chapters in the imperial chronicles.

3 Significance of Mughal Imperial Sieges

Sieges occupied a vital position within the diverse forms that Mughal warfare took in early modern South Asia. They were especially frequent in the broken plateaus, ravines, defiles, forests, and hills of Central and Western India, as well as the Deccan. The primary reason for this was that this terrain offers plenty of naturally defensible spots. Over the centuries, local powers of these regions fortified these spots in a bid to exploit the geographical peculiarities for their own strategic benefits. In the face of advancing Mughal armies seeking military engagements, these local powers would often prefer to withdraw into these forts and make a determined stand there. Consequently, Mughal imperial expansion into these regions entailed a large number of sieges. While battles and skirmishes were undoubtedly not entirely absent, conquest – and later territorial control – largely revolved around the control of fortified strongholds. Two of the sieges that this chapter studies – Chitor and Ranthambhor – unfolded under similar circumstances in the forested highlands of Western and Central India and exemplify this process.

19 Mukhia, *Historians and Historiography*, pp. 41–88; Nizami, *On History and Historians*, pp. 141–160; Rizvi, *Religious and Intellectual History*, pp. 262–277; M. Athar Ali, "The Evolution of the Perception of India: Akbar and Abu'l Fazl," *Social Scientist* 24, no. 1–3 (1996), 80–88.

But even outside this vast region, fortress warfare was quite common. The Himalayan foothills in the northern reaches of South Asia – where the siege of Mankot took place – comprised another belt of rugged terrain which saw several big sieges. Finally, the sieges for Patna and Surat exemplify the contests over strategically located strongholds in yet other regions. It was either the difficulty in taking these strongholds or the sheer political and strategic importance associated with them – or both – that drew Akbar to lead these sieges personally.

Mankot (1557) was the first siege of Akbar's career and was undertaken at a stage when he was still under the regency of Bairam Khan. Chased by Mughal troops, the Afghan chieftain Sikandar Sur had taken refuge in this fort in the Himalayan foothills near the Punjab Plains of North India. Sikandar was a member of the aristocracy of the Afghan sultanate that Mughal armies had defeated on the plains of Panipat the year before. This siege represented an important phase in the struggle of the emergent Mughal Empire against the increasingly-cornered Afghan aristocracy – one that would continue for decades to come. It was a keenly contested siege during which the garrison defended the fort valiantly with artillery and matchlocks. They sued for peace after almost six months, only after they were seriously threatened by rising desertions and a scarcity of food. Sikandar was pardoned by the young emperor and admitted into the Mughal officialdom.[20]

The next two imperial sieges did not take place till a decade later. Chitor (1567–68) and Ranthambhor (1569) represented Akbar's two most significant conquests against the Rajput warrior community of Western India. At Chitor, the Mughal army fought practically without any artillery. While the garrison relied heavily on their guns and cannons, the besiegers threw their weight behind their mines and saps. After six months, the decisive blow to the fortifications was dealt from the sap by means of a direct co-ordinated physical assault. The fort was stormed and both the garrison and the common populace were massacred.[21] In Ranthambhor, on the other hand, Akbar's army managed not only to transport artillery to the site of the siege, but also to mount it on a

20 *Akbar-nāma*, ed. Rahim, 2:50–52, 56–60; *Akbarnama*, trans. Beveridge, 2:79–81 and 87–91; *Ṭabaqāt-i Akbarī*, ed. De, 2:133–134; *Ṭabaqāt-i Akbarī*, trans. De, 2:221–223; *Ta'rīkh-i Akbarī*, trans. Ahmad, pp. 78–80; Abdul Qadir Badaoni, *Muntakhabu-t-Tawārīkh by 'Abdu-l-Qādir ibn-i-Mulūk Shāh known as al-Badāoni*, trans. W. H. Lowe, 3 vols (Delhi, 1986), 2:11–12; Mahomed Kasim Ferishta, *History of the Rise of the Mahomedan Power in India till the Year AD 1612 or Tarikh-i Firishta*, trans. John Briggs, 4 vols (Delhi, 1990), 2:116.

21 *Akbar-nāma*, ed. Rahim, 2:300–304 and 313–324; *Akbarnama*, trans. Beveridge, 2:441–444 and 464–477; *Ṭabaqāt-i Akbarī*, ed. De, 2:214–220; *Ṭabaqāt-i Akbarī*, trans. De, 2:341–348 and 352–355; *Ta'rīkh-i Akbarī*, ed. Nadwi, Ali, Arshi, pp. 109–115; *Ta'rīkh-i Akbarī*, trans. Ah-

neighbouring hill and deploy it successfully against the fortifications from there. Alarmed, the garrison surrendered. Their commander – Rai Surjan Hada of Bundi – was pardoned by the emperor and co-opted into the imperial body politic.[22]

The sieges of Surat (1572–73) and Patna (1574) signified a new wave of Mughal conquests that started in the early-1570s, with imperial armies marching beyond North India. The fort of Surat guarded a flourishing port on South Asia's western seaboard. The contest over it marked the climax of Akbar's conquest of Gujarat. Here the garrison was well-equipped with both handguns and artillery, including some of Ottoman origin.[23] Braving heavy firing, the Mughals had their mines and saps pushed forward. They also managed to erect an artificial elevation opposite to the fort and station some of their gunners on it. The combined pressure created by the firing by these gunners and the advance of the siegeworks forced the garrison to submit. Although the commander of the garrison was punished and some of his soldiers were imprisoned, the rest were pardoned.[24]

On the other hand, the siege of Patna signified the beginning a Mughal push towards the east. The fort stood on the riparian highway of the Ganga, whose east-west course drains the vast North Indian plains. With the siege in progress, Akbar sailed down from Agra to take charge of the operation himself. Once there, he primarily focused his attention on subduing the neighbouring fort of Hajipur, garrisoned by the same adversary. With Mughal forces storming Hajipur, the garrison of Patna relented. As its leader – the Afghan chieftain Daud

---

mad, pp. 148–153; *Muntakhabu-t-Tawārīkh*, trans. Lowe, 2:105–108; *Tarikh-i Firishta*, trans. Briggs, 2:229–232.

22 *Akbar-nāma*, ed. Rahim, 2:333–339; *Akbarnama*, trans. Beveridge, 2:489–496; *Ṭabaqāt-i Akbarī*, ed. De, 2:223–225; *Ṭabaqāt-i Akbarī*, trans. De, 2:352–355; *Ta'rīkh-i Akbarī*, ed. Nadwi, Ali, Arshi, pp. 115–121; *Ta'rīkh-i Akbarī*, trans. Ahmad, pp. 154–159; *Muntakhabu-t-Tawārīkh*, trans. Lowe, 2:110–111; *Tarikh-i Firishta*, trans. Briggs, 2:232–233.

23 Over the first half of the 16th century, the Sultanate of Gujarat cooperated first with the Mamluk Sultanate of Egypt and then – after the demise of the latter in 1517 – with the Ottoman Empire in fighting the increasing Portuguese presence in the Indian Ocean. Ottoman gunpowder weaponry and military personnel reached Gujarat as a part of these collaborations. Giancarlo Casale, *The Ottoman Age of Exploration* (New York, 2010), pp. 23, 47–48, 54–56; Andrew McGregor, *A Military History of Modern Egypt: From the Ottoman Conquest to the Ramadan War* (Westport, Connecticut, London, 2006), pp. 19–22.

24 *Akbar-nāma*, ed. Rahim, 3:18 and 27–30; *Akbarnama*, trans. Beveridge, 3:25–26 and 37–41; *Ṭabaqāt-i Akbarī*, ed. De, 2:241 and 246–248; *Ṭabaqāt-i Akbarī*, trans. De, 2:376–377 and 383–387; *Ta'rīkh-i Akbarī*, ed. Nadwi, Ali, Arshi, pp. 163–164 and 168–169; *Ta'rīkh-i Akbarī*, trans. Ahmad, pp. 194 and 197–198; *Muntakhabu-t-Tawārīkh*, trans. Lowe, 2:147–149; *Tarikh-i Firishta*, trans. Briggs, 2:238–239.

Khan Karrani – fled further eastwards into Bengal, Akbar's army entered Patna and took control of it.[25]

These five sieges played a key role in shaping Akbar's own political career.[26] His rule followed a dark period in Mughal history when an Afghan political resurgence in North India had forced his father Humayun (r. 1530–1540, 1555–1556) into abandoning his own empire and seeking temporary political asylum with his imperial rival – Shah Tahmasp, the Safavid monarch of Iran. Subsequently, he had returned to North India to reclaim his dominions. However, following his premature death in 1556, Akbar succeeded him at the tender age of fourteen. Over the next two decades, he led several military expeditions from the front. One could surmise that this was perhaps a conscious effort on the part of the young emperor to establish his own status as a strong and autonomous ruler, assert his own independence, and earn the respect of his aristocracy. His success in these expeditions eventually did help him create an image of a strong warrior and an able leader of men for himself. It was during this early phase that he also led his troops to victory in the five sieges under focus. The fact that each of these sieges was also politically and militarily crucial to the process of Mughal territorial expansion must have lent additional significance to Akbar's contribution. He did not lead any other siege personally in the remaining three decades of his rule. This shows that while campaigns went on without a break, the emperor no longer felt politically pressurised to demonstrate his own military acumen all the time.

The three biographies also forward their own reasons of why these five imperial sieges were important. First, invoking the image of the Mughal emperor as the rightful ruler of the entire world, they argue that by making a stand against the divinely-blessed Mughal sovereign, the rulers of these forts had committed the serious mistake of rebelling against divine will. So it was necessary for Akbar to take up leadership of the sieges in order to teach these rebels a lesson. Much of the language that justifies imperial intervention in these sieges plays on this idea of rebellion and punishment. Arif Qandahari's justification of Akbar's involvement in the siege of Ranthambhor is a case in point. He writes that by defying imperial overtures for a peaceful submission, the Rajput ruler

25 *Akbar-nāma*, ed. Rahim, 3:82–83, 85–90, and 92–101; *Akbarnama*, trans. Beveridge, 3:114–116, 120, 122–127, and 129–142; *Ṭabaqāt-i Akbarī*, ed. De, 2:284–292, and 297–298; *Ṭabaqāt-i Akbarī*, trans. De, 2:434–436, 438–445, and 452–453; *Ta'rīkh-i Akbarī*, ed. Nadwi, Ali, Arshi, pp. 189–192; *Ta'rīkh-i Akbarī*, trans. Ahmad, pp. 222–224; *Muntakhabu-t-Tawārīkh*, trans. Lowe, 2:176–184; *Tarikh-i Firishta*, trans. Briggs, 2:244–245.

26 For the best recent biography of Akbar, see Andre Wink, *Makers of the Muslim World: Akbar* (Oxford, 2009).

of the fort had "unleashed the fire of his rebellion (*ātish-i baghī*) which has blazed forth". Then, he explains that this incident "forced the imperial dignity to quench that fire by means of the water of the sword (*ba-āb-i ḥusām*), and that the wrath of turbulence and disobedience be extinguished by the lances of the victorious army".[27]

Second, the chroniclers, especially Abul Fazl, also conceptualized these imperial sieges as divine design – opportunities for the public demonstration of the greatness of the Mughal emperor. One of the arguments made is that by accomplishing with ease tasks generally acknowledged to be extremely difficult – and even unachievable, Akbar demonstrated how limitless his capabilities were and how he was the eternal recipient of divine blessings. Narrating Akbar's success in subduing the fort of Surat, for instance, Abul Fazl observes that while even "the sagacious would not have imagined [this] possible even after years of siege", Akbar had managed it within a remarkably short span of one month and seventeen days only.[28] For the biographers, therefore, these sieges served as some sort of case studies for illustrating the larger point about the greatness of Akbar.

Finally, Abul Fazl also projects these imperial sieges as occasions for the emperor to examine and assess the qualities of his commanders and soldiers. A passage from *Akbar-nāma* explains that the emperor often took charge of sieges so that "the gold encrusted copper may be placed in the dissolving crucible and the coin of the realm purified, and that the testing may be carried out to the uttermost".[29] This underlines how such events were also conceptualised as opportunities for the emperor to test the degrees of the loyalty of his employees, grade them according to their bravery, and separate incapability and unworthiness from real mettle. For Abul Fazl, this supposed ability of Akbar's to perform these functions derived from – and demonstrated – his elevated status and his possession of divine wisdom.

All of this made the imperial siege infinitely important to the official Mughal political paradigm. Far more being just a mundane military exercise, it became inherently tied up with various political rituals and diverse performative aspects of Mughal sovereignty. In the long run, it formed an integral part of the project of memorialisation of its greatness and achievements. On that note, let us proceed to examine the actual narratives of the imperial sieges.

27 *Ta'rīkh-i Akbarī*, ed. Nadwi, Ali, Arshi, p. 116; *Ta'rīkh-i Akbarī*, trans. Ahmad, p. 154.
28 *Akbar-nāma*, ed. Rahim, 3:28; *Akbarnama*, trans. Beveridge, 3:39.
29 *Akbar-nāma*, ed. Rahim, 3:28; *Akbarnama*, trans. Beveridge, 3:39.

4 The Fort as Impregnability Personified

It needs to be pointed out at the very outset that in describing these imperial sieges, all the three biographies make recurrent use of a narrative pattern and certain literary conventions. In the matter-of-fact narration of Nizamuddin Ahmad, these conventions appear in the most rudimentary and understated form. Arif Qandahari presents them with unabashed flamboyance, while the most complex formulations are made by Abul Fazl. But the differences in their style of narration notwithstanding, the pervasive presence of this pattern is unmistakable. This will become clear as we look at these narratives more closely in what follows.

The first piece in this pattern comprises the presentation of a magnified image of the impregnability of the fort in question. This would be done by using well-known literary tropes and by weaving them into a secular description of the physical features of the structure. Arif Qandahari's description of the fort of Ranthambhor is a case in point. He writes:

> The ramparts of the impregnable and tall fort were immovable and solid like high mountains (*koh-i shāmikh*), and its moat was wide and deep like a serpent-infested sea (*baḥr-i maḥyāt*), so that even fast falcons could not fly past it and the ray of vision could not travel from its base to its apex.[30]

Here, sentences describing the actual physical features of the fort – like its height and its moat – are carefully intertwined with hyperbolic similes that deliberately play up its impregnability. Similarly, describing the same fort, Abul Fazl exclaims that "[T]he fort is very lofty and strong, so that the lasso of the imagination cannot reach its [elevated] battlements, nor the catapult of the fancy be effectual against its high walls."[31] In the same vein, Arif Qandahari uses several similes and hyperboles to describe the fort of Chitor. He writes:

> The ring of the silvery moon is like an anklet at the feet of the fortifications, and Zohra, like an ear-ring, shines like a resplendent jewel on its tower, as if it were an ear. And the peak of its hill challenges the sword of Bahram, and the stones of its high abode rise till the feet of the Saturn.[32]

---

30 *Ta'rīkh-i Akbarī*, ed. Nadwi, Ali, Arshi, p. 117. Translation mine.

31 "*'wa īn qalā' īst dar ghā'ib-i rif'at wa raṣānat ki kamand-i andīsha ba-kungura-i irtifā-i u na-rasīda wa manjanīq-i wahm ba-dīwār-i taṣā'ud-i u kār-gar na-*uftāda'." *Akbar-nāma*, ed. Rahim, 2:335; *Akbarnama*, trans. Beveridge, 2:490.

32 "*Dar pāy-i bāra-ash chūn khalkhāl numaid, wa gosh-wāra-i Zohra dar gosh-i burj-ash chūn dāna-i durukhshanda. Tegh-i koh-ash bā tegh-i Bahram zabān-āwarī kunad wa az āsitān-i*

He concludes this graphic description by stating that no king of the world had ever been able to subdue this fortress. Through the similes of the crescent-shaped moon decorating the foot of the fort and the star Zohra adorning its ear, the image that Qandahari portrays is clearly that of a person. He seems to project the lofty stones that rise up to 'the feet of Saturn' as this person's head and the stones that are said to challenge the sword of Bahram as their weapon. What this amounts to, then, is the personification of the fort as an unassailable adversary. Both the chroniclers refer to Perso-Islamic mythology and cosmology to underline this point about unassailability.

Written in hindsight, with the knowledge of the eventual victory of Mughal forces in all these five imperial sieges, the objective of using these exaggerated images of the fort defenses was to create a befitting backdrop against which the greatness of the emperor could be etched. In the light of such an image of invincibility, the achievements of the emperor and his army were expected to appear even more pronounced. Abul Fazl's comment that Akbar took the "lofty fort (*qalʿā-i sipihr*)" of Ranthambhor in only a month – as opposed to the early-fourteenth century-sultan Alauddin Khalji, who took a whole year to complete the same task – demonstrates this succinctly.[33] Akbar's accomplishment of such seemingly difficult tasks thus became a direct way of demonstrating his greatness and charisma. Abul Fazl put this idea into words when he wrote in the context of the siege of Chitor:

> A lofty genius (*himmat-i ʿālī*) is the key to difficult enterprises (*miftāḥ-i maʿāqid-i maqṣūd*), and a lamp of the mystery of destiny (*miṣbāḥ-i makāmin-i taqdīr*). Especially, whenever such a lord of fortune put his heart into an arduous task, he easily accomplishes it in spite of the apprehension of worldlings.[34]

Having created such a premise, the rest of the narratives of the sieges then effectively became a way of explaining the workings of this "lofty genius" and recording how he subdued his apparently invincible adversary.

---

buland-ash pāy-i Kaiwān ba-sang dar āid." *Taʾrīkh-i Akbarī*, ed. Nadwi, Ali, Arshi, p. 110. Translation mine.

33 *Akbar-nāma*, ed. Rahim, 2:338–339; *Akbarnama*, trans. Beveridge, 2:495.

34 *Akbar-nāma*, ed. Rahim, 2:319; *Akbarnama*, trans. Beveridge, 2:471.

5 The Fort as the Imperial Adversary

Next, the chronicles describe the first encounter between Akbar and the forts. This was a defining moment, usually projected as some sort of a one-to-one confrontation between the emperor – whose greatness was about to unfold both in the real world and the narrative space – and the fortress – whose impregnability had already been recorded by the chroniclers. As Akbar's gaze would fall on his adversary and seek to assess their strengths and weaknesses, the fort would seem to stand opposite to him defiantly, as if returning the imperial gaze. The moment – and description – of this gaze was usually followed by Akbar recognising the fort as a worthy adversary and resolving to subdue it.

With the benefit of the knowledge that the Mughals would eventually win the sieges, the narrators probably found it worthwhile to project these moments as the beginning of the demonstration of imperial charisma. It is in Abul Fazl's narrative that these moments attain the greatest theatricality, although Arif Qandahari too seldom held his pen back. Describing this moment of encounter in the context of Chitor, Abul Fazl writes:

> [Akbar] arrived at the outskirts of the fort and pitched his camp. At this time there was a great storm of wind (*shiddat-i ʿawāṣif*), accompanied by thunder and lightning, so that the earth was shaken. But after an hour, the sky became clear (*hawā ṣāf shud*), the world was revealed (*ʿālam inkishāf yāft*) and the fortress appeared in the distance (*wa qalʿā az dūr namūdār gasht*).[35]

This description of darkness withdrawing itself for the fort to be revealed has almost the effect of curtains being removed off a stage at the outset of a dramatic performance. The above passage is followed by the verse quoted in the previous section describing the strength of the fort in the most hyperbolic terms. In the context of the siege of Ranthambhor, the same author narrates how Akbar positioned himself on top of a hill and took a thorough account of the adversary at hand. The moment of the imperial gaze was followed by a moment of resolution. Abul Fazl writes, "[H]e brought the figure of its conquest into the mirror of his imagination (*ṣūrat-i fatḥ -i ān rā dar āyina-i khayāl dar āwarda*) and tightened the straps of resolution for its capture (*niṭāq-i himmat bar taskhīr-i ān bastand*)".[36] The second moment is also said to have contained Akbar's public declaration of this resolution. Abul Fazl quotes Akbar as having

35 *Akbar-nāma*, ed. Rahim, 2:314; *Akbarnama*, trans. Beveridge, 2:464–465.

36 *Akbar-nāma*, ed. Rahim, 2:335; *Akbarnama*, trans. Beveridge, 2:491.

announced that "[B]y the favour of God the Creator/ I shall cast this fort to the ground".[37]

6 The Fort as the Site of Operations

The next part of the narratives comprises the description of the actual prosecution of the siege. While sieges indeed required a massive amount of well-planned and coordinated work by a vast body of soldiers and labourers, the way these activities are narrated in the three imperial biographies has a couple of specific purposes behind it. First, through a specific kind of description of the prosecution of the siege, the narrators seem to tell us that in taming the wild strength of the fort, what the emperor resorted to was method. By heading a systematic approach comprising elaborate planning, measurement, organisation, and skillful execution, Akbar is presented as having thrown his stock behind a proper methodical approach in taming the wild.

In the context of the siege of Chitor, Abul Fazl describes in details how Akbar went around the fort, how his surveyors measured the circumference of the structure, how he sent his commanders to raid the neighbouring areas to plunder his adversary's territory, and so on.[38] The value of such a methodical approach is underlined by contrasting it against the failure of more sporadic and spontaneous ones. To emphasize this point Abul Fazl writes that while Akbar went about the daunting task at hand in such a systematic fashion, some of his soldiers – who did not seem to grasp the necessity of his methodical approach – rushed to the outer ramparts of Chitor and tried to climb their way past them. Quickly pointing out that these efforts failed, the chronicler explains that "terrestrials cannot reach celestials (*zamīnīyān rā dast ba-āsmānīyān namī-rasad*)".[39] The implication is clearly that to subdue one celestial (the fort), it requires another (Akbar). This entire episode presents the emperor as the fountainhead of 'true' knowledge – only he was able to make a correct assessment of the strength of the fort. Consequently, he alone knew that nothing less than a patient, methodical technique could bring about the discomfiture of his mighty adversary. Following the above failure of the poorly coordinated attempts at escalade by the Mughal soldiers, Akbar is described to have taken the siege back on track by ordering the selection of proper spots for

---

37 *"ba-taufīq-i yazdān-i jān-āfrīn/ dar andāzam īn ḥiṣn-rā bar zamīn." Akbar-nāma*, ed. Rahim, 2:335; *Akbarnama*, trans. Beveridge, 2:491.

38 *Ṭabaqāt-i Akbarī*, ed. De, 2:216–217; *Ṭabaqāt-i Akbarī*, trans. De, 2:343–344.

39 *Akbar-nāma*, ed. Rahim, 2:315; *Akbarnama*, trans. Beveridge, 2:466.

the construction of batteries and the building of mines and saps.[40] This valorisation of a methodical approach as the only way to achieve victory is done even more explicitly later. Blaming the poor calculation and haste of some Mughal officers in causing an accident in the mines, Abul Fazl writes that in matters of subduing mighty fortresses, haste usually was of no avail. What was necessary was "patience (*ṣabr*) and planning (*sar-anjāmī*)".[41] The entire process of prosecution of the siege is thus projected as a case of the emperor educating the lesser mortals around him about the virtues of a patient and methodical effort by enlightening them with his divinely inspired knowledge about how to do things right.

In South Asia, more often than not forts are located on extremely inaccessible terrain. Besieging them consequently required armies to engage with this difficult physical geography as a part of their military endeavour. Led by the emperor in these five sieges, Mughal armies needed to plan the execution of the sieges carefully, construct elaborate mines and saps, and sometimes even exploit natural elevations to their advantage. Attributing all this to the emperor allowed the chronicler to project the emperor as a figure who was capable of altering and ordering nature itself, in a bid to fulfill his divine destiny of keeping his military juggernaut rolling for the benefit of humanity.[42]

In case of the other sieges also, the point about the impregnability of the fort and the first encounter between the fort and the emperor is followed by a detailed account of the actual siege operations. In fact, the planning and execution of these operations was narrated as the main substance of the great struggle between the fort and Akbar.[43] In fact, the narratives of almost all these sieges across the three chronicles are framed in a language of combat between two individuals; the work of an entire army engaged in a tedious military operation against a fort is memorialised as a tussle between two larger than life figures, both towering over everybody else.

In describing the combat between the garrison and the besiegers, the Mughal chroniclers attribute a great amount of bravery and valour to their adversaries. This was a convenient way for justifying the emperor's personal

---

40 *Akbar-nāma*, ed. Rahim, 2:318; *Akbarnama*, trans. Beveridge, 2:470.

41 *Akbar-nāma*, ed. Rahim, 2:318; *Akbarnama*, trans. Beveridge, 2:469–470.

42 What gets majorly eclipsed in such imperial narratives is the role of military engineers in Mughal employment. A prime example of such people was Muhammad Qasim Khan, Akbar's *Mir Barr wa Bahr*. He was the man in charge of building the siegeworks in the sieges of Chitor, Ranthambhor, and Surat. For a detailed analysis of Qasim Khan's career as a military engineer in Akbar's army, see Pratyay Nath, "Building the Empire: Military Infrastructure and the Career of Muhammad Qasim Khan in Mughal North India," *Proceedings of the Indian History Congress* 75 (2014), 270–274.

43 For Mankot, see *Akbar-nāma*, ed. Rahim, 2:58; *Akbarnama*, trans. Beveridge, 2:89.

involvement in the siege. To the same end, the chroniclers additionally highlight how keenly contested the military contest was. In describing the storm of Chitor, for example, Nizamuddin Ahmad slips into verse that uses graphic imagery comparing the Mughal army to a whirlwind that blew its mountain-like adversaries away like straws:

> Making ready for a war of victorious faith,
> The mountain of iron moved to the sea,
> The army like a whirl-wind came to that land,
> That its stones were blown like grass away.[44]

In the context of Chitor, Arif Qandahari writes:

> There was fierce fighting (*nīrān qitāl*) day and night. The speedy arrows were constantly hitting the target from bows and the stones hurled from catapults were all presenting the scene of the Day of Judgement. The noise produced did frighten even the bravest men, but it further encouraged the zeal of bravery among them. The sound of guns seemed to remind the sound of angel Israphil's horn while the noise of battle seemed to have filled or gripped the whole world (*g̲h̲aug̲h̲ā'-i rastā-k̲h̲ez dar jahān mī-āndāk̲h̲t*).[45]

Once again, the amplification of the contested quality of combat through the evocation of mythological motifs like the Day of Judgement and the horn of Israfil – which announces this day – sought to magnify the achievements of the Mughal army.

7 The Fort as the Site of Imperial Theatre

In the second half of the 16th century, a whole range of courtly rituals – together with the mythology surrounding Mughal kingship – highlighted the role of Akbar as a performer of various miracles and charismatic acts. For instance, Abul Fazl narrates how the emperor once cured his leopard-keeper by simply

---

44 "*'Hama sāz 'azā' karda muhaiyā/ rawān shud koh-i āhan, suwī daryā/ dar āmad bād īn lashkar, dar ān k̲h̲āk/ ki sangash ham gurezān shud chu k̲h̲āshāk.*'" *Ṭabaqāt-i Akbarī*, ed. De, 2:219; *Ṭabaqāt-i Akbarī*, trans. De, 2:347.

45 *Ta'rīk̲h̲-i Akbarī*, ed. Nadwi, Ali, Arshi, p. 111; *Ta'rīk̲h̲-i Akbarī*, trans. Ahmad, p. 149.

breathing on him.[46] The performance of such miracles constituted a vital element of Akbar's kingship. It represented one of the Sufistic qualities Mughal sovereignty had imbibed by this time.

Alongside this ritualistic theatrics in the real world, the ceaseless performance that formed an integral part of Mughal sovereignty also played itself out in the pages of the imperial biographies. The long duration and the challenging circumstances of imperial sieges gave Akbar's biographers the chance to emphasize this dimension of his rule through the narration of different types of imperial performance. Instances of Akbar performing superhuman miracles by the grace of god would regularly be interwoven into the narratives of the sieges. Once again, it is *Akbar-nāma* that furnishes us with the most fabulous examples of these miracles. One recurrent theme is how Akbar – fortified as he was by divine blessing – would never be harmed by garrison-fire. Describing the siege of Surat, for example, Abul Fazl writes:

> [B]alls from cannon and culverins came several times into the holy quarters, but by the Divine protection they did no harm (*az ṣiyānat-i īzidī āsīb na-rasīd*). As the station was very near to the fort, H.M. at the request of his officers moved to a place near the Gopi Tank. That, too, was near the fort, but it was screened by forest, and uneven ground. Here, too, cannon balls reached the bounds of the quarters, but the Divine protection did its work (*ṣiyānat-i ghaibī kār-i khūd mī-kard*).[47]

Abul Fazl narrates a similar incident in the context of the siege of Patna. He then notes how "[T]he fact of his being guarded by the King of the world (*nigāh-bānī-i khidev-i hastī*) and the amount of his reliance upon Him (*pāya-i tawakkul-i wālā*) was impressed upon friends and strangers."[48] Narrating the siege of Chitor, he writes that Akbar was so sure about this divine protection that he would pace around slowly in the face of garrison fire; yet, while shots and bullets would strike down people around him, he would remain unharmed. Reflecting on this, the biographer writes that such miraculous incidents served

46 *Akbar-nāma*, ed. Rahim, 3:212; *Akbarnama*, trans. Beveridge, 3:298–299. One is immediately reminded of Marc Bloch's discussion of the beliefs surrounding the healing power of the touch of the medieval English and French monarchs. Marc Bloch, *The Royal Touch: Monarchy and Miracles in France and England*, English trans. J. E. Anderson (New York, 1989).

47 *Akbar-nāma*, ed. Rahim, 3:18; *Akbarnama*, trans. Beveridge, 3:25–26.

48 *Akbar-nāma*, ed. Rahim, 3:98. Translation mine.

as 'guidance to the simple' and enhanced 'the devotion of the loyal' towards the sovereign.[49]

Once again, the narration of such incidents gave the chroniclers the ideal opportunity for projecting the image of Akbar as a divinely-protected sovereign. Abul Fazl describes Akbar as an inhabitant of the "fort of divine care (*ḥiṣār-i ḥiẓānat-i īzidī*)".[50] By virtue of the divine mandate, not only was Akbar described to have been able to protect himself from harm at all times, but also extend this protection to all those who would choose to be loyal to him. Like many others, this quality of Akbar's too is given a strong corporeal dimension. Above all, it was the body of the emperor that was projected as the source of supernatural powers. While the limitless kindness or wrath of the emperor could spread anywhere unhindered, physical proximity seemed to be beneficial in the transmission of certain qualities. Thus, in course of the imperial sieges, the people near the emperor are described to have evaded death even after being shot through their armours, as the bullets would tend to cool down and render themselves harmless before reaching the flesh.[51]

There was another way Akbar's biographers highlight his divine connection – describing how he would pray for heavenly assistance in favour of his army and ascribing to these prayers crucial importance while narrating the outcome of the sieges. For instance, Arif Qandahari describes how the emperor prayed for heavenly intervention in favour of his army at Chitor.[52] In the context of the siege of Patna, Abul Fazl writes that Akbar "asked for assistance from the hidden armies".[53] The fact that the Mughals eventually won all these sieges allowed the chroniclers to use this literary trope to underline that however much Mughal soldiers might have exerted themselves towards the subjugation of these forts, it was only through divine intervention on behalf of the Mughals – prayed for and secured by Akbar – that victory could be achieved. This was one of the ways the chroniclers appropriated all the credit for the success of the siege and ascribed it to the emperor, thereby projecting him as an intercessor between divinity and mankind.

Narratives of imperial sieges were also used to demonstrate Akbar's ability of predicting the future. An episode from Abul Fazl's narrative of the siege of Chitor bears evidence to this. Describing an accident in a Mughal mine that killed hundreds of Mughal soldiers, he points out that Akbar had already

---

49 *Akbar-nāma*, ed. Rahim, 2:319; *Akbarnama*, trans. Beveridge, 2:470–471.

50 *Akbar-nāma*, ed. Rahim, 3:98. Translation mine. Beveridge renders *ḥiẓānat* as "protection." But perhaps 'care' translates it more accurately. *Akbarnama*, trans. Beveridge, 3:137.

51 *Akbar-nāma*, ed. Rahim, 2:319; *Akbarnama*, trans. Beveridge, 2:471.

52 *Ta'rīkh-i Akbarī*, Persian text, p. 112; *Ta'rīkh-i Akbarī*, trans. Ahmad, p. 149.

53 *Akbar-nāma*, ed. Rahim, 3:90; *Akbarnama*, trans. Beveridge, 3:127.

foreseen how and why this mishap would happen. Hence, he was not shocked when the incident was reported to him.[54] In the same vein, during his first encounter with the fort of Patna, Akbar is said to have realized that the key to the conquest of that fort lay in gaining control over the neighbouring fort of Hajipur.[55] Akbar is repeatedly said to have possessed knowledge and wisdom unknown to others, principally owing to his divine connection. The imperial siege was thereby made to emerge as a site for the emperor enlightening other lesser mortals that surrounded him and sharing with them the benefits of his farsight.

Alongside these more passive ways in which Akbar is said to have contributed to these sieges, there were also more direct ones. There are several examples of Akbar performing outstanding military feats in the course of these sieges. Abul Fazl narrates in the context of the siege of Chitor that Akbar "killed many of the noted members of the garrison and sent them to the sleep of annihilation".[56] Two examples can be cited here. First, he describes Akbar once visiting one battery where Mughal soldiers were exchanging gunfire with the garrison. One of the matchlockmen of the garrison was displaying particular skill and marksmanship. As his shot injured a Mughal commander by the name of Jalal Khan, Akbar decided to avenge him. Abul Fazl quotes Akbar as having said, "[W]e shall this very instant take revenge for you on his gun".[57] Then, he is said to have aimed his gun at the spot on the fortification from where the adversary's gun was projecting and shot the marksman dead with a single bullet. It also turned out later that the victim of the emperor's wrath was none other than the chief of the musketeers of the garrison.[58] Abul Fazl carries on, saying that the emperor's next feat was to shoot down Jaimal, the very leader of the garrison, during a decisive struggle on the eve of the storm of the fort.[59] This is supposed to have created tremendous confusion and panic among the garrison. They stopped engaging the besiegers, burnt their women in ritual fire to supposedly protect their honour, and prepared to make a final stand.[60] The narration of this immensely theatrical event – the leader of the

---

54 The biographer, however, is silent about why the emperor let this tragedy happen, given that he had prior knowledge of it. One can only speculate about whether allowing the incidence of this human tragedy was his way of imparting a lesson in the value of the proper methodical approach to his soldiers. *Akbar-nāma*, ed. Rahim, 2:317; *Akbarnama*, trans. Beveridge, 2:468–469.

55 *Akbar-nāma*, ed. Rahim, 3:95; *Akbarnama*, trans. Beveridge, 3:135.

56 *Akbar-nāma*, ed. Rahim, 2:319; *Akbarnama*, trans. Beveridge, 2:470.

57 *Akbar-nāma*, ed. Rahim, 2:319; *Akbarnama*, trans. Beveridge, 2:470.

58 *Akbar-nāma*, ed. Rahim, 2:318–319; *Akbarnama*, trans. Beveridge, 2:470.

59 *Ṭabaqāt-i Akbarī*, ed. De, 2:218; *Ṭabaqāt-i Akbarī*, trans. De, 2:346.

60 *Ṭabaqāt-i Akbarī*, ed. De, 2:218; *Ṭabaqāt-i Akbarī*, trans. De, 2:346.

besieging army shooting down the leader of the garrison and thereby deciding the fate of the long-drawn siege – allowed the narrator yet another way to showcase the young emperor's military prowess and his ability to force a resolution when events seemed to have reached a stalemate. Whatever military feats our narratives claim he performed are said to have been made possible by his being the recipient of unlimited divine favour. This is made explicit by a sentence in Arif Qandahari's text, which reads, "[Akbar] performed feats of such amazing courage which could only be performed by the grace of God".[61]

Finally, the military genius and leadership capabilities of the emperor were emphasized by ascribing to him various important decisions taken by his army during the course of these sieges. The deadlock of the six months-long siege of Chitor, for example, is said to have been finally broken when the Mughal army focused their attention on specific spots of the fortification and tried to physically take them down. This worked, and the besiegers eventually stormed the fort. Narrating this episode, Arif Qandahari says that it was none other than the emperor who came up with the idea that the Mughal "army should launch a sudden attack on the fort from all sides".[62] In the same vein, Abul Fazl says that during the siege of Patna, it was Akbar who came up with the crucial suggestion that the Mughal army should first focus their efforts to capture the neighbouring fort of Hajipur, which was supplying the fort of Patna with its provisions.[63] Once again, this decided the outcome of the siege. The fall of Hajipur disheartened the garrison of Patna, forcing Daud Khan Karrani – the Afghan chief leading the defence of Patna – to flee to Bengal and leave the fort for the Mughal army to take.

8 The Fort Defeated

Since the Mughals emerged successful in all the five sieges in question, each of them presented the imperial chroniclers with the opportunity of projecting Akbar as the decider of the fate of the garrison at the end of the siege. He was free to either choose to display his kindness and magnanimity, and forgive them – as was the case in Surat or Ranthambhor; or he could punish them, as was the case at Chitor. The biographies project either decision as completely justified and as the need of the hour. When the garrison was pardoned, the biographers reason that these soldiers had realized their mistake, had repented

61 *Taʾrīkh-i Akbarī*, ed. Nadwi, Ali, Arshi, p. 113; *Taʾrīkh-i Akbarī*, trans. Ahmad, p. 150.
62 *Taʾrīkh-i Akbarī*, ed. Nadwi, Ali, Arshi, p. 112; *Taʾrīkh-i Akbarī*, trans. Ahmad, p. 149.
63 *Akbar-nāma*, ed. Rahim, 3:83; *Akbarnama*, trans. Beveridge, 3:115–116.

opposing the imperial forces for so long, and had finally chosen the virtuous path of loyalty to the Mughal emperor. For instance, Abul Fazl narrates that "the boundless ocean of his [Akbar's] benevolence was put in motion (*daryā-i bī-karān-i ʿāṭifat-ash ba-josh āmad*)" by the surrender of the garrison of Surat.[64] Justifying Akbar's subsequent pardoning of the garrison, Nizamuddin Ahmad quotes Akbar as having said:

> For evil to return evil,
> To outward seeming, might be wise;
> But those who to the soul have reached,
> Have evil seen, and good have done.[65]

Here the narrator highlights the greatness of the Mughal emperor by creating a binary. He implies that on the one hand there are common people who appreciate only the face value of things, while on the other there are exceptional men like Akbar, who can see right through everything and comprehend their inner meaning. Hence, unlike commoners, Akbar does not retaliate against evil with evil, but neutralises it with generosity and forgiveness. Similarly, justifying Akbar's decision to forgive the Afghan garrison of Mankot, Abul Fazl observes how the emperor "regarded the sweets of forgiveness as greater than those of revenge".[66]

By way of describing Akbar forgiving these garrisons, the chroniclers create an occasion for celebrating these positive qualities of his. What is repeatedly stressed is that the Mughal emperor was not an unreasonable rapacious conqueror. In certain cases, the chroniclers mention the meting out of symbolic punishments. Abul Fazl, for example, narrates that following the submission of the garrison of Surat, most of the garrison was left unharmed. Only their leader and some of his associates were punished.[67] This reinforces the image of Akbar as a just ruler, since he was shown to be capable of identifying the real culprits and punishing them specifically. It also serves to underline the idea that just as the Mughal emperor did not indiscriminately use violence like a madman, his commitment to the dispensation of justice was also not affected by

---

64 *Akbar-nāma*, ed. Rahim, 3:28; *Akbarnama*, trans. Beveridge, 3:40.

65 "*badī rā mukāfāt kardan badī/ bar ahl-ṣūrat būd ba-khurdī/ ba-maʿnī kasānī ki pai burda-and/ badī dīda wa neko'ī karda-and.*" *Ṭabaqāt-i Akbarī*, ed. De, 2:248; *Ṭabaqāt-i Akbarī*, trans. De, 2:386.

66 *Akbar-nāma*, ed. Rahim, 2:59; *Akbarnama*, trans. Beveridge, 2:91.

67 *Akbar-nāma*, ed. Rahim, 3:29; *Akbarnama*, trans. Beveridge, 3:40.

his kindness and his ability to overlook offenses. He could use violence when there was need for it.[68]

Finally, in the case of Chitor all three texts say that Akbar ordered a general massacre (*qatl-i ʿāmm*) of the garrison as well as the general population. Here, the chroniclers go to great lengths to justify this spectacular act of violence. They do so by reminding the reader that these people had resisted the forces of the just emperor with great viciousness and evil. Abul Fazl elaborates how such a massacre was not an unreasonable act of vengeance but exactly what justice demanded. He reminds his readers that when Alauddin Khalji – the early-thirteenth – century Sultan of Delhi – had seized the fort, no such general massacre had been necessary since the common people had not joined the garrison in resisting his army back then.[69]

Another important element of the narration of these post-victory activities of the emperor was the description of his dealings with the common populace. Barring exceptional cases like Chitor, our texts take great care to underline that the victorious Mughal armies left the civilians unharmed. Doing this was necessary for the projection of a benevolent image of Mughal rule, one that was committed to the well-being of the entire human race. Exemplifying this, Nizamuddin Ahmad describes how Akbar interacted with the common people of Patna in the aftermath of the siege and "gave assurance of safety to high and low".[70]

Finally, the chronicles narrate the process of the ritual incorporation of the vanquished forts into the empire. The first stage of this would comprise the emperor making a dramatic entry into the fort. Describing this moment in the aftermath of the siege of Chitor, for instance, Nizamuddin Ahmad writes:

> The whole of that night was passed in war and slaughter; and in the morning, which was the dawn of the daily-increasing grandeur, the fort was conquered. His Majesty mounted on an elephant, with all his devoted servants attending on his victorious stirrups, on foot, entered the fort.[71]

---

68 Ali Anooshahr has argued that the ability to use calculated violence was one of the markers of warrior masculinity in the early Mughal empire. It is beyond the scope of the present chapter, but it will be interesting to explore elsewhere the inter-relationship between the ideals of masculinity, rulership, and the use of violence in Akbar's empire. Ali Anooshahr, "The King Who Would Be Man: The Gender Roles of the Warrior King in Early Mughal History," *Journal of the Royal Asiatic Society* 18, no. 3 (2008), 327–340.

69 *Akbar-nāma*, ed. Rahim, 2:322–323; *Akbarnama*, trans. Beveridge, 2:475.

70 *Ṭabaqāt-i Akbarī*, ed. De, 2:294; *Ṭabaqāt-i Akbarī*, trans. De, 2:447.

71 *Ṭabaqāt-i Akbarī*, ed. De, 2:219; *Ṭabaqāt-i Akbarī*, trans. De, 2:347.

Akbar is said to have gone around the fort, surveying his most recent acquisition, and taking stock of everything. Narrating a similar episode after the fall of Surat, the same author writes:

> On the following day (the emperor) went to inspect the fort, and having gone inside the citadel, after much consideration and examination, he gave orders to the servants of the threshold about the repairs of the fort and its improvement.[72]

Akbar discovered a lot of artillery of Ottoman origin in the fort of Surat. Deeming its presence there unnecessary in terms of defense of the fort, he ordered some of these to be transported to the Mughal capital.[73] Aside from the military value of the cannons, they must also have held immense significance as trophies of war for the Mughals, not only because it belonged to their vanquished Gujarati adversaries, but probably also because they had a link to the rival empire of the Ottomans. Descriptions of the aftermath of the other sieges follow a similar pattern. They usually close by narrating the assignment a Mughal commander to the charge of the fort.

Such surveys – followed by an elaborate narration of orders issued towards repairs, alterations, and assignment of the fort – are projected by the texts as ways in which the emperor would order and discipline his new conquest. This description of a ritualistic disciplining would bring to a full circle the story of Akbar taming the wild adversary that the fort once represented, one which had begun in the literary space with the theatrical description of the emperor's first encounter with the fort. It would also perfectly bring to a conclusion the earlier description of the methodical approach adopted by Akbar to subdue the fort.

9 Conclusion

This essay has highlighted the various literary strategies Mughal chroniclers deployed in the late-16th century in order to describe imperial sieges led by Akbar. I have argued that the literary description of these sieges was used by these chroniclers as a site for extolling the greatness of Akbar as a universal

---

72 Here "servants of the threshold" refers to Mughal soldiers. *Ṭabaqāt-i Akbarī*, ed. De, 2:249; *Ṭabaqāt-i Akbarī*, trans. De, 2:387.

73 *Ṭabaqāt-i Akbarī*, ed. De, 2:249; *Ṭabaqāt-i Akbarī*, trans. De, 2:387–388; *Akbar-nāma*, ed. Rahim, 3:29; *Akbarnama*, trans. Beveridge, 3:40–41.

emperor. While the three texts under focus vary from each other on several counts – including their literary style and the immediate context within which they were produced, they share the agenda of glorifying imperial rule using similar arguments and a conviction about the universal nature of Mughal sovereignty. Their narratives of these sieges follow a similar sequential pattern. This points to the emergence of certain literary conventions within Mughal official circles by the late-16th century. The narratives usually begin by highlighting the invincibility of the forts in question. Next, by projecting them as mighty imperial adversaries, the chroniclers cast the narrative of the imperial sieges as descriptions of their individual tussles against the Mughal emperor. Following this, the chroniclers narrate the various activities of the Mughal army and the emperor on the one hand and the resistance put up by the garrison of the fort on the other. The language of these parts of the narratives highlights the ability of the emperor to perform miracles, foresee the future, and enlighten everybody around him about the merits of a thorough and methodical approach in subduing the fort. Finally, the chroniclers narrate the moment of the fall of the fort, the administration of justice to the garrison by the emperor, and the ritual incorporation of the fort into the body politic of the empire. The language used by the chroniclers throughout these steps was aimed at projecting Akbar as a just, deserving, and divinely mandated ruler of all of humanity.

This increasing thrust on universal sovereignty was arguably the most significant direction Mughal imperial ideology took under Akbar in the second half of the 16th century.[74] Timurid ideals of statecraft aside, it also came to accommodate other ideological influences that shaped the emerging political discourse under Akbar.[75] All of this contributed towards the emergence of a complex and sophisticated *Akbari* imperial ideology, whose cornerstone was a vigorous claim to universal sovereignty. Composed in the late-16th century – the same period when this ideology reached a stage of maturity, all the three chronicles under focus represent a growing conviction in this claim among the

74 Under Babur (r. 1526–1530) and Humayun, Mughal ideological vocabulary had still been firmly couched in the Timurid discourse of power and empire. It was under Akbar that Mughal ruling ideology really came of its own as an independent entity. Timurid political philosophy continued to have a significant position in it. For a discussion on the sustained importance of Timurid dynastic memory and ideological influence on generations of Mughals, see Lisa Balabanlilar, *Imperial Memory*.

75 These included Ibn al-Arabi's ideas about *waḥdat al-wujūd* (the unity of being) and *insān-i kāmil* (the perfect man), Brahmanical religious and philosophical elements, and a reinvigorated thrust on Nasirean political ethics. For more, see Mukhia, *The Mughals of India*, pp. 14–71; Muzaffar Alam, "The Mughals, the Sufi Shaikhs and the Formation of the Akbari Dispensation," *Modern Asian Studies* 43, no. 1 (2009), 135–174.

imperial ranks. This conviction strengthened under Akbar's successors in the 17th century and found many more new avenues of expression.[76]

All this gives us an opportunity to rethink the relationship between warfare and Mughal empire-building. At a time when the universalist Mughal imperial discourse was taking shape under Akbar, literary chronicles – represented here by the three texts under focus – served as one of the prime sites for the elaboration of this evolving ideology and an occasion for celebrating the emergent imperial cult. Imperial wars – exemplified here by the five sieges – provided these chroniclers with ample material for memorialising the martial glory of the empire and extolling the greatness of the emperor. It is primarily for this reason that narratives of wars comprise one of the most important subjects of Mughal chronicles. This means that not only was warfare important to the process of Mughal empire-building as a tool for territorial expansion and control, but also significant in providing the empire with substantial material to validate its claims to universal sovereignty within shared literary spaces. In turn, this pushes us to acknowledge that the importance of war spilled well beyond the domain of battles, technologies, and army organisation – the three registers conventionally used to analyse Mughal war-making. Warfare shared a much more complex and pervasive relationship with the broader processes of Mughal state-formation and empire-building. Through its literary analysis of Mughal narratives of Akbar's sieges, the present essay has made an effort towards probing this relatively unexplored relationship.

Bibliography

Printed Sources

Ahmad, Khwaja Nizamuddin, *Ṭabaqāt-i Akbarī*, Persian text ed. Brajendranath De, 3 vols (Calcutta, 1931); *The Ṭabaqāt-i Akbarī of Khwājah Niẓāmuddīn Aḥmad*, English

76 Akbar's son Jahangir presided over the production of allegorical political paintings that portrayed him as the just, semi-divine, all-conquering lord of time and space. Koch, "The Symbolic Possession of the World;" Koch, "How the Mughal *Padshahs* Referenced Iran;" Ebba Koch, "The Mughal Emperor as Solomon, Majnun, and Orpheus, or the Album as a Think Tank for Allegory," *Muqarnas* 27 (2010), 277–311; Ramaswamy, "Conceit of the Globe." This tradition was also continued – although in a more circumscribed way – under his son and successor, Shah Jahan (r. 1627–1658). Allusions to universal sovereignty were also inscribed in the imperial built spaces that came up in this century, including the spectacular Taj Mahal and Shahjahanabad complexes in Agra and Delhi respectively. See Ebba Koch, "The Taj Mahal: Architecture, Symbolism, and Urban Significance," *Muqarnas* 22 (2005), 128–149; Ebba Koch, *Mughal Art and Imperial Ideology: Collected Essays* (New Delhi, 2001); Stephen P. Blake, *Shahjahanabad: The Sovereign City in Mughal India, 1639–1739* (New York, 1991).

trans. Brajendranath De, revised, edited and completed by Baini Prasad, 3 vols (Delhi, 1992).

Badaoni, Abdul Qadir, *Muntakhabu-t-Tawārīkh by 'Abdu-l-Qādir ibn-i-Mulūk Shāh known as al-Badāoni*, English trans. W. H. Lowe (Delhi, 1986).

Fazl, Abul, *Akbar-nāma*, Persian text ed. Maulawi Abdur Rahim, 3 vols (Calcutta, 1876); *The Akbarnama of Abu'l Fazl*, English trans. H. Beveridge, 3 vols (Calcutta, 1904).

Ferishta, Mahomed Kasim, *History of the Rise of the Mahomedan Power in India till the Year AD 1612 or Tarikh-i Firishta*, English trans. John Briggs, 4 vols (Delhi, 1990).

Nathan, Mirza, *Bahāristān-i Ghā'ibī*, transcribed copy of the original Persian manuscript preserved in Bibliotheque Nationale, Paris, JS 60–62, Jadunath Sarkar Collection, National Library, Kolkata; *Bahāristān-i Ghaybī: A History of the Mughal Wars in Assam, Cooch Behar, Bengal, Bihar and Orissa during the Reigns of Jahangir and Shah Jahan,* 2 vols, English trans. M. I. Borah (Guwahati, 1992).

Qandahari, Muhammad Arif, *Ta'rīkh-i Akbarī*, Persian text ed. Haji Syed Muinuddin Nadwi, Syed Azhar Ali, Imtiaz Ali Arshi (Rampur, 1962); *Ta'rīkh-i Akbarī*, English trans. Tasneem Ahmad (Delhi, 1993).

Tusi, Nasiruddin, *Akhlāq-i Nāṣirī*, Persian text ed. Mojtaba Minavi and Ali Riza Haidari (Tehran, 1976); *The Nasirean Ethics*, English trans. G. M. Wickens (London, 1964).

Secondary Sources

Alam, Muzaffar, "Akhlaqi Norms and Mughal Governance," in *The Making of Indo-Persian Culture: Indian and French Studies*, ed. Muzaffar Alam et al. (New Delhi, 2000), pp. 67–95.

Alam, Muzaffar, *The Languages of Political Islam in India, c. 1200–1800* (Delhi, 2004).

Alam, Muzaffar, "The Mughals, the Sufi Shaikhs and the Formation of the Akbari Dispensation," *Modern Asian Studies* 43, no. 1 (2009), 135–174.

Alam, Muzaffar, "The Pursuit of Persian: Language in Mughal Politics," *Modern Asian Studies* 32, no. 2 (1998), 317–349.

Ali, M. Athar, *The Apparatus of Empire: Awards of Ranks, Offices and Titles to the Mughal Nobility 1574–1658* (New Delhi, 1985).

Ali, M. Athar, "The Evolution of the Perception of India: Akbar and Abu'l Fazl," *Social Scientist* 24, no. 1–3 (1996), 80–88.

Anooshahr, Ali, "The King Who Would Be Man: The Gender Roles of the Warrior King in Early Mughal History," *Journal of the Royal Asiatic Society* 18, no. 3 (2008), 327–340.

Aziz, Abdul, *The Mansabdari System and the Mughal Army* (Lahore, 1945).

Balabanlilar, Lisa, *Imperial Memory and Dynastic Politics in Early Modern South and Central Asia* (New Delhi, 2013).

Balabanlilar, Lisa, "Lords of the Auspicious Conjunction: Turco-Mongol Imperial Identity on the Subcontinent," *Journal of World History* 18, no. 1 (2007), 2–39.

Bang, Peter Fibiger; Kołodziejczyk, Dariusz, eds., *Universal Empire: A Comparative Approach to Imperial Culture and Representation in Eurasian History* (Cambridge, UK, 2012).

Begum, Ayesha, "Mughal Fort Architecture in Bengal with an Introduction to Some Important River Forts," *Journal of Asiatic Society of Bangladesh* 47, no. 1 (2002), 1–24.

Blake, Stephen P., *Shahjahanabad: The Sovereign City in Mughal India, 1639–1739* (New York, 1991).

Bloch, Marc, *The Royal Touch: Monarchy and Miracles in France and England*, English trans. J. E. Anderson (New York, 1989).

Casale, Giancarlo, *The Ottoman Age of Exploration* (New York, 2010).

Crossley, Pamela Kyle, *A Translucent Mirror: History and Identity in Qing Imperial Identity* (Berkeley, Los Angeles, London, 1999).

Deloche, Jean, *Studies on Fortification in India* (Pondicherry, 2007).

Eaton, Richard M.; Wagoner, Philip B., *Power, Memory, Architecture: Contested Sites on India's Deccan Plateau, c. 1300–1600* (New Delhi, 2014).

Garza, Andrew de la, *The Mughal Empire at War: Babur, Akbar and the Indian Military Revolution, 1500–1605* (London, New York, 2016).

Gordon, Stuart, "Forts and Social Control in the Maratha State," *Modern Asian Studies* 13 (1979), 1–17.

Gommans, Jos, *Mughal Warfare: Indian Frontiers and High Roads to Empire, 1500–1700* (London, New York, 2002).

Habib, Irfan, "Akbar and Technology," *Social Scientist* 20, no. 9–10 (1992), 3–15.

Hunt, Margaret R., "The 1689 Mughal Siege of East India Company Bombay: Crisis and Historical Erasure," *History Workshop Journal* 84, no. 1 (2017), 149–169.

Irvine, William, *The Army of the Indian Moghuls* (Delhi, 2004).

Khan, Iqbal Ghani, "Metallurgy in Medieval India – The Case of Iron Cannons," *Proceedings of the Indian History Congress* 45 (1984), 464–471.

Khan, Iqtidar Alam, "Early Use of Cannon and Musket in India, AD 1442–1526," *Journal of Economic and Social History of the Orient* 24, no. 2 (1981), 146–164.

Khan, Iqtidar Alam, *Gunpowder and Firearms: Warfare in Medieval India* (New Delhi, 2004).

Khan, Iqtidar Alam, "Origin and Development of Gunpowder Technology in India, AD 1250–1500," *The Indian Historical Review* 4, no. 1 (1977), 20–29.

Kinra, Rajeev, *Writing Self, Writing Empire: Chandar Bhan Brahman and the Cultural World of the Indo-Persian State Secretary* (Oakland, 2015).

Koch, Ebba, "How the Mughal *Padshahs* Referenced Iran in Their Visual Construction of Universal Rule," in *Universal Empire: A Comparative Approach to Imperial Culture and Representation in Eurasian History*, ed. Peter Fibiger Bang, Dariusz Kołodziejczyk (Cambridge, UK, 2012), pp. 194–209.

Koch, Ebba, *Mughal Art and Imperial Ideology: Collected Essays* (New Delhi, 2001).

Koch, Ebba, "The Mughal Emperor as Solomon, Majnun, and Orpheus, or the Album as a Think Tank for Allegory," *Muqarnas* 27 (2010), 277–311.

Koch, Ebba, "The Taj Mahal: Architecture, Symbolism, and Urban Significance", *Muqarnas* 22 (2005), 128–149.

Koch, Ebba, "The Symbolic Possession of the World: European Cartography in Mughal Allegory and History Painting," *Journal of the Economic and Social History of the Orient* 55, no. 2–3 (2012), 547–580.

McGregor, Andrew, *A Military History of Modern Egypt: From the Ottoman Conquest to the Ramadan War* (Westport, Connecticut, London, 2006).

Moin, A. Azfar, *The Millennial Sovereign: Sacred Kingship and Sainthood in Islam* (New York, 2012).

Mukhia, Harbans, *Historians and Historiography during the Reign of Akbar* (New Delhi, 1976).

Mukhia, Harbans, *The Mughals of India* (Malden, Oxford, Carlton, 2004).

Nath, Pratyay, "Building the Empire: Military Infrastructure and the Career of Muhammad Qasim Khan in Mughal North India," *Proceedings of the Indian History Congress* 75 (2014), 270–274.

Nath, Pratyay, "Siege Warfare in Mughal India, 1519–1538," in *Warfare and Politics in South Asia from Ancient to Modern Times*, ed. Kaushik Roy (New Delhi, 2011), pp. 121–144.

Nath, Pratyay, "Through the Lens of War: Akbar's Sieges (1567–69) and Mughal Empire-Building in Early Modern North India," *South Asia: Journal of South Asian Studies* 41, no. 2 (2018), 245–258.

Nizami, Khaliq Ahmad, *On History and Historians of Medieval India* (New Delhi, 1983).

Pagden, Anthony, *Lords of All the World: Ideologies of Empire in Spain, Britain and France c. 1500–c. 1800* (New Haven, London, 1995).

Pant, G. N., *Mughal Weapons in the Baburnama* (Delhi, 1989).

Ramaswamy, Sumathi, "Conceit of the Globe in Mughal Visual Practice," *Comparative Studies in Society and History* 49, no. 4 (2007), 751–782.

Rizvi, S. A. A., *Religious and Intellectual History of the Muslims in Akbar's Reign, with Special Reference to Abul Fazl* (New Delhi, 1975).

Roy, Kaushik, *India's Historic Battles: From Alexander the Great to Kargil* (New Delhi, 2004).

Roy, Kaushik, *Military Transition in Early Modern Asia, 1400–1750: Cavalry, Guns, Government and Ships* (London, New York, 2014).

Sarkar, Jadunath, *Military History of India* (Delhi, 1970).

Sharma, Sandhya, *Literature, Culture and History in Mughal North India, 1550–1800*, (Delhi, 2011).

Streusand, Douglas, *The Formation of the Mughal Empire* (New Delhi, 1990).

Talbot, Cynthia, "Justifying Defeat: A Rajput Perspective on the Age of Akbar," *Journal of the Economic and Social History of the Orient* 55, no. 2–3 (2012), 329–368.

Truschke, Audrey, "Setting the Record Wrong: A Sanskrit Vision of Mughal Conquests," *South Asian History and Culture* 3, no. 3 (2012), 373–396.

Wink, Andre, *Makers of the Muslim World: Akbar* (Oxford, 2009).

CHAPTER 8

Siege Warfare in Verse and Prose: the Ottoman Conquest of Kamianets-Podilsky (Kamaniçe), 1672

Kahraman Şakul

1 Introduction[1]

In June 1672, the Ottoman imperial army marched into western Ukraine in support of the pro-Ottoman Cossack Hetman Petro Doroshenko (1666–1676) against his former Polish-Lithuanian suzerain and captured the unassailable fortress of Kamianets-Podilsky. This chapter presents a threefold analysis of this siege based on Ottoman literary production. The political and ideological background of the Ottoman campaign is followed by a day-to-day reconstruction of the siege that deals with technical and literary aspects and incorporates military details, religious and ideological components, and Ottoman forms of cultural representations.

2 The Ideological Context

This was the first campaign in Europe led by a sultan since the failed Polish (Khotyn/Hotin) Campaign of 1621. Sultan Mehmed IV (1648–1687) took the imperial family with him on the campaign, including his mother, son (the future sultan Mustafa II), chief consort and two sisters. The long march was full of occasions for displaying imperial pomp and grandeur, with royal hunts along the way, and the Sultan and his family riding in silver-plated coaches.[2] But, the campaign was distinctive for other reasons.

1 This research is sponsored by TÜBİTAK (The Scientific and Technological Research Council of Turkey) as part of the ARDEB 1001 Research Projects Funding Program (119K056 Ottoman Siege Warfare in the Seventeenth Century). I am grateful to Gábor Ágoston, Caroline Finkel, Anke Fischer-Kattner, Dariusz Kołodziejczyk, Rhoads Murphey, Victor Ostapchuk, and Jamel Ostwald for their valuable suggestions and criticisms. I am indebted to Andriy Zhyvachivskyi and Volodymyr Pivtorak for translations from Polish and locating the fortress plans. Lastly, I owe special thanks to Fatma Aladağ and Zeynep Eroğlu for the preparation of the maps and plan of the siege.

2 Halime Doğru, *Lehistan'da Bir Osmanlı Sultanı: IV. Mehmed'in Kamaniçe-Hotin Seferleri ve Bir Masraf Defteri* (İstanbul, 2005), pp. 11–19.

 | DOI:10.1163/9789004395695_009

Located 100 km southeast of Lviv and 200 km southeast of Cracow, Kamianets, center of the historic region of Podolia, was seen as a northern extension of the centuries-long struggle between Islam and Christianity in the Mediterranean. It was a key segment of the *antimurale Christianitatis*, a Polish bulwark against Ottoman 'barbarism' as was La Valetta in Malta. By contrast, the Ottomans labeled it the "key to northern Europe."[3]

The fall of formidable fortresses such as Uyvar [Hung.: Érsekújvár] (1663), Candia [Tur.: Kandiye] (1669), and Kamianets (1672) one after another were glorified in campaign books (*fetihname*) describing these sieges in both technical and poetic terms with the intention of praising their specific vizier patron. The colorful description of the siege of Kamianets by the poet Yusuf Nabi, secretary of the vizierial council of Musahip ("the Favorite") Mustafa Pasha, who had the honor of commanding the right flank as the second vizier in rank, is one of the most prominent works in this genre. His elaborate account of the campaign made full use of metaphors inspired by Persian mythology and Islamic tradition in narrating the technical details of the siege. Indeed, this work launched him upon an illustrious career.[4] The most detailed account of the siege with a tactical-strategic focus was written by Hacı Ali Efendi, a man experienced in siege warfare who participated on the left flank; he was the chief secretary of Kara ("the Black") Mustafa Pasha, the third vizier in rank. Hacı Ali's account is a eulogy that presents Kara Mustafa as a frontline soldier whom the rank and file adored as one of their comrades. Abdi Pasha, the seventh vizier and head of the chancery, participated in the siege under Fazıl ["the Wise"] Ahmed Pasha and produced a third eyewitness account in his chronicle.[5] By contrast, Silahdar ("the Arms-Bearer") Mehmed, who stayed behind in Babadağ where queen-mother Hatice Turhan Sultan encamped, relied on Hacı Ali and Abdi Pasha in his account of the siege in his chronicle with some observations of his own.[6]

3 Hacı Ali Efendi, *Tarih-i Kamaniçe (Tahlil ve Metin)*, ed. Ayşe Hande Can (Ankara, 2007), p. 83. Another transcription is available in Musa Taçkın, "Ali Efendi ve Tarih-i Kamaniçe Adlı Eseri (Metin ve Tahlil)" (MA thesis, Marmara University, 2004).

4 Şerife Yalçınkaya, *Nabi'nin Nasirliği ve Kamaniçe Fetih-namesi* (*Tenkitli Metin-İnceleme*) (İzmir, 2012). One of Nabi's verses was inscribed on the main gate of Kamianets, see Tülay Artan, "Osmanlı Ordusunun Sefere Çıkış Alayı (1672): Osmanlı Kadimciliği mi, Püritenlik Gösterisi mi – The Departure Procession of 1672: Ottoman Antiquarianism or a Puritan Statement," in *Uzak Komşu Yakın Anılar: Türkiye-Polonya İlişkilerinin 600 Yılı – Distant Neighbour Close Memories: 600 Years Of Turkish-Polish Relations*, ed. Ayşe Anadol (Istanbul, 2014), pp. 60–77, here: p. 61.

5 Fahri Çetin Derin, "Abdurrahman Abdi Paşa Vekayinamesi: Tahlil ve Metin Tenkidi" (Ph.D. thesis, Istanbul University, 1993).

6 Nazire Karaçay Türkal, "Silahdar Fındıklılı Mehmed Ağa Zeyl-i Fezleke (1065–22 Ca. 1106/1654–7 Şubat 1695)" (Ph.D thesis, Marmara University, 2012).

Technical and literary sources are compatible in all these sources that usually combined verse and prose, conforming to the usual Ottoman literary conventions. They all took advantage of the power of metaphor, simile, and alliteration in narrating the military-technical details to varying degrees. The metaphor of 'unconquerable maiden,' for instance, pervades the works that celebrated the conquest of the impregnable fortress of Kamianets.[7] There was a kernel of truth in this gendered language since it was only captured twice, in 1393 and 1672, despite numerous sieges. An impressed Nabi described the fortifications in verse: "never saw the eye of the heavens such a strong and extraordinary castle, since this wheel of destiny began rotating around the world." In his world of symbols, the winding riverbed surrounding the town and the fortress was a formidable dragon curled up to protect its dearest treasure. Alternatively, it was a blessed moonlike halo around the waist of a maiden. For Nabi the domes of the churches in the town were so tall that they reached up to the feet of Jesus Christ residing in the fourth heaven.[8] One should not dismiss elaborate descriptions of the siege as examples of oriental exaggeration. Highlighting odd technical aspects through literary devices was a convenient means of familiarizing the Ottoman literati with siegecraft in such eulogies.

These eye-witness accounts reflect a certain religious-ideological view that represented the sultan as a strong man who liberated himself from the power of royal women and moved his court to Edirne, the gate of holy war (*gaza*), in order to fight the Christian enemy.[9] This is best illustrated in the magnificent military procession held on May 7, 1672 in Edirne that officially launched the campaign. The parade highlighted Ottoman antiquarianism and religious revivalism[10] popular among the Ottoman ruling elite that pursued expansionist policies. Guided by Vani Mehmed Efendi, the preacher of the sultan, the revivalist *Kadızadeli* movement viewed the 'infidels' along the borders of the empire and 'irreligious innovations' (e.g., Sufi rituals, tobacco, coffee and so on) as the gravest threat to the state and religion. Sources on the Ottoman siege of

---

7 Muhammet Kuzubaş, "17. Yüzyıl Şairi Mezaki'ye Göre Lehistan Savaşı," *Turkish Studies/ Türkoloji Dergisi* 1, no. 2 (2006), 213–225, here: p. 220; Türkal, "Silahdar," p. 610; Mark Baer, *Honored by the Glory of Islam: Conversion and Conquest in Ottoman Europe* (Oxford, 2008), p. 166.

8 Yalçınkaya, *Nabi*, pp. 75–79, 231–234. Nabi's description of the fortress was repeated by others, see *Çehrin Seferi* in Lubomyr Andrij Hajda, "Two Ottoman Gazanames Concerning the Chyhyryn Campaign of 1678" (Ph.D. thesis, Harvard University, 1984), p. 68 and Filiz Duman, "Bir Gazavatname Türü Olarak Hüseyin Behçeti'nin Miracü'z-Zafer Adlı Zafernamesi (H.1088/M.1678)" (MA thesis, Süleyman Demirel Üniversitesi, 2010), pp. 143–144.

9 Baer, *Honored by the Glory of Islam*, pp. 139–146. Baer contradicts the traditional understanding of Mehmed IV as a weak sultan fond of hunting.

10 Artan, "The Departure Procession of 1672," pp. 60–77.

Kamianets mentioned Vani Mehmed on many occasions in recognition of his spiritual role with the sultan.[11] Aspiration to the military victories and religious simplicity of early Ottoman centuries was evident in the contrast between the modest costumes of the dignitaries and the splendid armor of their household troops. The message was clear: all wealth should be spent on the holy war. While the Turkish accounts praised the vivid religiosity and military expertise by reference to Persian mythology, Orientalist Antoine Galland, an eye-witness, described Sultan Mehmed IV (r. 1648–1687) as the "Turkish Mars." The new image of the sultan was also replicated in European prints that featured a victorious Mehmed IV against a background of citadels and battles after the 1660s.[12]

3 The Intercultural Context

Although their emphasis is on the religiosity and martial skills of the sultan and commanders, Ottoman accounts about the campaign incorporate ethnographic representation of troops – friends and foe alike – based on ethnicity as well as rivalry between several branches of military service. They presented the Ottoman army as an assortment of imperial subjects of all ethnic backgrounds with different levels of military competence. By the same token the enemy force, composed of *Leh* (Poles), *Nemçe* (Slavic *Nemets* meaning German speakers), and *Rus* (Ruthenians: old name for Ukrainians), was a worthy foe to defeat comparable to 'Franks' (Catholic powers).

This chapter also addresses the question of cultural perspective of representation of the Ottoman siege of Kamianets. Narration of the phases of a proper siege and comparisons between different fortresses indicate that there was a cross-cultural debate over the ways of besieging a fortress. Ottoman literary production suggests that siege warfare was a familiar topic for the Ottoman reading public that took pride in the conquest of impregnable strongholds of Kamianets-Podilsky, Uyvar, the *Ak Tabya* [white redoubt, i.e., St Andrew's bastion] of Candia in Crete (remained under Ottoman siege from 1648–1669), and Baghdad (1638).

11 Marinos Sariyannis, "The Kadızadeli Movement As a Social and Political Phenomenon: The Rise of a 'Mercantile Ethic,'" in *Political Initiatives "From the Bottom-Up" in the Ottoman Empire*, ed. Antonis Anastosopoulos (Rethymno, 2012), pp. 263–289; Baer, *Honored by the Glory of Islam*, pp. 150–151, 165.

12 Antoine Galland, *Journal d'Antoine Galland Pendant son Séjour à Constantinople, 1672–1673*, vol. 1, ed. Charles Schefer (Cambridge, UK, 2012), p. 137; Baer, *Honored by the Glory of Islam*, pp. 148–149.

4 The Technical Context

Literature on Ottoman warfare leaves the impression that the Ottoman conquest of this distant fortress was a foregone conclusion of a fairly straightforward campaign since the entire siege lasted ten days. This is not surprising, for the Ottoman military victories were by and large based on their mastery of siege warfare.[13] We will offer a detailed description of the siege based on Ottoman sources in order to reveal the technical complications. This description will also demonstrate how incorrect it is to insist on a diffusionist model in which military innovations spread in waves from the first-tier innovators in Western Europe, to second-tier exporters, and to third-tier imitators such as the Ottomans.[14] What we need, instead, is a geostrategic view that compares the Ottoman Empire as a dominant power with its immediate rivals in the theatres where they confronted one another.[15] One such theatre was Podolia, where they scored a major success against the Polish-Lithuanian Commonwealth in the era of imperial resurgence under the Köprülü grand viziers.

5 The Ottomans' New Northern Policy

Coined the 'Second Golden Age,' the period of 1661–1683 was marked by the expansionist policies of three grand viziers from the Köprülü family (of Albanian origin) who held the office without interruption throughout this period. Köprülü grand viziers Mehmed Pasha (1656–61), his son Fazıl Ahmed (1661–76), and son-in-law Kara Mustafa (1676–83) were the real decision makers in this period and, after the conquest of Crete, they renewed the territorial expansion in their attempt to control the Ottoman 'Northern Quadrant:' "a vast arc of territory stretching from Slovakia to Podolia and across the Pontic Steppe as far

13 Between 1521 and 1566 only 13 castles in Hungary resisted Ottoman siege for more than ten days, and only nine for more than 20 days. See Gábor Ágoston, "Habsburgs and Ottomans: Defense, Military Change and Shifts in Power," *The Turkish Studies Association Bulletin* 22, no. 1 (Spring 1998), 126–141.

14 For this model see, Keith Krause, *Arms and the State: Patterns of Military Production And Trade* (Cambridge, UK, 1992); for criticisms see, Nicola Di Cosmo, "Did Guns Matter? Firearms and the Qing Formation," in *The Qing Formation in World-Historical Time*, ed. Lynn Struve (Cambridge, MA, 2004), pp. 121–166; Gábor Ágoston, "Behind the Turkish War Machine: Gunpowder, Technology and Munitions Industry in the Ottoman Empire, 1450–1700," in *The Heirs of Archimedes: Science and the Art of War through the Age of Enlightenment*, ed. Brett D. Steele, Tamera Dorland (Cambridge, MA, 2002), pp. 101–137.

15 For such an approach see, David Nicolle, *Ottoman Fortifications, 1300–1710* (Oxford, 2010) and *The Frontiers of the Ottoman Empire*, ed. Andrew Peacock (Oxford, 2009).

as the lower Volga."[16] While the Ottoman conquest of Kamianets-Podilsky (1672) and of Chyhyryn (Çehrin) (1677–78) served this policy, it was inevitably doomed to failure after the unsuccessful siege of Vienna (1683).

The main objective of the traditional Ottoman northern policy was the security of the Black Sea, which supplied Istanbul with grains, slaves, and furs among other things. Its safeguarding was ensured by preserving an equilibrium between Muscovy and Poland-Lithuania, monitoring the steppes through the empire's Crimean Tatar vassals, and maintaining strong garrisons in strategic coastal areas.[17] This was soon to change in the Köprülü era due to developments in the Ukraine.

The time of the Köprülüs roughly corresponded to the period known in Ukrainian historiography as The Ruin (1657–1686) characterized by endemic warfare and civil strife after the death of Bohdan Khmelnytsky, the hetman of the Zaporozhian Cossacks, and of the Cossack State (1648–1657) he founded.[18] The divided Cossack leadership often fought one another in order to reunite the different regions of Khmelnytsky's state and, in the struggle to preserve autonomy against the encroachments of Poland-Lithuania and Muscovy, shifted its allegiance between the former and the latter. The Ottoman Sultan stood out as a third alternative who could potentially protect the Cossack autonomy after the treaty of Andrusovo (1667).[19] The treaty endangered the security of the Ottomans' northern frontier by dividing the Ukraine along the Dnieper into a Polish Right-Bank in the west and Muscovite Left-Bank in the east while

16 Colin Heywood, "The Shifting Chronology of the Chyhyryn (Çehrin) Campaign (1089/1678) Acording to the Ottoman Literary Sources, and the Problem of the Ottoman Calendar," in *The Ottoman Empire: Myths, Realities and 'Black Holes' (Contributions in Honour of Colin Imber)*, ed. Eugenia Kermeli, Oktay Özel (Istanbul, 2006), pp. 283–295, here: p. 286.

17 Halil İnalcık, "The Origins of the Ottoman-Russian Rivalry and the Don-Volga Canal 1569," *Les annales de l'Université d'Ankara* 1 (1946–1947), 47–106 (esp. 50–55); Peter Bartly, "17. Yüzyılda ve 18. Yüzyılın İlk Yarısında Kazak Devleti ve Osmanlı İmparatorluğu," *İlmi Araştırmalar* 6 (1998), 302–329, here: p. 301 [transl. by Kemal Beydilli of "Der Kosakenstaat und das Osmanische Reich im 17. und in der ersten Hälfte des 18. Jahrhunderts," *Südost-Forschungen* XXXIII (1974), 166–194].

18 Known under several names, e.g., the Hetmanate, the Cossack State encompassed the right and left banks of the Dnieper, and Zaporozhia, Paul Robert Magosci, *A History of Ukraine* (Toronto, 1996), p. 231.

19 Victor Ostapchuk, "Cossack Ukraine In and Out of Ottoman Orbit, 1648–1681," in *The European Tributary States of the Ottoman Empire in the Sixteenth and Seventeenth Centuries*, ed. Gábor Kármán, Lovro Kunčević (Leiden, 2013), pp. 123-152, here: pp. 139–140; Dariusz Kołodziejczyk, "Ottoman Podillja: The Eyalet of Kam'janec,' 1672–1699," *Harvard Ukrainian Studies* 16, no. 1/2 (June 1992), 87–101, here: pp. 90–91.

placing Zaporozhia under dual Polish-Muscovite overlordship as a buffer against the Crimean Tatars.[20]

Another factor that prompted the Ottomans towards a new northern policy was an urge to bring the unruly Ottoman vassal states of Transylvania, Wallachia and Moldavia under greater central control. These states had joined forces with Sweden against Poland-Lithuania in the 1650s despite Ottoman objections. The Köprülüs hoped to transform these buffer vassal states into an interior zone by surrounding them with the new provinces of Yanova (Ineu), Varad (Oradea), and Kamaniçe (Kamianets-Podilsky) while the Crimean Tatars were likely to come under closer scrutiny in this process as well.[21]

The fortress of Kamianets came on the agenda in the negotiations in 1648–53 when Khmelnytsky, who offered vassalage to the Sublime Porte (the Ottoman central government), banned naval raids by the Zaporozhian Cossacks in 1648 that had even hit Istanbul itself in the early seventeenth century and proposed to hand over Kamianets to the Ottomans. In return, the Porte, could provide military support, though only limited due to domestic turmoil, through Tatar proxies. Yet, Tatars often changed sides in Polish-Cossack clashes in order to prevent one side from gaining too much power.[22]

Another prospect for seizing Kamianets was on the horizon when Hetman Doroshenko sought recognition from the Porte as the sole hetman of all Cossacks and an Ottoman vassal (1669) after reuniting both banks of the Dnieper under his authority. In the words of the celebrated poet Yusuf Nabi, referring to previous hostile relations between the Ukrainian Cossacks and the Porte, "this hetman of the Right-Bank" was granted the status of "the thornbush attached to the walls of the moist rose garden of Islam."[23] However, this was a fleeting success due to the interferences of Poland-Lithuania, Muscovy, and the Crimean Khanate. On the eve of the Ottoman campaign to Kamianets, the Polish-Lithuanian Commonwealth condemned him as a rebel and invaded the Right Bank.[24] Meanwhile, the Zeydi Shi'is of Yemen blockaded the pilgrim routes

20 Magosci, *A History of Ukraine*, pp. 215–216, 225–227; Orest Subtelny, *Ukraine: A History* (Toronto, 2009), pp. 146–147; Ostapchuk, "Cossack Ukraine," pp. 125–127, 139–143; Kołodziejczyk, "Ottoman Podillja," pp. 87–90.

21 See İbrahim Metin Kunt, "17. Yüzyılda Osmanlı Kuzey Politikası Üzerine Bir Yorum," *Boğaziçi Üniversitesi Dergisi* 4–5 (1976–1977), 111–116.

22 Subtelny, *Ukraine: A History*, p. 147.

23 İlhan Genç, "Nabi'nin Fetihname-i Kamaniçe'deki Naşirliğinin Estetik Boyutu," in *Nabi Sempozyum Bildirileri*, ed. Ali Bakkal (Ankara, 2014), pp. 379–401, here: p. 398.

24 Dariusz Kołodziejczyk, *The Ottoman Survey Register of Podolia (ca. 1681): Defter-i Mufassal-i Eyalet-i Kamaniçe* (Boston, 2004), pp. 2–9; Dariusz Kołodziejczyk, *Ottoman-Polish Diplomatic Relations (15th–18th Century): An Annotated Edition of 'Ahdnames and Other Documents* (Leiden, 2000), pp. 144–146.

and Sultan Mehmed IV was about to leave Edirne for Yemen to lead his first campaign. When Doroshenko requested military aid in 1671, he instead shifted his attention to the urgent situation in Poland-Lithuania.[25]

6 The Siege of Kamianets

The fortress of Kamianets-Podilsky (*Stary Zamek*: the old bastion/citadel) was surrounded by the deep gorge of the river Smotryc and connected to the town (the medieval castle) by a bridge over the moat. In the 1630s a 'new bastion' (Pol.: *Nowy Zamek*; Turk.: *tabya-yı kebir*) in the shape of a Dutch hornwork was added to the western side of the fortress.[26] Evliya Çelebi, the famous Ottoman traveler of the seventeenth century, had seen Kamianets fortress several years before the Ottoman conquest, and commended in his usual literary technique of exaggeration that one could not encounter the likes of such a stronghold either in the country of the Czechs, Swedes, Flemings or Germans. In his view, any prospective Ottoman conquest would have to starve the garrison into surrender in a dogged blockade by a numerous army, while the Tatars would have to fend off any relief army.[27]

Numbering about 100,000 men, the Ottoman army in 1672 was commanded by Fazıl Ahmed Pasha rather than the sultan. It was an imperial assortment of Tatars, Cossacks, Moldavians, Wallachians, Kurds, Arabs and several French renegade specialists.[28] The legendary Janissary infantry probably comprised only a quarter of the army, but supplied most of the trench troops. By contrast, the prebendal provincial light cavalry (timariots from the Balkans and Anatolia) were mostly relegated to auxiliary services as they were substituted by the infantry mercenaries enrolled in the pasha households in this century.[29]

The arduous march of the Ottoman army from Edirne took almost three months to complete, with 52 way-stations punctuating the route, and 75 days of rest [see Map of Itinerary]. Heavy rains, swamps, the need to construct bridges, and the delayed arrivals of provincial contingents were the usual problems of any military march. River crossings were always a dangerous enterprise

25 Türkal, "Silahdar," pp. 586–587.

26 Kołodziejczyk, "Ottoman Podillja," pp. 91–92.

27 *Evliya Çelebi Seyahatnamesi*, 5 (Istanbul, 2007), ed. Yücel Dağlı, Seyit Ali Kahraman, İbrahim Sezgin, p. 70.

28 İnbaşı, *Ukrayna'da Osmanlılar*, pp. 115–135; for a criticism of the exaggerated role of European renegades in the Ottoman military see Gábor Ágoston, *Guns for the Sultan: Military Power and Weapons Industry in the Ottoman Empire* (Cambridge, UK, 2005), pp. 44–47.

29 Özgür Kolçak, "XVII. Yüzyıl Askeri Gelişimi ve Osmanlılar: 1660–64 Osmanlı-Avusturya Savaşları" (Ph.D thesis, Istanbul University, 2012), pp. 177–178, 186.

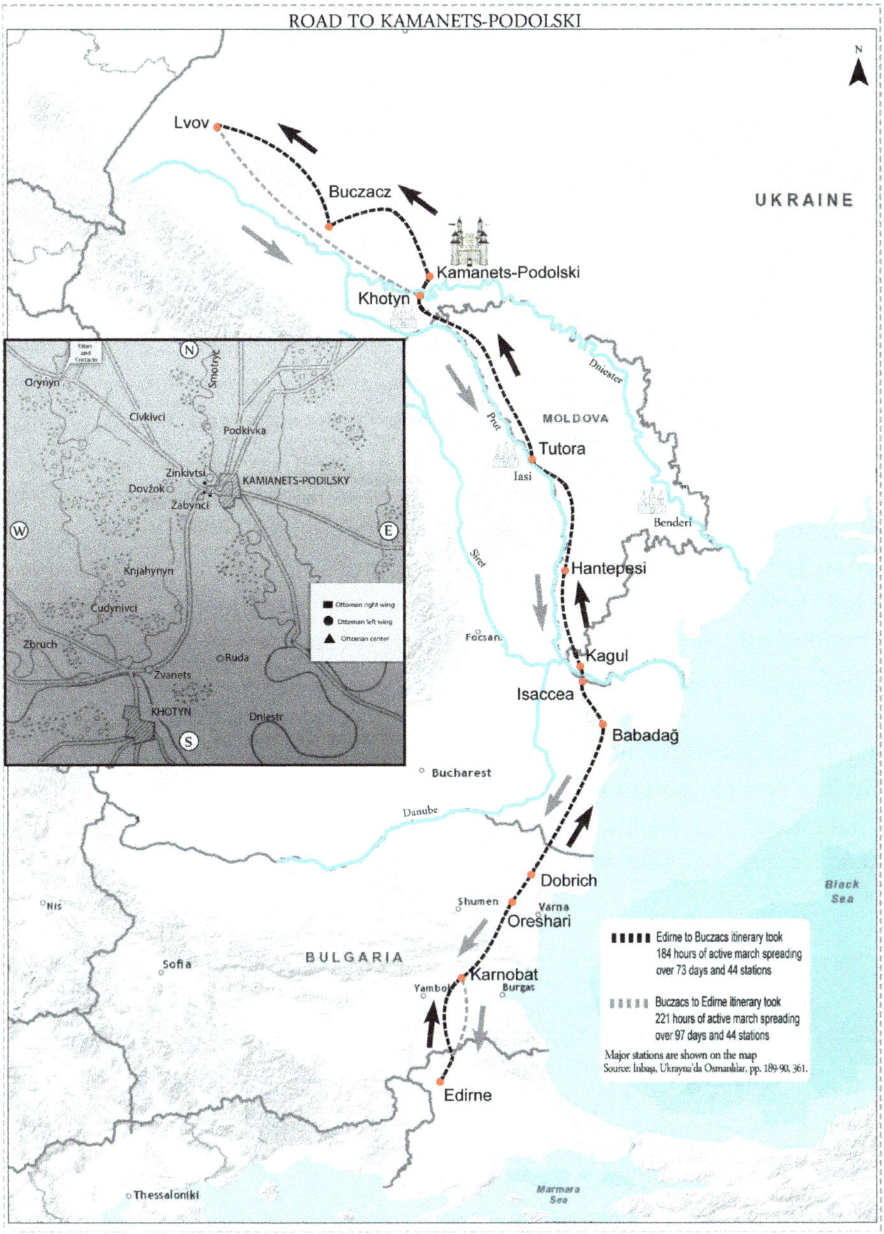

FIGURE 8.1 Itinerary: Road to Kamanets-Podolski

SOURCE: THE MAP OF THIS ITINERARY IS ADAPTED FROM MEHMET İNBAŞI, *UKRAYNA'DA OSMANLILAR*, 189–90, 361. THE INSERTION IN THE MAP IS ADAPTED FROM JANUZS WOLINSKI, "OBLOZENIE KAMIENCA W 1672 ROKU," *Z DZIEJÓW WOJEN POLSKO-TURECKICH* (WARSAW, 1983), P. 37.

in the Ottoman campaigns in Europe due to several reasons: Besides the constant threat of attack, the nature of rivers usually precluded the possibility of fording and required the throwing of pontoon bridges that were not always match for unpredictable floods. Thus, by tradition the sultan paid the customary campaign bonus to 34,825 men of the standing army before crossing over the Dniester into enemy territory and organized a feast once across to celebrate the dangerous feat.[30] With a view to emphasize the popular concerns over the campaign, Galland recorded rumors circulating in the streets of Istanbul by the end of August that the Poles had ambushed the Ottoman army upon its crossing the Dniester to the Polish side. Similar false reports from Warsaw even reached London in a matter of two weeks. These reports claimed that Ottomans could not pass over the river due to high waters and had abandoned the siege.[31] The Ottomans were fortunate that their river crossing on August 12 – 14 over pontoon bridges was not blocked by a Polish vanguard. Nor was a war of attrition, as suggested by Evliya, necessary for victory because of the political strife reigning in the Polish-Lithuanian Commonwealth. The pro-Habsburg King Michal of Poland-Lithuania had ignored the warnings of the imminent Ottoman attack voiced by the pro-French Great Hetman Jan Sobieski and failed to reinforce the fortress in advance.[32]

6.1 *Preparations for the Siege*[33]

Once across the Dniester the army encamped near Zvanets on August 15, four hours away from Kamianets to the south. Kara Mustafa Pasha, with around 5000 troops, had gone ahead of the army on August 14, in order to carry out

30 For a detailed account of the march see Mehmet İnbaşı, *Ukrayna'da Osmanlılar: Kamaniçe Seferi ve Organizasyonu (1672)* (İstanbul, 2004), pp. 55–115; Palmira Brummett, "The River Crossing: Breaking Points (Metaphorical or Real) in Ottoman Mutiny," in *Mutiny and Rebellion in the Ottoman Empire*, ed. Jane Hathaway (Wisconsin, 2004), pp. 45–60; Türkal, "Silahdar", p. 603.

31 Galland's entry is dated August 31, five days after the London news that relied on a letter dated August 26 and arrived from Warsaw, see Galland, *Journal d'Antoine Galland*, p. 199; *London Gazette*, issue 710, September 5–9, 1672, p. 1.

32 Kołodziejczyk, "Ottoman Podillja," p. 90.

33 The following description of the Ottoman siege is inspired by Ostwald's description of the siege of Ath (1697) – Vauban's 'model siege' –, see Jamel Ostwald, *Vauban Under Siege: Engineering Efficiency and Martial Vigor in the War of the Spanish Succession* (Leiden, 2007), pp. 22–46; on Ottoman siege technique, see Vernon J. Parry, "Hisar," *Encyclopedia of Islam* [2nd ed.], vol. 3 (Leiden, 1971), pp. 469–483 [under the subtitle "v. – Ottoman Empire", pp. 476–481] and Rhoads Murphey, *Ottoman Warfare, 1500–1700* (New Jersey 1999), pp. 105–132.

reconnaissance.[34] His examination of the site from the north was especially fruitful as the sloping land on this side provided a clear view of the fortress, the new bastion and the town itself. When the reconnaissance troops came too close to the Russian Gate of the town and to the Nowy Zamek itself, the defenders fired large shots of 10 to 15 kg, hitting a horse in the foot. This required the slaughtering of the horse on the spot for a spontaneous feast in Tatar culture, to the astonishment of the Ottoman dignitaries. Kara Mustafa took the defenders' cannonballs as examples of the enemies' ordnance to his meeting with Mehmed IV, where they decided to besiege the hornwork from both sides.[35]

On August 15 the sultan organized the customary banquet in the new army camp; this was a crucial occasion to boost the morale of the troops after the long march. The 'Tatar looting' of the food prepared for the Khan's contingent in the banquet was another spectacle custom as related in the accounts of Ottoman military campaigns. Accounts of the siege described in wonder the alacrity these 'noble savages' displayed in carrying off the food, causing a violent brawl among 1000 Tatars over the 'spoils:' 500 trays of food, as well as 300 sheep and 100 cattle cooked on skewers. The next day the siege train reached the camp near Zvanets, while Doroshenko received robes of honor from the Sultan in front of his 500 Cossacks.[36]

6.2 *Investing the Fortress*

In compliance with the agreed siege plan, the Sultan encamped on the wooded high plateau to the west of the citadel and town on August 17 after arriving in a grand military procession within full view of the defenders. In no single campaign had there been such a vast assemblage of troops since the campaign of Baghdad – the yardstick used to describe the size of the Ottoman army in any given campaign. The garrison troops, taken by surprise, demolished the mills by the river to deny any facilities to the besiegers and began preparations to resist the siege. While scattered fire did not dissuade the Ottomans from watering their horses in the river, the bombshells fired from Nowy Zamek almost hit the sultan's tent in the woods when he was receiving the grand vizier. Hacı Ali praised the sultan, who had no prior experience of war, for keeping calm at this

34 Kara Mustafa, as the commander-in-chief, also surveyed Chyhyryn in 1678 ahead of the army, see *Çehrin Seferi* in Hajda, "Two Ottoman Gazanames," p. 78.

35 Can, ed., *Kamaniçe*, p. 43; Derin, "Abdi Paşa," p. 336; Türkal, "Silahdar," pp. 610–611; Januzs Wolinski, "Oblozenie Kamienca w 1672 roku," *Z Dziejów Wojen Polsko-Tureckich* (Warsaw, 1983), pp. 21–50 (see p. 39) [reprint: *Zeszyty Naukowe Wojskowej Akademii Politycznej, Seria historyczna*, no. 14 [44] (1966), 21–55].

36 Can, ed., *Kamaniçe*, pp. 44–45; Derin, "Abdi Paşa," p. 337; Türkal, "Silahdar," pp. 611–613.

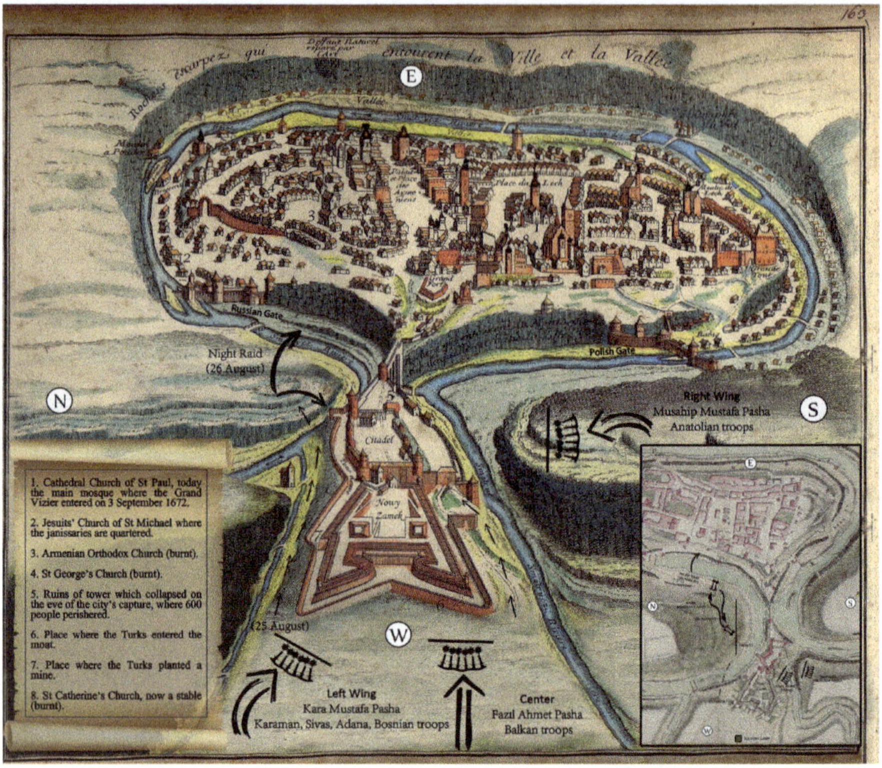

FIGURE 8.2 Plan of the Ottoman Siege of Kamianets-Podilsky (1672)
SOURCE: THE MAP THAT SHOWS THE SIEGE PLAN AND ITS LEGEND ARE ADAPTED FROM THE CITY PLAN DRAWN BY NICOLAS DE FER IN 1691 IN PARIS. NOTABLY, THIS SEEMS TO BE BASED ON THE MAP DRAWN BY CYPRIAN TOMASZEWICZ IN 1672, BUT IT EXHIBITS A LEFT-RIGHT REVERSAL THAT IS COMMONPLACE IN THE ERA. THE INSERTION IN THE MAP IS ADAPTED FROM THE MAP ENTITLED "KAMINIECK PODOLSKI, CAPITALE DU PALATINATE DE LA HAUTE PODOLIE [...]," ACCESSIBLE AT: <http://gallica.bnf.fr/ark:/12148/btv1b8439930x> (ACCESSED 23 OCTOBER 2018)

critical moment. He implied that the sultan showed strong leadership by not moving his tent out of range of the guns. Upon the defenders' refusal of the formal offer that they surrender, the Ottoman troops began to move into their assigned positions.[37]

Fazıl Ahmed endorsed the siege plan previously proposed by Kara Mustafa after a final reconnaissance [see the siege plan]. The Ottoman forces spread along a front that ran 7.5 kilometers from north to west. Accordingly, Kara Mustafa commanded the left flank, composed of troops from the Anatolian

37 Can, ed., *Kamaniçe*, pp. 48–49; Derin, "Abdi Paşa," p. 338; Türkal, "Silahdar," p. 614.

provinces of Karaman, Adana and Sivas, and set out to establish the first parallel (*metris*: lateral trench) to approach Nowy Zamek from the northwest. The forces under the command of the grand vizier, together with the Balkan troops and the bulk of the Janissaries, were deployed in the west forming the center in front of the Sultan's camp spreading along a line of three kilometers opposite the hornwork. Musahip Mustafa Pasha occupied the high ground to the southeast of the fortress, on the right flank, his force also being composed of Anatolian troops. Janissaries were assigned to all three sides as the elite among the trench soldiers, while the timariots and household troops of the pashas were relegated to such auxiliary duties as weaving gabions and dragging an unspecified number of light guns, mortars and six heavy cannons to each army group. The main base of the Tatar and Cossack contingents was Orynyn (NW) but they were on constant patrols to fend off a possible relief force.[38]

The meagre forces of the defending garrison were concentrated in Nowy Zamek. Captain Myśliszewski[39] led the defense in the north of the bastion with 700 men under his command. A tiny force of roughly 50 men was responsible for the defense of the citadel. The defenders also took some measure to protect the two important gates to the town: the Russian Gate, and the Polish Gate, which was more vulnerable than the former. Lack of provisions was as acute a problem as was the shortage of men.[40]

All these routine siege works started at nighttime, with the result that there were fewer casualties in the first phase of the siege than in previous sieges.[41] The defenders, on the other hand, lit torches and shot solid balls and bombshells at moving trains of cannon causing no appreciable damage. Those in the army camp surmised from the redirection of the garrison's fire on the Janissaries that the latter had begun entering the trenches in the first parallel.[42] Opening trenches was the very first task undertaken in an Ottoman siege, even before the siege train had completely arrived, whereas the emergent cult of military engineering in Europe advised against such premature trenchwork.

38 Can, ed., *Kamaniçe*, p. 50; Türkal, "Silahdar," p. 617; Wolinski, "Oblozenie Kamienca," pp. 35, 40.

39 Ottoman sources misidentified Myśliszewski as the commander of the garrison since he carried out negotiations in person on several occasions, see İnbaşı, *Kamaniçe*, p. 63 and Wolinski, "Oblozenie Kamienca," p. 36.

40 Wolinski, "Oblozenie Kamienca," pp. 36, 38; the bishop of Kamianets-Podilsky communicated on July 10 to the king that the fortress "was so ill provided with all necessaries required in a Siege, that it could not be expected it should hold out long," *London Gazette*, issue 700, August 1–5, 1672, p. 1.

41 Janissaries entered the trenches by daytime in the siege of Chyhyryn, see *Ahval-i İcmal-i Sefer-i Çehrin* in Hajda, "Two Ottoman Gazanames," p. 224.

42 Can, ed., *Kamaniçe*, p. 50.

Furthermore, it seems the Ottomans established the first parallel closer to the fortress, within gun range, than was the theoretical norm in Europe. Polish sources, however, claimed that the Ottoman gunners opened fire from a distance of 2–3 kilometers.[43]

On August 18, the Ottomans were busy reinforcing the parallels by means of raised firing platforms in the form of high earthen ramparts (*tabiye, domuz damı*) and constructing upright gabions as a screen to conceal their gun positions. Since the 25-year long siege of Candia, Ottoman custom required commanders of the vizieral rank to stay in these redoubts to motivate the frontline volunteers known as *serdengeçti* (i.e., 'forlorn hopes' or 'daredevil') in their dangerous task of driving forward the zigzag saps (*sıçan yolları*: approach trenches).[44]

A brief digression is warranted to contextualize the Ottoman siege tactics in the contemporary art of siege. The diffusionist model credits French military architect Sébastien le Prestre de Vauban with introducing the three-parallel approach and the elevated fire platform (*cavalier de tranchée*) based on his observations on the Ottoman siege of Candia. Although the Ottomans seem to have employed "a simplified three-parallel approach" improved with the addition of redoubts at each end of the parallels to provide better protection for the attacker, it was the Europeans who systematized the application of these innovations.[45] On the other hand, new research suggests that zigzag, trench cavaliers, and parallels had already been present in various forms also in Europe and Vauban may have equally been inspired by these European precedents.[46] Therefore, there is no need to argue for a single source of diffusion for these innovations. Although the Ottomans had no equivalent of Vauban, they also systematized earlier practices and incorporated novelties in their siegecraft thanks to the existence of the Corps of Miners and Sappers as an independent brigade.[47] This military service produced institutional memory based on artisanal practice. It is remarkable that Nabi imagined the trenches as 'schools' in which the troops could study the difficult subject of war. Ottoman trenches,

43 Ostwald, *Vauban Under Siege*, pp. 235–240; Parry, "Hisar," p. 479; Wolinski, "Oblozenie Kamienca," p. 40.

44 This was also the case at Chyhyryn *Ahval-i İcmal-i Sefer-i Çehrin* in Hajda, "Two Ottoman Gazanames," p. 224.

45 John A. Lynn, *Giant of the Grand Siècle: The French Army, 1610–1715* (Cambridge, UK, 1997), pp. 569–571; Gábor Ágoston, "Erken Modern Devirde Avrupa ve Osmanlı İmparatorluğunda Bilgi, Teknoloji ve Savaş Uğraşı," in *Osmanlı'da Ateşli Silahlar ve Askeri Devrim Tartışmaları*, ed. and trans. Kahraman Şakul (Istanbul, 2017), pp. 101-115, here: pp. 107–108.

46 Ostwald, *Vauban Under Siege*, p. 53.

47 Ágoston, *Guns for the Sultan*, pp. 40–42; Vauban's proposal for a separate corps of sappers and miners was found too costly, see Ostwald, *Vauban Under Siege*, pp. 166–168.

represented by Nabi as "eloquent imprints the sappers engraved on the page of the soil by pickax and shovel" were described in Polish sources as wide enough to accommodate the Ottoman cavalry.[48] This contradicts the Western accounts of the Ottoman trenchworks in the siege of Vienna (1683). Ottoman saps in 1683 tended to be sinuous, deep and long as opposed to rectilinear European zigzags to better protect the soldiers in the trenches.[49]

Nabi uses to great effect the typical Ottoman prose conventions to aestheticize the daily routines of the siege which, otherwise, is brim with fatigue and hardship. In his poetic imagination, the noise of the hammers and nails used to pitch thousands of tents was to the besieged like the noise of the mosquito entering Nimrod's brain as divine retribution, who then had his head banged with a stone all the time to stop the pain.[50] Digging a trench meant digging a grave for the enemy to whom the Janissary morticians fit shrouds by using their muskets as iron measuring rods. They were also conjurors who juggled cups and balls using their muskets and lead balls. With sleight of hand, the Janissaries made balls vanish into the musket and reappear in the chest of the enemy.[51]

Rugged terrain was the major obstacle to transporting the sluggish siege train in the first days of the siege preparations, and contrary to Ottoman expectations, only two siege guns were in position by midnight (August 18/19) in the center and on the left flank. These guns concentrated their fire on the bastions of Nowy Zamek till midmorning, while the Janissaries advanced the approach trenches towards the ditch.[52]

6.3 *First Negotiations*

On August 19, the garrison sent envoys to Fazıl Ahmed requesting the opening of negotiations in the morning. Suspecting that this was a trick to gain precious time he demanded the garrison raise a white flag as a precondition for the ceasefire. This demand was accepted only in late afternoon upon which the Ottoman envoys dispatched to the fortress repeated the previous offer of safe conduct in return for unconditional surrender (*vire*). The negotiations failed as

48 Yalçınkaya, *Nabi*, p. 237; Wolinski, "Oblozenie Kamienca," p. 41.

49 Ostwald, *Vauban Under Siege*, pp. 58–61; Parry, "Hisar" p. 479; Whether the sheer duration of the siege of Vienna accounted for a deviation from the normal Ottoman practice – if there was any – needs further research. Ottoman diggers dug in a sitting position rather than on their knees, digging narrower mine galleries with lower ceilings so as to increase the impact of the explosion, see Ágoston, *Guns for the Sultan*, pp. 40–42. It is likely that Ottoman sappers also dug more narrowly, as did the miners.

50 Yalçınkaya, *Nabi*, pp. 75–79, 231–234.

51 Yalçınkaya, *Nabi*, pp. 73, 84, 86, 230, 236, 237, and 240.

52 Can, ed., *Kamaniçe*, p. 51; Türkal, "Silahdar," p. 616.

the garrison flatly rejected the Ottoman proposal.[53] Both the rank-and-file in the Ottoman camp and the inhabitants of the town misinterpreted the white flag as the announcement of surrender so that a number of besiegers came too close to the ditch to enjoy the view to the dismay of the defenders. As the besieged garrison opened fire to prevent a possible storming of the Nowy Zamek, a feverish race back to the Ottoman trenches followed among the exposed men near the ditch who incurred some losses. To Nabi's mind, the besiegers hoped for the ripening of the fruit of peace on the sapling of the white flag but the sound of unexpected gunfire from the garrison alerted them to the danger just the same as archangel Raphael resurrected the dead by blowing his trumpet. Indeed, the men in the trenches grew furious and promised to avenge their loss caused by the sudden enemy fire.[54]

Notably, the military command of both sides selectively complained about the other for violating the ceasefire. Although it was not against the terms of negotiations, the besieged force interpreted the progress of the siege as Ottoman ill-will and abruptly sent back the Ottoman envoys. The Ottomans, conversely, charged the Poles with shooting their guns before their envoys could reach the Grand vizier's redoubt. Hacı Ali contrasted these imprudent Ottoman troops who went to the ditches unsuspectingly with those experienced men who benefitted from the ceasefire to advance the trenchwork. He noted that this could not have been accomplished otherwise, as the available labor force was to melt away in a matter of days because of heavy fighting.[55]

On the morning of August 20, Fazıl Ahmed was relieved by the imminent arrival of the remainder of his siege guns. But, Kara Mustafa had had to rush in with his household troops the previous night to drag the cannons to the siege park in the left flank, since his Anatolian Turks and Bosnians abandoned the siege guns in the open field because of the heavy garrison fire. In the absence of siege guns, there was no way to silence the defensive artillery, which had freely trained its fire on the trenches and tents of the left flank for the past two days. Hacı Ali as a veteran of various sieges attempted to exonerate his master Kara Mustafa from a possible charge of negligence by letting his artillery to be transported to the siege park within the range of the garrison. He argued that it would have taken three days at best to bring these heavy guns to the left flank over the broken terrain had it not been for the efforts of his master to drag his

53 Can, ed., *Kamaniçe*, pp. 52–53; Türkal, "Silahdar," p. 616; Abdi Pasha recounted that Fazıl Ahmed Pasha urged him to keep an account of the siege on this day, Derin, "Abdi Paşa," pp. 339–340.

54 Can, ed., *Kamaniçe*, p. 53; Yalçınkaya, *Nabi*, pp. 85–88, 90–93, and 238–242

55 Wolinski, "Oblozenie Kamienca," pp. 41–43; Can, ed., *Kamaniçe*, p. 53; Türkal, "Silahdar," p. 616; Derin, "Abdi Paşa," pp. 339–340.

cannons the previous night. These guns finally reached their assigned positions on the left flank by dawn on August 20.[56]

6.4 *Methodic Siege: Closing on the Ditch*

A religious component of Ottoman siege warfare was the traditional Muslim sheep sacrifice. The combination of religious ritual with technical expertise was a common practice and necessary to motivate the men. A sheep sacrifice ceremony held on the left flank at dawn on August 20 signaled the beginning of the methodic siege with the parallels and zigzag saps dug under enemy fire, guns deployed on the raised firing platforms, high ramparts screened by caltrop shields (*domuz damı*) for the protection of Janissary musketeers and bowmen built under the supervision of the pashas, and gabions woven. A competent sultan in military matters as he appeared in Abdi Pasha's chronicle, Mehmed IV urged Kara Mustafa to silence the garrison guns by artillery fire[57] so that the trench troops could fully engage the enemy with musket fire and arrows on the left flank. Polish sources pointed out that the Ottomans fired 400 projectiles and 100 grenades per diem starting by August 20.[58]

The Ottoman focus on the commanders, rather than the engineer, as the true expert in siegecraft was the same as in Europe.[59] Emphasis on the physical proximity of high-ranking commanders to the rank and file also shows close resemblance to western European sources on siegecraft. In the siege of Uyvar, Fazıl Ahmed wandered the flooded zigzags on a stormy night in water up to his knees, to motivate the sappers building the causeway over the moat.[60] He had narrowly escaped death before the walls of Candia in September 1667, when the trenches collapsed in a counter-mine attack at the time of his visit, killing a number of high dignitaries.[61]

Kara Mustafa spent the whole of August 20 in the trenches, distributing bonuses to the wounded and encouraging the sappers to drive forward. Hacı Ali portrayed his presence on the frontline as a cause for the rapid progress of the zigzag saps. While the defenders tried to stop the sappers by grenade fire, the Ottoman gunners on the left flank concentrated their fire on a protruding half-

---

56 Can, ed., *Kamaniçe*, p. 54; Türkal, "Silahdar," pp. 615–617.

57 Abdi Pasha was on a visit to the left flank when a messenger arrived with the sultan's order, see Derin, "Abdi Paşa," p. 341; Türkal, "Silahdar," p. 617.

58 Wolinski, "Oblozenie Kamienca," p. 43.

59 Ostwald, *Vauban Under Siege*, pp. 147–154.

60 Arslan Boyraz, "Köprülüzade Fazıl Ahmet Paşa Devrinde (1069–1080) Vukuatı Tarihi Transkripsiyon ve Değerlendirme" (MA thesis, Marmara Üniversitesi, 2002), p. 17.

61 Abubekir Sıddık Yücel, "Mühürdar Hasan Ağa'nın Cevahirü't-Tevarih'i" (Ph.D. thesis, Erciyes Üniversitesi, 1996), p. 357.

bastion of the hornwork, damaging a sentinel post and killing most of the defenders behind their gabion screens.[62]

The Ottoman center under Fazıl Ahmed also managed to bring the trenches close to the ditch after silencing most of the defender's guns, whom Hacı Ali praised for their determined resistance. According to him, the defenders opposite the right flank commanded by Musahip Mustafa put up an even more resolute fight. Warfare in this sector was largely a gun duel, since the right flank was on a high plateau separated from the fortress by a deep valley. Ottoman gunners targeted the round bastion that connected the citadel to the Nowy Zamek because most of the defenders were deployed in this bastion the gunfire of which covered a wide radius in all directions.[63]

Apparently, the Ottomans had an understanding of a hierarchy of national expertise in military affairs based on their long-term clashes with European and Asian powers. In the initial phases of the siege the besiegers on the right flank had attributed the lack of mortar fire from the garrison to the inferiority in this science of the Poles to the 'Franks' (Catholic powers of the Mediterranean). Therefore, the opening of mortar fire stunned the veterans of the siege of Candia, who were quick to realize that these bombshells flying over the Janissary trenches were as voluminous as those of the Venetians at Candia. Sentries were appointed to watch for the enemy bombshells as 'required by the custom' in Ottoman siegecraft. A frontline Janissary volunteer astonished his superiors by spying the bombshell magazine in the bastion from afar,[64] provoking the Head of the Bombardiers to successively shoot three large mortar shells (75 kg each). The explosion destroyed the vault of the afore-mentioned bastion where the Lutheran chapel that served as the ammunition store was located, and it caused a sharp tremor in the fortress and the town. Encouraged by the explosion, the Ottomans then trained their cannons and mortars on Nowy Zamek from all sides during the ensuing fire that lasted three hours. The defenders could not harass the Ottoman right flank with mortar fire anymore as they had lost all their grenades to the flames.[65]

On August 21, the war raged on, with the full engagement of gunners, musketeers and bombardiers. The firepower of the German musketeers, who were defending the round bastion opposite the left flank, was devastating throughout the day. They rained bullets down on the trenches and Kara Mustafa's redoubt from an organ gun situated behind the gabions. It had 15–20 barrels

62 Can, ed., *Kamaniçe*, pp. 55–56; Derin, "Abdi Paşa," p. 341.
63 Can, ed., *Kamaniçe*, p. 56; Türkal, "Silahdar," p. 618.
64 Türkal, "Silahdar," p. 618.
65 Can, ed., *Kamaniçe*, p. 57; Wolinski, "Oblozenie Kamienca," p. 43.

fixed in a row on a wooden stock, and fired bullets of 90–130 gr with a single lock mechanism. As it mowed down a good number of the entrenched Janissaries, Kara Mustafa charged the bombardiers to deal with this deadly weapon as well as the enemy mortar fire. In the presence of high-ranking commanders, a bombardier officer shot bombshells of 75 kg and 32 kg – the standard calibers issued in this siege – and hit the target. Hacı Ali observed that the damage was especially remarkable since the Germans were unsuspecting, as no mortar shells had been fired from the left flank until then. The German musketeers neither sighted the projectiles in daylight nor heard them until the explosion, owing to the high quality of the mortars. Kara Mustafa granted the officer, who thus proved his proficiency, a handsome bonus as related in Hacı Ali's straightforward technical analysis of the siege.[66]

On the night of August 21/22, Kara Mustafa moved two of the siege cannons down from the ramparts and placed them in forward positions to shoot at the barracks. The firing that began at dusk was detrimental for the German soldiers, whose "black hats burst into the sky like ravens" as the Ottoman guns battered their barracks. Due to high casualties among the Germans, the Poles and Ruthenians took the lead in the battle the next day.[67]

Ottoman ethnographic representation was judgmental about the martial qualities of their foes. There was a pattern to which groups the Ottomans considered worthy foes. Accordingly, Germans were the ablest in siege warfare (*kale cengi*), whereas the Ruthenians were considered cowards, shooting their pieces randomly behind the gabions. Poles, by contrast, tried to look different from the Ruthenians in their manners and uniforms, acting arrogantly towards the latter and never entering their redoubts. They, furthermore, forced the Ruthenians to repair the gabions as a sign of their superstitious claim to superiority over them and stood in the exposed field fearlessly to open protective fire. Such observations by Hacı Ali suggest that the Ottomans were aware of the tensions between the Poles and their Ruthenian subjects. In his account, Ottoman stratagems were to overcome the Polish stupid (*eblehane*) bravery of exposing themselves on the parapets. Agitated by this needless show of courage,

66 Can, ed., *Kamaniçe*, pp. 58–59; Türkal, "Silahdar," p. 618.

67 Baer mistook the *Rus* in Ottoman sources for modern Russians (Baer, *Honored by the Glory of Islam*, p. 164). Ottomans usually meant Ruthenians by the term 'Rus' as opposed to 'Moskof' (Muscovites), see Omeljan Pritsak, "Das erste Türkisch-Ukrainische Bündnis (1648)," *Oriens* 6 (1953), 266–298 (especially, pp. 296–298) [Turkish trans. by Kemal Beydilli: "İlk Türk-Ukrayna İttifakı (1648)," *İlmi Araştırmalar* 7 (1999), 256–283]; the population of Podolia (roughly 37,662) was Ruthenian/Ukrainian by over 90 percent in this era, Kołodziejczyk, *The Ottoman Survey Register of Podolia*, pp. 35, 37; after the conquest, the *Rus* in Kamianets offered their allegiance to Doroshenko "on account of ethnicity," see Can, ed., *Kamaniçe*, p. 84.

the Ottoman troops in the trenches spontaneously stopped firing at the Poles in order to lure them out from behind the gabion screen. The Poles, assumed that their firepower broke the spirit of 'the Turks' and boasted that their firepower proved superior to the Germans.' Kara Mustafa and troops in the distant trenches were quite puzzled by this strange sight, dispatching messengers to clarify the situation. The traditional Muslim cheer of 'God is Great,' and the following rain of bullets on the Poles who had indeed come out from behind the gabions, thrilled the men in the trenches. The foe was not to be seen again until late afternoon.[68]

Fazıl Ahmed inspected the flanks on August 22 and discussed the situation with Musahip Mustafa and Kara Mustafa. All three commanders were convinced that the defenders were losing the will to fight, as their projectiles were all aiming upwards, flying very high. Thus, they reconnoitered the fortress and the town from all sides to inspect the status of the defense.[69]

The capture of the covered way was the next phase in the siege. In European siege warfare of this era, most of the casualties suffered by a besieging army resulted from clashes over the covered way. Field generals usually resented the engineers, who advocated the mechanical attack of sapping forward, and preferred storming the covered way to avoid delay during a siege.[70] Ottoman literary production about the siege of Kamianets, however, does not suggest the existence of such a doctrinal clash between the cults of 'vigor' and 'efficiency.' Even the highest authorities had a firm belief in the wisdom of siege *à la sap*. In Nabi's poem, the representation of the common tactic to sap rather than storm had mystical connotations. The Janissaries were all followers of the fourteenth-century saint Hacı Bektaş Veli (Bektashi religious order) whose miraculous deeds involved transmuting a still wall into a horse that he then rode. Just as the still wall moved forward thanks to Hacı Bektaş so the front walls of the trenches were driven forward thanks to the relentless efforts of his Janissary followers.[71] According to Hacı Ali, the Ottoman siege plan was based on the prediction that the sappers would not reach the ditch in less than 20 days, despite the presence of the sultan in the camp. It was due to the efforts of the sappers, "the men who used stone and soil for a pillow and comforter", that it took only five days to reach this critical stage of the methodic siege.[72]

As in European siege warfare, the siege escalated into a heated battle after the sappers reached the ditch, which was in the night of August 22/23. Both

68 Can, ed., *Kamaniçe*, pp. 60–61; Türkal, "Silahdar," p. 619.

69 Can, ed., *Kamaniçe*, p. 61.

70 Ostwald, *Vauban Under Siege*, pp. 240–252.

71 Yalçınkaya, *Nabi*, pp. 73, 230.

72 Can, ed., *Kamaniçe*, pp. 61–62.

sides were ready for the collision on the following day as the viziers' redoubts were moved forward to the bank of the ditch and the defenders refilled their upright gabions throughout the night. The regular troops, who had been held back until the sappers reached the ditch, were ready to fully engage the defenders. The garrison desperately began an all-out fight to keep the Ottomans at bay by the ditch on August 23. Besides the intense fighting, heavy rain also frustrated the troops in the trenches on this bloody day. The Sultan set up a field hospital next to his tent for the wounded and distributed bonuses among them in order to demonstrate his compassion and generosity to the rank and file.[73]

Nabi represented this gruesome combat as an opportunity for the Muslim soldiers to watch the beautiful face of the beloved – the conquest and victory (*feth ü zafer*) – at the expense of pouring their rose-pink blood at her feet. The siege was a shadow theatre in which the enemy was the passive audience that had to watch the unfolding of the plot shown on the curtain – the walls of the fortress. Ottoman cannons were dragons in this show that sprayed with flame the embrasures that were the sinister eyes of the soulless bastions. Avian imagery is evident in Nabi's depiction of the siege: the sultan, whose nickname was "the hunter", was a falcon, a bird of prey. The redoubts erected by the parallels as the command posts of the pashas were the nests of hawks moving forward as the parallels closed on the ditch. By contrast, the fortress was "the nest of polytheist crows and kites."[74]

6.5 *Methodic Siege: Passage of the Ditch*

Notably, the counterscarp and the ditch were considerable obstacles for engineers engaged in siege warfare. On August 24, the besiegers were in control of the ditch of the Nowy Zamek which was dry as it was located on a steep ridge 100 metres above the riverbed. The passage of the dry ditch was only possible by heaping up fascines and bags on the sides for the protection of the sappers as they crawled across the floor of the ditch to the breach. It was necessary to dig tunnels known as descents from the steep ridge to the bottom of the counterscarp for bringing in troops for the imminent attack on the ditch. These protected spaces collected these troops for the second phase of the operation in the side of the ditch at two points, in the center and on the left flank.[75]

---

73 Can, ed., *Kamaniçe*, pp. 60–61; Derin, "Abdi Paşa," p. 341.

74 Yalçınkaya, *Nabi*, pp. 84–86, 89, 92–93, and 236–240, 242–243; see Baer, *Honored by the Glory of Islam*, p. 166 for the quotation and the avian imagery in Nabi's poem.

75 Christopher Duffy, *Fire and Stone: The Science of Fortress Warfare, 1660–1860* (London, 1996), pp. 178–180.

The relevant experts had already determined the spots for the descents of the ditch the previous day. The auxiliaries had readied the necessary timber, woolsacks and gabions. The Ottoman logistics service could have a rough estimate of the immense number of bags and sacks to be used in the descents since the timariots had to deliver their quotas assigned in proportion to their income levels.[76] Meanwhile, Ottoman gunners and musketeers were ordered to fire on Nowy Zamek in order to divert the garrison's attention away from the ditch. The Ottoman auxiliaries, however, were still struggling to rush the remaining guns and 110 metric tons of gunpowder (one fifth of the state's annual production) from Isaccea (Tur.: İsakçı) which was 600 km southeast of Kamianets.[77]

The frontline volunteers, who usually manned the zigzag saps, entered from the hole into the ditch and began crawling forward, sheltered by the woolsacks to the fore and sandbags and gabions at the sides. When the defenders in another bastion opposite of the left flank noticed the descent into the ditch and warned their fellows, a violent struggle broke out that involved throwing stones by hand as well as shooting bullets and arrows. Many Ottoman troops took to the hills to have a better view of the ongoing battle in the ditch while the roar of the struggle reached the sultan's tent.[78]

Turkish sources portrayed both Fazıl Ahmed and Kara Mustafa as attentive commanders who were ready on the spot, distributing spare weapons and occasionally shooting arrows at the defenders from the redoubts erected by the ditch. From the Ottoman point of view, the defenders put up a good fight out of desperation since they realized that resolute action to defend the ditch was their only remaining option. After reaching the walls of the round bastion, the frontline volunteers and miners propped up planks against the walls and improvised shelter for the covered mining operation to be carried out at ground level.[79] Underground mine attack was a common feature of Ottoman sieges since the fifteenth century, although mining required more time and effort than breaching artillery.[80]

76 Can, ed., *Kamaniçe*, pp. 62–63; each timariot delivered two sacks for every 1000 aspers (*akçe*) of their annual revenue at the siege of Uyvar, see Kolçak, "XVII. Yüzyıl Askeri Gelişimi ve Osmanlılar," p. 178–179.

77 Derin, "Abdi Paşa," p. 341.

78 Can, ed., *Kamaniçe*, pp. 63–64; Türkal, "Silahdar," p. 620; Wolinski, "Oblozenie Kamienca," p. 43.

79 Counter-mining was impossible, for the bastions were filled with earth, see Can, ed., *Kamaniçe*, p. 64.

80 At Rhodes and Uyvar, Ottomans applied the ancient technique of burning out the foundation of the fortress wall, see Boyraz, "Köprülüzade Fazıl Ahmet Paşa," p. 18 and Parry, "Hisar," p. 476. At Győr (1594), they made cave-like holes in the brick walls and drilled the

By late evening the two sides disengaged only for a brief period of time, during which the Ottomans heard 'some strange music' from the direction of the town. They assumed that it was a warning message for the town dwellers to prepare for a last stand. The defenders resumed the fight after midnight from the Russian and Polish gates, throwing torches and grenades into the ditch until dusk.[81]

6.6 *Capture of the Nowy Zamek*

On August 25, the Balkan troops – who were skillful besiegers according to Hacı Ali Efendi – were ready by daybreak to storm Nowy Zamek at the center commanded by Fazıl Ahmed. On the left flank, however, the commander of the Janissary frontline volunteers hesitated to storm the bastion facing them as he suspected a ruse. According to Hacı Ali, his master embarrassed the reluctant commander by volunteering to climb to the top of the parapet, which was a thinly veiled charge of cowardice. The defenders, it transpired, had retreated to the citadel before dawn following the frantic nighttime battle and had pulled up the drawbridge. Around 2000 men were up on the walls in a matter of minutes, and calls to prayer were heard from Nowy Zamek, symbolizing the conversion of the 'infidel' space into a Muslim space.[82]

The defenders had actually decided to retreat into the citadel by August 23 in order to shorten the line of defense and prevent the underground attacks, for the citadel was built on solid rock. The rearguard defending the bridge repelled the first wave of Ottoman attackers with grapeshot and completely retreated to the citadel by daylight under the lethal fire of the Ottoman right flank.[83]

The grand vizier and Kara Mustafa examined Nowy Zamek with delight on the very same day. This inspection convinced the Ottoman engineers of the superiority of the hornwork to Ottoman bastions with the sole exception of the *Ak tabya* of Candia. Especially impressive were the vaulted domes, giant arches, and the open courtyard surrounded by magnificent towers at four corners. Opposite the Ottoman right flank were the earthworks in the courtyard which were, in Hacı Ali's words, as high as the Montenegro – a conventional simile in Ottoman literature. The depth and width of the ditch was also astonishing; it was 'a steel ditch surrounding an iron bastion.' In short, all were cheer-

foundation of the wall from inside so as to fill up the cavities with gunpowder, see Tibor Szalontay, "The Art of War During the Ottoman-Habsburg Long War (1593–1606) According to Narrative Source" (Ph.D. thesis, University of Toronto, 2004), p. 124.

81 Can, ed., *Kamaniçe*, p. 65; Türkal, "Silahdar," p. 621.

82 Can, ed., *Kamaniçe*, pp. 66–67; Baer, *Honored by the Glory of Islam*, pp. 175–176.

83 Can, ed., *Kamaniçe*, p. 68; Wolinski, "Oblozenie Kamienca," pp. 45–47.

ful by the end of the inspection tour that the hornwork had fallen in just six days.[84]

It should be noted that fortifications and siegecraft became part of cross-cultural debates with mutual claims to superiority. Foreign observers of Ottoman military architecture in this era asserted that the Ottomans had actually captured most of their modern fortresses from their Christian rivals.[85] The Ottomans resented such European denigration of their knowledge of fortification design. Evliya Çelebi, for instance, responded to these critics by referring to fortresses like Bender (Bendery, Moldova), Segedin (Szeged, Hungary), Özi (Ochakiv, Ukraine), and Avlonya (Vlorës, Albania) all of which were substantial Ottoman-built structures.[86] However, the Ottomans also appreciated the expertise of their European foes in the art of fortification as suggested above; praising a conquered European bastion after all was just another form of self-glorification.

6.7 *Final Phase: Mining the Citadel and Battering the Town*

After the inspection of the hornwork, Fazıl Ahmed ordered the repositioning of the siege guns to target the citadel, which was, however, out of view of the left flank. Kara Mustafa therefore decided to move the trenches and guns northwards across the river, to the valley between the citadel and the town in order to engage the defenders in the citadel and the town simultaneously. He set up a small field tent within range of the garrison guns, so as to be close to his Janissaries whom he sent out to dig open trenches between the citadel and the river. This dictated that the long side of the trenches had to run parallel to the walls of the citadel, thereby exposing the Janissary diggers to enemy fire. Solid balls of all sorts landed around his tent, injuring members of his household. Casualties in a single day in these trenches amounted to the overall casualty rates of the past 6 days since the parallel was quite shallow at least in the initial phase of the work.[87]

While two of the siege guns that were repositioned behind the trenches on the left flank, began battering the town so as to strike fear into the civilians as a military necessity,[88] the remaining two tried to silence the garrison artillery.

84 Can, ed., *Kamaniçe*, pp. 66–67; Türkal, "Silahdar," p. 622.

85 Kevin Andrews, *Castles of the Morea* (New Jersey, 1953), p. 231; Simon Pepper, "Ottoman Military Architecture in the Early Gunpowder Era: A Reassessment," in *City Walls: The Urban Enceinte in Global Perspective*, ed. James D. Tracy (Cambridge, UK, 2000), pp. 282–316.

86 *Evliya Çelebi Seyahatnamesi*, 5: 65; *Evliya Çelebi Seyahatnamesi*, 8, ed. Yücel Dağlı, Seyit Ali Kahraman, Robert Dankoff (Istanbul, 2003), p. 312.

87 Can, ed., *Kamaniçe*, pp. 69–70.

88 For Ottoman false night attacks at Candia, see "Hisar," p. 479.

In haste to save their own lives, the Janissaries did a month's work in a day, bringing the lines close to the lower course of the river (NW) and to the citadel (SE) by the evening of August 25. The defenders in the town lit torches and threw bundles of scrap smeared with tar to illuminate the ditch against a night attack since the musket fire from the trenches continued all night long. At his usual evening inspection of the trench troops, the gunners convinced Kara Mustafa to position all four guns to bear on the well-lit town. Gunners fired two successive salvos in the direction of the Russian and Polish gates, hitting the churches of St. Catherine and St. George as well as the Armenian Cathedral and compelling the defenders to put out the torches. After being rewarded generously, they took two of the pieces back to their previous positions before dusk while the bombardiers prepared the mortar shells to be used in the upcoming attack on the citadel.[89]

A Tatar prisoner who hurled himself down from the citadel into the river the next day provided valuable information to Kara Mustafa about the situation inside. He reported that the defenders had not anticipated that Kara Mustafa would attack the town and the citadel simultaneously from the northwest. According to him the bombardment of the town at nighttime had caused more consternation than the loss of Nowy Zamek.[90]

Hacı Ali credited Fazıl Ahmed with ordering the mining of one of the great bastions of the citadel on August 26, saying that he had acquired expert knowledge of siege warfare at the sieges of Candia and Uyvar. This was apparently the round "bastion adjacent to the citadel and visible between the two towers", which defended the way by the bridge to the citadel.[91] The legion of miners who set out to work under shelter at ground level at the base of the walls suffered considerable losses, although the gunners attempted to divert the defenders' attention elsewhere by deceptive fire.[92]

When the mine was ready to explode, Ottoman troops were pulled back from its vicinity, which the garrison mistook for exhaustion; therefore, they played musical instruments in celebration. The Head of the Miners "accompanied the merriment with his own instrument and exploded the mine" as described sarcastically by Hacı Ali. The citadel and the town remained wreathed

89 Can, ed., *Kamaniçe*, p. 71, Türkal, "Silahdar," p. 623; Mandzy, "Entrepôt of the Ukrainian Steppe Frontier," p. 299; Wolinski, "Oblozenie Kamienca," p. 44.

90 He could only tell what he had overheard in *Rus* language in Turkish through a Tatar interpreter and strange body language. See Can, ed., *Kamaniçe*, pp. 72–73; Türkal, "Silahdar," p. 624.

91 Defterdar Sarı Mehmed Paşa, *Zübde-i Vekayiat*, ed. Abdülkadir Özcan (Ankara, 1995), p. 29; Wolinski, "Oblozenie Kamienca," p. 47.

92 Can, ed., *Kamaniçe*, pp. 71–72; Derin, "Abdi Paşa," p. 341.

in clouds of smoke for the next two hours, during which the gunners trained their pieces on the citadel from all sides in support of the infantry charge. However, the bastion was double-walled and only the outer wall collapsed. In addition, earth inside the bastion filled the breaches and eliminated any chance of storming it. The defenders cut down the exposed Ottoman troops massed in front of the tower with musket and grenade fire.[93]

In his elaborate account of the final storming of the Nowy Zamek, Nabi saw chambers of powder deposited beneath the bastions as 'treasuries' – homophone of chamber in Arabic (*hazine*) – that split the fortification in two like a burning heart split the chest of a lover. The capture of the fortress was the end of the army's longing for the beloved.[94]

Hacı Ali described the 'outlandish' military culture of the Tatars and Cossacks[95] through ridicule and wit. He alleged that these poor men watched the battle in awe from the hills to the north and mistook the explosion for an earthquake, screaming out, "*Tengri* hit the infidel." Ferocious Muslim war hurrah "*Allah, Allah*" in his narrative poses a stark contrast with Tatar's "*Tengri*" – the Turkic/shamanic sky god and the non-religious synonym for God. Tatars and Cossacks condemned the use of mortar and mine upon learning the truth.[96] This story illustrates the Ottomans' traditional opinion of the Tatar 'vassals' as an unfit force in siege warfare if no patrolling operations were involved. The Tatars had their own concerns about the Ottoman 'allies,' so much so that Khan Selim Giray had even maintained a spy – Pavel the Cossack – in the Ottoman headquarters at the siege of Uyvar, who would complain to his Habsburg interrogators about Fazıl Ahmed's condescending behavior towards the Tatars.[97] Selim Giray acted somewhat independently in this siege as well by encouraging Myśliszewski as early as August 16 to enter into negotiations with the sultan through his own intermediacy.[98]

Meanwhile the left flank once again was short of its two siege guns for a good part of the day, as it had taken a considerable time to move them across

---

93 Can, ed., *Kamaniçe*, p. 73; Türkal, "Silahdar," p. 624; Derin, "Abdi Paşa," pp. 341–342; Raşid Mehmed Efendi, *Tarih-i Raşid ve Zeyli*, 1, ed. Abdülkadir Özcan, Yunus Uğur, Baki Çakır, A. Zeki İzgöer (Istanbul, 2013), p. 163.

94 Yalçınkaya, *Nabi*, pp. 91, 245.

95 L. J. D. Collins, "The Military Organization and Tactics of the Crimean Tatars, 16th–17th Centuries," in *War, Technology and Society in the Middle East*, ed. Vernon J. Parry, Malcolm E. Yapp (London, 1975), pp. 257–276.

96 Can, ed., *Kamaniçe*, pp. 68, 74.

97 Kolçak, "XVII. Yüzyıl Askeri Gelişimi ve Osmanlılar," pp. 205–210; in the siege of Vienna, Kara Mustafa despised the Tatars in his letter to the Khan as "eaters of rotten horse meat," see Cevat Üstün, *1683 Viyana Seferi* (Ankara, 1941), p. 86.

98 Turkish sources are silent on this event. Wolinski, "Oblozenie Kamienca," p. 40.

the river and over the hills before silencing the light artillery deployed in the town. As a result of the failure to storm the citadel, Mehmed IV sent messengers to his commanders in order to rekindle their enthusiasm, predicting that the citadel would fall in a few days.[99] This reveals the sultan's role in the whole campaign; his was not an expert opinion but a divine guidance embodied in his persona as the head of the dynasty. He boosted the overall morale by his presence at the siege, distributing bonuses, and tending the wounded.

At the night of August 26, around 500 men, including many of Kara Mustafa's Croatian mercenaries[100], volunteered for a night raid into the town. Kara Mustafa yielded only reluctantly as he deemed it unnecessary to storm the town itself. Hacı Ali seems to have underlined his master's wisdom by mentioning his concerns about the repulsed raid. Armenians in the town would later testify to the panic reigning there that night, loud organ music and cries reaching the army camp. Hacı Ali fancied that the organ – a symbol of the Church – was begging Muslims to end the raid. In a frenzy, the town-dwellers lit up the town, making it easy prey for the Ottoman gunners. Splinters of rock poured down on the people as the artillery projectiles continued to hit the stone houses for hours after the night raid.[101]

Meanwhile, the Ottoman center readied a second mine beneath the ruined tower and the wall for a surprise dawn attack on August 27.[102] An incident in the last phase of the siege sheds light on the competition between different branches of service in the Ottoman army. The Ottoman elite traditionally favored the Balkan contingents over those of Anatolia in terms of martial qualities.[103] The Anatolian troops, jealous of the former, who had been the first to enter the Nowy Zamek, seized the gun batteries on the right and left flanks in order to demonstrate that they were just as competent in siege warfare. They even seized the cannons, confident of their capability that they allegedly gained at the siege of Candia. The indiscriminate fire they opened on the town and citadel before dusk puzzled the dignitaries. Unaware of the incident, many commanders believed the gunners were making an effort to show their gratitude for the handsome bonuses they had previously received. All in all, the rank and file rejoiced at the sound of roaring guns on the eve of the planned assault so that the commanders decided to drop the matter. Early in the morn-

99 Can, ed., *Kamaniçe*, p. 75.

100 Croatian mercenaries in the private armies of many Ottoman pashas in the seventeenth century, see Kolçak, "XVII. Yüzyıl Askeri Gelişimi ve Osmanlılar," pp. 123, 141, 153, 341

101 Can, ed., *Kamaniçe*, pp. 75–76.

102 Özcan, ed., *Zübde-i Vekayiat*, p. 29.

103 For a similar observation based on Evliya see Kolçak, "XVII. Yüzyıl Askeri Gelişimi ve Osmanlılar," pp. 177–178.

ing a white flag could be seen flying opposite the left flank, and the siege effectively came to a conclusion.[104]

6.8 *Capitulation of Kamianets-Podilsky*

That evening a similar incident of fierce rivalry also occurred among the besieged. After a Janissary contingent entered the citadel, the tower used as the ammunition store exploded and caused much confusion on both sides. The Ottomans initially thought the defenders had planned a ruse in order to demolish the citadel with the Janissaries inside, whereas the Polish authorities blamed it on the Ottomans at first.[105] The Polish envoys then assured the Ottomans that the explosion was an accident caused by a brawl between the Polish guardians and the 'greedy' Germans who had tried to pilfer powder from the stores. As a precaution, the Ottoman soldiers spent the entire night in the trenches and were only allowed to retreat to the camp on August 28.[106] Meanwhile the shortage of provisions in the camp was a matter of some concern, but it turned out that fresh supplies were due to arrive that day. It is worth mentioning that this is one of the rare occasions when Turkish sources mention logistical problems in the campaign, which indicates that Ottoman logistics usually functioned smoothly. The keys of the fortress were delivered the next morning by a delegation composed of twenty individuals including nobles as well as priests.[107]

Hacı Ali boasted that, following the successful Ottoman siege of Kamianets, all the Armenian and Jewish townspeople as well as a majority of the Ruthenians chose to remain. According to him, "the Jews greeted the Ottomans saying, 'We know how sweet is life in the shadow of Islam.'"[108] Although Hacı Ali maintained that all the Poles wanted to leave, the 1681 Ottoman survey register shows that there still remained a sizable Polish population in the city. In fact,

---

104 Polish sources claimed that a man flew the white flag on his own initiative, see Wolinski, "Oblozenie Kamienca," p. 47; Can, ed., *Kamaniçe*, pp. 77–78.

105 For historical precedents of such traps in the Long War see Szalontay, "Ottoman-Habsburg Long War," pp. 119–121; for a similar accident at Uyvar, see Kolçak, "xvii. Yüzyıl Askeri Gelişimi ve Osmanlılar," pp. 217–218.

106 Can, ed., *Kamaniçe*, pp. 80–81; Derin, "Abdi Paşa," p. 343; Wolinski, "Oblozenie Kamienca," p. 49.

107 On logistics see İnbaşı, *Ukrayna'da Osmanlılar*, pp. 217–290; Potocki, the foreman (*starosta*) of Kamianets, and Bishop Lanckoronski were in the delegation, Wolinski, "Oblozenie Kamienca," p. 49; Can, ed., *Kamaniçe*, pp. 81–82; Derin, "Abdi Paşa," p. 343.

108 As quoted in Kołodziejczyk, *Ottoman Survey Register of Podolia*, p. 36; Armenians and Jews were reluctant to participate in the defense because of their trade contacts with the Ottoman Empire, Wolinski, "Oblozenie Kamienca," p. 38.

many residents did not wish to surrender and protested the decision of the town authorities.[109]

On August 30, a group of 1,500 people, including soldiers and inhabitants of the town, left under Ottoman escort with 'marks of honor' for a destination of their choice. First to go were the civilians and the injured, who, in a wretched condition, left with 300 wagons provided by the Ottomans. The pikemen and German musketeers (360 in total) led by the fortress commander paraded through the Ottoman regiments standing on the sides – or, in the words of Hacı Ali, "overwhelmed by the fully equipped Ottoman regiments, they stared around like bewildered oxen as they went."[110] Fazıl Ahmed had let the defeated garrison of Uyvar march out with 'marks of honor' with a grain of sarcasm, telling them, "leave [the fortress in parade] as you wish if you are not ashamed of yourselves for doing it!"[111] Such parading was an unjustified defiance of the victor by an already defeated foe rather than self-appreciation of brave resistance by a worthy foe. Therefore, so-called 'marks of honor' could only be 'marks of shame' from an Ottoman point of view.

The procession was followed by celebration of the victory in the evening in the army camp. Lit by candles, the camp resounded with a salute of three salvoes of artillery and muskets. Doroshenko and his Cossacks also organized their own *feu de joie* in the celebrations that went for two days as required by custom. The sultan had the biggest share in the victory in Hacı Ali's account: his presence as 'the heart and soul of the universe' in the army camp, his determination to pitch his tent that represented the state within gun range, and his verbal decree that heralded the impending victory were indispensable to the conquest. This victory, according to Hacı Ali, was a demonstration of the world-famous Ottoman expertise in siegecraft to "soldiers of various nationalities" who were rather spectators in the show staged by the Ottomans.[112]

News of the victory reached Istanbul on September 11 and official celebrations were held as far away as the province of Baghdad for three days. It is worth mentioning that the *London Gazette* declared in a postscript dated Sep-

109 Mandzy, "Entrepôt of the Ukrainian Steppe Frontier," p. 93; a 30-day period of grace was granted for absent townsmen to reclaim their property in the town, see Can, ed., *Kamaniçe*, p. 82; the twenty-seven years long Ottoman administration facilitated the "Ukrainization" of the city through immigration from the suburbs and also proved beneficial for the Jews, see Kołodziejczyk, *Ottoman Survey Register of Podolia*, pp. 32–37.

110 Can, ed., *Kamaniçe*, p. 82.

111 Boyraz, "Köprülüzade Fazıl Ahmet Paşa," p. 19.

112 Can, ed., *Kamaniçe*, pp. 83–85; Derin, "Abdi Paşa," pp. 344–345; *Çehrin Seferi* in Hajda, "Two Ottoman Gazanames," p. 99.

tember 26, "We seem as yet unwilling to give any credit to the report of the taking of *Caminiec*, it being a place of the greatest strength in all these parts."[113]

Mehmed IV entered the town on the first Friday after the conquest (September 2) so that he could perform the Friday sermon in the main cathedral, once this and other churches were cleansed of the icons and 'idols,' 'signs of polytheism and sacrilege' from the Ottoman point of view. Another symbolic act of purification of the town involved exhumation of thousands of corpses buried inside over centuries only to be thrown onto dunghills; this was to be reciprocated by the Poles after retaking the town in 1699.[114]

7 The Aftermath

Warfare ravaged Ukraine for three more years until both sides reached a final settlement in the treaty of Zhuravne [Pol.: Zurawno] (1676). The treaty recognized the Ottoman territorial gains, but dropped the matter of tribute imposed on Warsaw in the previous treaty of Buchach (18 October 1672). As it made no clause for a unified Cossack state under his leadership, Doroshenko felt abandoned by the Porte as well as his Cossacks who revolted against his pro-Ottoman policy in the war. He handed over the strategic fortress of Chyhyryn (Ukr.: Chigirin) to the Muscovites and went into exile in Moscow (1676), precipitating the Ottoman-Muscovite War of 1678–81.[115]

It is questionable whether the Ottomans could have colonized and reshaped the new province of Kamianets as a regular administrative unit ruled from the center so as to both provide regional security and also enhance the provisioning of the imperial seat at Istanbul. Due to political instability, Ottoman rule in the province was restricted to the fortress itself for 19 of the 27 years they held

113 Galland, *Journal d'Antoine Galland*, pp. 198–205; *London Gazette*, issue 715, p.1; confirmation came on September 17 that Kamianets had surrendered due to lack of provisions; the loss of the fortress "is of so great consequence to Poland, that we are yet unwilling to believe it." *London Gazette*, issue 716, p.1.

114 Derin, "Abdi Paşa," p. 345; Türkal, "Silahdar," p. 629; this is corroborated by the Ruthenian Eyewitness Chronicle: "All the dead had been dug up from the tombs and graves and taken away from the city, and the holy images removed from the Catholic and Orthodox churches had been laid in the mud on the streets upon which the Turk [i.e., the sultan] and his servant, unfaithful hetman Dorošenko, entered Kam'ianec," quoted in Kołodziejczyk, *Ottoman Survey Register of Podolia*, p. 51; Baer, *Honored by the Glory of Islam*, pp. 173–174.

115 Ostapchuk, "Cossack Ukraine," pp. 145–147; Alan W. Fisher, "Ottoman Kamanets-Podolsk," *Journal of Turkish Studies* 7 (1984), 55–83, here: pp. 55–57; Kołodziejczyk, *Ottoman-Polish Diplomatic Relations*, pp. 148–151.

it, whereas poverty in the region meant that the strong garrison (200 guns and 6000 men) had to be supplied from Moldavia.[116] A certain Hasan who was on garrison duty in Kamianets described the difficult garrison life in 1673 as ten times worse than that experienced in the twenty-five year long siege of Candia because of the hoarding of the provisions by the army top-brass.[117] With hindsight, the new northern policy of the Köprülü period only weakened Poland-Lithuania relative to Russia and the Austrian Habsburgs. The Ottomans, on the other hand, invested in the nearby fortress of Khotyn as its northern bulwark after losing Kamianets in 1699.[118]

8 Conclusion

The analysis of the Ottoman literary production on the siege of Kamianets illustrates the political, religious, cultural, and technical aspects of Ottoman siege warfare. Turkish sources obviously represented Sultan Mehmed IV and the high dignitaries as the champions of Islam. They customarily focused on the deeds of the commanders at the expense of the rank and file. Our major source on the siege, Hacı Ali, who wrote the most complete Turkish account of the operations in terms of technical aspects, is biased towards his master Kara Mustafa. He viewed him as the strong-willed person pushing the entire siege forward thanks to his work ethic, religiosity, and technical expertise.

An overview of the stages of the siege revealed that Ottoman and European ways of siegecraft did not constitute binary oppositions regardless of a number of differences in command structure and technical aspects. It appears that mining and sapping was the favorite tactics in Ottoman sieges by this period as opposed to breaching and storming. The existence of a separate corps of miners and sappers in the Ottoman army since the fifteenth century may have resulted in the accumulation and systematization of knowledge in a professional discipline over the course of time. Although the Turkish sources credited the commanders with the victory, they were also sympathetic to the hardships of

116 Some local peasants had immediately converted to Islam presumably to evade certain taxes, see Kołodziejczyk, *Ottoman Survey Register of Podolia*, p. 37.

117 Orhan Şaik Gökyay, "Kamaniçe Muhafızlarının Çektiği," *İ.Ü.E.F. Tarih Dergisi* 32 (1979), 281–300. For the poem see pp. 295–299 and 981–983 [facsimile]; Rhoads Murphey, "Forms of Differentiation and Expression of Individualism in Ottoman Society," *Turcica* 34 (2002), 135–170, here: p. 163.

118 Kołodziejczyk, "Ottoman Podillja," pp. 90, 94–96, and 99. Kołodziejczyk argues that the Ottoman conquest was a preventive measure in securing the northern Black Sea region rather than a preliminary stage of a new expansion in the region, Kołodziejczyk, *The Ottoman Survey Register of Podolia*, p. 5.

the professional miners, sappers, and the troops – 'the men who used stone and soil for a pillow and comforter.' They paid special attention to the siege technicians and gunners and usually mentioned by name to those rewarded generously for their precise shooting. Ottoman siegecraft also diverged from its European counterpart. The Ottoman concept of honorable surrender was not likely to embrace the exhibition of marks of honor in a final parade although the Ottomans granted the surrendering foe the courtesy of marching out bearing arms.

Most of the elements in Ottoman and European siegecraft, however, were similar. These included basic siege methods such as the three-parallel approach and the equipment. In addition, the army top-brass was expected to stay close to their men in the trenches night and day. Although Ottoman sources suggest that the siege of Kamianets was untypically short, sieges elsewhere also tended to be shorter over the course of the 'long seventeenth century:' half of the sieges during the War of the Spanish Succession lasted two weeks or less.[119]

A number of outstanding victories at difficult siege operations such as Baghdad, Candia, and Uyvar armed the Ottoman literati with the necessary technical information on siegecraft and enabled them to compare the siege of Kamianets with the previous examples. Such a comparative outlook was not restricted to the technical aspects of sieges, but also involved military ethnographies. Based on experience, the Ottomans developed their own *scala militia* in siege warfare, in which the Germans and Franks ranked first to be followed by Poles and Ruthenians among the European rivals. As for the friendly troops, the Tatars appeared to be the exotic albeit incompetent allies in siege wars, whereas the Muslim troops from the Balkans were found superior to those from Anatolia.

Short Ottoman rule in Kamianets (1672–1699) left the wrong impression on the modern researcher that it was a short-term military adventure with no appreciable impact in Ottoman and modern Turkish culture. The fortress was the first thing in the open steppe that caught the eyes of Hacı Ali Agha, the Ottoman envoy to Warsaw in 1755, in Poland-Lithuania. He fancied that the fortress in the horizon lowered its head in sorrow as if it were sighing quietly to itself, for it was ceded to the infidels. Then, he continued with a story he was told of how in the post-Ottoman era "the minaret of the town" was saved miraculously when the priests who wanted to destroy it perished at the attempt "by the will

119 Ostwald, *Vauban Under Siege*, pp. 325–340.

of God."[120] The minaret at the entrance to the SS. Peter and Paul Cathedral (Catholic) with the shining statue of St. Mary erected on its top is an object of wonder for every visitor. Many Turks are imbued with deep nostalgia for the glorious past when they visit the site that was once the sultan's mosque (*cami-i hüdavendigar*). A remote Ottoman siege in eastern Europe still resonates through literature and shapes our view of the past as revealed in *Fire in the Steppe* by Henryk Sienkiewicz and Orhan Pamuk's *The White Castle*, which actually was the fortress of *Kamaniçe*.[121]

Bibliography

Printed Sources

Can, Ayşe Hande, ed., *Hacı Ali Efendi, Tarih-i Kamaniçe (Tahlil ve Metin)* (Ankara, 2007).

Dağlı, Yücel; Kahraman, Seyit Ali; Sezgin, İbrahim, eds., *Evliya Çelebi Seyahatnamesi*, 5 (Istanbul, 2007).

Dağlı, Yücel; Kahraman, Seyit Ali; Dankoff, Robert, eds., *Evliya Çelebi Seyahatnamesi*, 8 (Istanbul, 2003).

Galland, Antoine, *Journal d'Antoine Galland Pendant son Séjour à Constantinople, 1672–1673*, vol. 1, ed. Charles Schefer (Cambridge, UK, 2012).

Kołodziejczyk, Dariusz, *The Ottoman Survey Register of Podolia (ca. 1681): Defter-i Mufassal-i Eyalet-i Kamaniçe* (Boston, 2004).

Kołodziejczyk, Dariusz, *Ottoman-Polish Diplomatic Relations (15th–18th Century): An Annotated Edition of 'Ahdnames and Other Documents* (Leiden, 2000).

London Gazette

Özcan, Abdülkadir, ed., *Defterdar Sarı Mehmed Paşa, Zübde-i Vekayiat* (Ankara, 1995).

Özcan, Abdülkadir; Uğur, Yunus; Çakır, Baki; İzgöer, A. Zeki, eds., *Raşid Mehmed Efendi, Tarih-i Raşid ve Zeyli*, 1 (İstanbul, 2013).

Topaktaş, Hacer, ed., *Lehistan'da Bir Osmanlı Sefiri: Ziştovili Hacı Ali Ağa'nın Lehistan Elçiliği ve Sefaretnamesi (1755)* (Ankara, 2015).

Yalçınkaya, Şerife, *Nabi'nin Nasirliği ve Kamaniçe Fetih-namesi (Tenkitli Metin-İnceleme)* (İzmir, 2012).

120 Hacer Topaktaş, ed., *Lehistan'da Bir Osmanlı Sefiri: Ziştovili Hacı Ali Ağa'nın Lehistan Elçiliği ve Sefaretnamesi (1755)* (Ankara, 2015), p. 72.

121 The protagonist of Sienkiewicz' novel was Michal Wołodyjowski – the national hero who died in the explosion in the fortress on the day of surrender as described above, see *Fire in the Steppe* (Boston, 1917); Orhan Pamuk, *The White Castle* (Manchester, 1990).

Secondary Sources

Ágoston, Gábor, *Guns for the Sultan: Military Power and Weapons Industry in the Ottoman Empire* (Cambridge, UK, 2005).

Ágoston, Gábor, "Behind the Turkish War Machine: Gunpowder, Technology and Munitions Industry in the Ottoman Empire, 1450–1700," in *The Heirs of Archimedes: Science and the Art of War through the Age of Enlightenment*, ed. B. Steele, T. Dorland (Cambridge, MA, 2002), pp. 101–137.

Ágoston, Gábor, "Erken Modern Devirde Avrupa ve Osmanlı İmparatorluğunda Bilgi, Teknoloji ve Savaş Uğraşı," *Osmanlı'da Ateşli Silahlar ve Askeri Devrim Tartışmaları*, ed. and trans. Kahraman Şakul (İstanbul, 2017), pp. 101–115.

Ágoston, Gábor, "Habsburgs and Ottomans: Defense, Military Change and Shifts in Power," in *Ottomans, Hungarians, and Habsburgs in Central Europe: The Military Confines in the Era of Ottoman Conquest*, ed. Gézá Dávid, Pál Fodor (Leiden, 2000), pp. 126–141.

Andrews, Kevin, *Castles of the Morea* (New Jersey, 1953).

Artan, Tülay, "Osmanlı Ordusunun Sefere Çıkış Alayı (1672): Osmanlı Kadimciliği mi, Püritenlik Gösterisi mi – The Departure Procession of 1672: Ottoman Antiquarianism or a Puritan Statement," in *Uzak Komşu Yakın Anılar: Türkiye-Polonya İlişkilerinin 600 Yılı – Distant Neighbour Close Memories: 600 Years Of Turkish-Polish Relations*, ed. Ayşe Anadol (Istanbul, 2014), pp. 60–77.

Baer, Mark, *Honored by the Glory of Islam: Conversion and Conquest in Ottoman Europe* (Oxford, 2008).

Bartly, Peter, "17. Yüzyılda ve 18. Yüzyılın İlk Yarısında Kazak Devleti ve Osmanlı İmparatorluğu," *İlmi Araştırmalar* 6 (1998), [transl. by Kemal Beydilli of "Der Kosakenstaat und das Osmanische Reich im 17. und in der ersten Hälfte des 18. Jahrhundert," *Südost-Forschungen* XXXIII (1974), 166–194], 302–329.

Boyraz, Arslan, "Köprülüzade Fazıl Ahmet Paşa Devrinde (1069–1080) Vukuatı Tarihi Transkripsiyon ve Değerlendirme" (MA thesis, Marmara Üniversitesi, 2002).

Brummett, Palmira, "The River Crossing: Breaking Points (Metaphorical or Real) in Ottoman Mutiny," in *Mutiny and Rebellion in the Ottoman Empire*, ed. Jane Hathaway (Wisconsin, 2004), pp. 45–60.

Collins, L. J. D., "The Military Organization and Tactics of the Crimean Tatars, 16th–17th Centuries," in *War, Technology and Society in the Middle East*, ed Vernon J. Parry, Malcolm E. Yapp (London, 1975), pp. 257–276.

Cosmo, Nicola Di, "Did Guns Matter? Firearms and the Qing Formation," in *The Qing Formation in World-Historical Time*, ed. Lynn Struve (Cambridge, MA, 2004), pp. 121–166.

Derin, Fahri Çetin, "Abdurrahman Abdi Paşa Vekayinamesi: Tahlil ve Metin Tenkidi" (Ph.D. thesis, Istanbul University, 1993).

Doğru, Halime, *Lehistan'da Bir Osmanlı Sultanı: IV. Mehmed'in Kamaniçe-Hotin Seferleri ve Bir Masraf Defteri* (İstanbul, 2005).

Duffy, Christopher, *Fire and Stone: The Science of Fortress Warfare, 1660–1860* (London, 1996).

Duman, Filiz, "Bir Gazavatname Türü Olarak Hüseyin Behçeti'nin Miracü'z-Zafer Adlı Zafer-namesi (H.1088/M.1678)" (MA thesis, Süleyman Demirel Üniversitesi, 2010).

Fisher, Alan W., "Ottoman Kamanets-Podolsk," *Journal of Turkish Studies* 7 (1984), [special issue: Pierre Oberling, ed., *Turks, Hungarians and Kipchaks: A Festschrift in Honor of Tibor Halasi-Kun*], pp. 55–83.

Genç, İlhan, "Nabi'nin Fetihname-i Kamaniçe'deki Naşirliğinin Estetik Boyutu," in *Nabi Sempozyum Bildirileri*, ed. Ali Bakkal (Ankara, 2014), pp. 379–401.

Gökyay, Orhan Şaik, "Kamaniçe Muhafızlarının Çektiği," *İ.Ü.E.F. Tarih Dergisi* 32 (1979), 281–300.

Hajda, Lubomyr Andrij, "Two Ottoman Gazanames Concerning the Chyhyryn Campaign of 1678" (Ph.D. thesis, Harvard University, 1984).

Heywood, Colin, "The Shifting Chronology of the Chyhyryn (Çehrin) Campaign (1089/1678) Acording to the Ottoman Literary Sources, and the Problem of the Ottoman Calendar," *The Ottoman Empire: Myths, Realities and 'Black Holes' (Contributions in Honour of Colin Imber)*, ed. Eugenia Kermeli, Oktay Özel (İstanbul, 2006), pp. 283–295.

İnalcık, Halil, "The Origins of the Ottoman-Russian Rivalry and the Don-Volga Canal 1569," *Les annales de l'Université d'Ankara* 1 (1946–1947), 47–106.

İnbaşı, Mehmet, *Ukrayna'da Osmanlılar: Kamaniçe Seferi ve Organizasyonu (1672)* (İstanbul, 2004).

Karaçay Türkal, Nazire, "Silahdar Fındıklılı Mehmed Ağa Zeyl-i Fezleke (1065–22 Ca. 1106/1654 – 7 Şubat 1695)" (Ph.D thesis, Marmara University, 2012).

Kolçak, Özgür, "XVII. Yüzyıl Askeri Gelişimi ve Osmanlılar: 1660–64 Osmanlı-Avusturya Savaşları" (Ph.D thesis, Istanbul University, 2012).

Kołodziejczyk, Dariusz, "Ottoman Podillja: The Eyalet of Kam'janec', 1672–1699," *Harvard Ukrainian Studies* 16, no. 1/2 (1992), 87–101.

Krause, Keith, *Arms and the State: Patterns of Military Production and Trade* (Cambridge, UK, 1992).

Kunt, İ. Metin, "17. Yüzyılda Osmanlı Kuzey Politikası Üzerine Bir Yorum," *Boğaziçi Üniversitesi Dergisi* 4–5 (1976–1977), 111–116.

Kuzubaş, Muhammet, "17. Yüzyıl Şairi Mezaki'ye Göre Lehistan Savaşı," *Turkish Studies/Türkoloji Dergisi* 1, no. 2 (2006), 213–225.

Lynn, John A., *Giant of the Grand Siècle: The French Army, 1610–1715* (Cambridge, UK, 1997).

Magosci, Paul Robert, *A History of Ukraine* (Toronto, 1996).

Mandzy, Adrian, "Entrepôt of the Ukrainian Steppe Frontier: An Urban History of Early Modern Kamianets-Podilsky, Origins to 1672" (Ph.D. thesis, York University, 1998).

Murphey, Rhoads, *Ottoman Warfare, 1500–1700* (New Jersey, 1999).

Murphey, Rhoads, "Forms of Differentiation and Expression of Individualism in Ottoman Society," *Turcica* 34 (2002), 135–170.

Nicolle, David, *Ottoman Fortifications, 1300–1710* (Oxford, 2010).

Ostapchuk, Victor, "Cossack Ukraine In and Out of Ottoman Orbit, 1648–1681," in *The European Tributary States of the Ottoman Empire in the Sixteenth and Seventeenth Centuries*, ed. Gábor Kármán, Lovro Kunčević (Leiden, 2013), pp. 123–152.

Ostwald, Jamel, *Vauban Under Siege: Engineering Efficiency and Martial Vigor in the War of the Spanish Succession* (Leiden, 2007).

Parry, Vernon J., "Hisar," *Encyclopedia of Islam* [2nd ed.], vol. 3 (Leiden, 1971), pp. 469–483 [under the subtitle "v. – Ottoman Empire," pp. 476–481].

Peacock, Andrew, ed., *The Frontiers of the Ottoman World* (Oxford, 2009).

Pepper, Simon, "Ottoman Military Architecture in the Early Gunpowder Era: A Reassessment," in *City Walls: The Urban Enceinte in Global Perspective*, ed. James D. Tracy (Cambridge, UK, 2000), pp. 282–316.

Pritsak, Omeljan, "Das erste Türkisch-Ukrainische Bündnis (1648)," *Oriens* 6 (1953), 266–298 [Turkish trans. by Kemal Beydilli: "İlk Türk-Ukrayna İttifakı (1648)," *İlmi Araştırmalar* 7 (1999), 256–283.]

Sariyannis, Marinos, "The Kadızadeli Movement As a Social and Political Phenomenon: The Rise of a 'Mercantile Ethic,'" in *Political Initiatives "From the Bottom-Up" in the Ottoman Empire*, ed. Antonis Anastosopoulos (Rethymno, 2012), pp. 263–289.

Subtelny, Orest, *Ukraine: A History* (Toronto, 2009).

Szalontay, Tibor, "The Art of War During the Ottoman-Habsburg Long War (1593–1606) According to Narrative Source" (Ph.D. thesis, University of Toronto, 2004).

Taçkın, Musa, "Ali Efendi ve Tarih-i Kamaniçe Adlı Eseri (Metin ve Tahlil)" (MA thesis, Marmara University, 2004).

Üstün, Cevat, *1683 Viyana Seferi* (Ankara, 1941).

Wolinski, Januzs, "Oblozenie Kamienca w 1672 roku," *Z Dziejów Wojen Polsko-Tureckich* (Warsaw, 1983), pp. 21–50 [reprint: "Zeszyty Naukowe Wojskowej Akademii Politycznej," *Seria historyczna*, no. 14 [44] (1966), 21–55].

Yalçınkaya, Şerife, *Nabi'nin Nasirliği ve Kamaniçe Fetih-namesi* (*Tenkitli Metin-İnceleme*) (İzmir, 2012).

Yücel, Abubekir Sıddık, "Mühürdar Hasan Ağa'nın Cevahirü't-Tevarih'i" (Ph.D. thesis, Erciyes Üniversitesi, 1996).

PART 3

Definitions

CHAPTER 9

Siegecraft in Ming and Qing China

Tonio Andrade

Today we know much about the siege in Western Europe but little about sieges and fortification in the rest of the world. Our lack of knowledge is particularly striking for China, where the gun was invented. We know that in Europe, as siege guns became more effective in the course of the 1400s, warmakers began building new types of fortification, so that by the early 1500s the various experiments had crystalized into a durable and long-lasting model. The artillery fortress, also known as the *trace italienne*, revolutionized European warfare and may even have helped further European expansion overseas. But if gunpowder artillery led to such monumental changes in Western Europe, why don't we see similar developments in China? And, if it is indeed true that China saw no revolution in siegecraft similar to that of Europe, were Chinese military forces at a disadvantage vis-à-vis European ones?

This chapter attempts to provide preliminary answers to these questions. Until now, our understanding of siege warfare has been overwhelmingly dominated by Western historiographies and sources. It is vital to redress that balance, and I hope in this article to help show, by means of specific examples, that to truly understand the siege – even the European siege – we need to widen our source base to include non-Western perspectives.

Despite the fact that the gun was invented in what is today China during the 1200s, warmakers in East Asia did not generally use the new weapon to bombard walls, preferring to aim it at living creatures: men, horses, elephants. Thus, whereas firearms (i.e., handheld guns) became a key part of Chinese armies by the mid-1300s, artillery did not. This is a clear contrast to the West, where it was the other way around: firearms did not play a major role on European battlefields until the mid- to late 1400s, whereas European siege artillery became significant by the end of the 1300s.[1] The reason for this divergence is probably not

1 On huge guns in Europe, see Kelly DeVries, "The Technology of Gunpowder Weaponry in Western Europe during the Hundred Years' War," in *XXII. Kongress der internationalen Kommission für Militärgeschichte* (Vienna, 1997), pp. 285–299; Kelly DeVries, Robert D. Smith, *Medieval Military Technology* (2nd ed., Toronto, 2012); Kelly DeVries, Robert D. Smith, *The Artillery of the Dukes of Burgundy, 1363–1477* (Woodbridge, 2005); Robert D. Smith, "Artillery and the Hundred Years War: Myth and Interpretation," in *Arms, Armies and Fortification in the Hundred Years War*, ed. Anne Curry, Michael Hughes (Woodbridge, 1994), pp. 151–160; Robert

 | DOI:10.1163/9789004395695_010

to be sought in China's military culture, despite Sun Zi's famous dictum that attacking towns is the last resort of warmakers.[2] To explain the divergence we should instead look at wall-building practices. Before 1500, walls in Western Europe tended to be thin and brittle relative to the massive and absorbent walls of China. So starting in the late 1300s, European warmakers found it expedient to invest the huge sums necessary to build and operate siege artillery, whereas Chinese warmakers did not. Their pre-existing walls were, as we'll see, built very differently and were virtually impregnable to the fire of fourteenth-, fifteenth- and early sixteenth-century artillery.

As a result, Chinese focused on making smaller guns until at least the mid-sixteenth century. Thereafter, however, practices on both sides of the Eurasian supercontinent began to converge. In the early 1500s, people in East Asia began adopting and adapting European guns of all types and sizes, and in China this process has been referred to as Sino-Western fusion or hybridization, reflecting the fact that new designs were nativized and melded with traditional Chinese guns.[3] Chinese military adoption and innovation sped up still further in the 1600s, especially between 1620 and 1683, when massive wars convulsed continental East Asia. During this period, armies in China were increasingly equipped with siege artillery. These cannons, known in China as "red barbarian cannon" (紅夷炮), after the hair color of the Dutch and English who brought them to China's shores, were unprecedentedly effective, even against the thick walls of China.[4] Indeed, it is quite intriguing that the *trace italienne* walls that

---

D. Smith, "All Manner of Peeces: Artillery in the Late Medieval Period," *Royal Armouries Yearbook* 7 (2002), 130–138; Robert D. Smith, *Rewriting the History of Gunpowder* (Sundby Lolland, 2010); Clifford Rogers, "The Military Revolutions of the Hundred Years War," *The Journal of Military History* 57, no. 2 (1993), 241–278; Clifford Rogers, "'Military Revolutions' and 'Revolutions in Military Affairs:' A Historian's Perspective," in *Toward a Revolution in Military Affairs? Defense and Security at the Dawn of the Twenty-First Century*, ed. T. Gongora, H. von Riekhoff (Westport, 2000), pp. 21–35; Clifford J. Rogers, "The Artillery and Artillery Fortress Revolutions Revisited," in *Artillerie et fortification: 1200–1600*, ed. Nicolas Prouteau, Emmanuel de Crouy-Chanel, Nicolas Facherre (Rennes, 2011), pp. 75–80; Bert S. Hall, *Weapons and Warfare in Renaissance Europe: Gunpowder, Technology, and Tactics* (Baltimore, 1997). On early Chinese guns, see Tonio Andrade, "Late Medieval Divergences: Comparative Perspectives on Early Gunpowder Warfare in Europe and China," *Journal of Medieval Military History* 13 (2015), 247–276; and Tonio Andrade, *The Gunpowder Age: China, Military Innovation, and the Rise of the West in World History* (Princeton, 2016).

2 Sun Zi famously said, "The best strategy is to attack plans; the next best is to attack alliances; the next best is to attack armed forces; and the worst is to attack walls." Sun Zi 孫子, *Bing Fa* 兵法, 3:3.

3 Wang Zhaochun 王兆春, *Shi jie huo qi shi* 世界火器史 (Beijing, 2007), pp. 159–181. See also Li Yue, *Ming dai*.

4 It is important to note that the Iberians also introduced large muzzle-loaders, at around the same time (circa 1600), and Chinese officials contracted Chinese artisans who had worked in

stimulated the development of seventeenth-century European siege artillery were in many ways quite similar to the traditional walls of China: thick, earthen-cored, sloped.

Yet the *trace italienne* was not just effective because of its thick walls. It was designed from the ground up as a gun fortress, with geometrical principles of defense that included the angled bastion and a variety of ingenious outworks. It stimulated not just the development of powerful siege artillery, but also a series of innovations in siegecraft, including advances in siegeworks and counter-siegeworks, in trench warfare, in logistics, and in siege tactics and planning. By the mid- to late-1600s, these practices were so refined and systematized that a well-planned and well-provisioned siege could quickly and straightforwardly reduce the most powerful *trace italienne* fortifications.[5] Perhaps guns and fortresses are the most significant aspects of the military revolution, but siegecraft itself was far from trivial.

To what extent did East Asian warmakers import – or develop – Western-style fortification and siegecraft? Although research on this topic is in its infancy, it does seem that Europeans maintained an advantage. Recent work shows that although many Chinese did indeed adopt *trace-italienne* designs, those designs did not spread widely. Moreover, by examining seventeenth-century sieges in which the forces of China attacked Western artillery fortresses, we can infer that when it came to siegecraft, the armed forces of China were less effective than those of Europe, at least when they had to grapple with *trace italienne*-style defenses.

1 The Culture of Wall Building in China and Europe

Walls were highly significant in China.[6] The main character for "city" (城) is also the main character for "(defensive) wall", and any significant Chinese settlement with an administrative function (most Chinese towns and cities did have administrative functions) tended to be surrounded by wall complexes

foundries in Macau and Manila to help them cast large cannons, even as they did their best to acquire English and Dutch cannons. Moreover, there were other earlier episodes of cannon adoption, including large muzzle loaders during the mid-1500s. On that, see Zheng Cheng 鄭誠, "Fa gong kao: 16 shi ji chuan Hua de Ou shi qian zhuang huo pao ji qi yan bian" 發熕考—16 世紀傳華的歐式前裝火炮及其演變, *Zi ran ke xue shi yan jiu* 自然科學史研究 32, no. 4 (2013), 504–522.

5 Jamel Ostwald, *Vauban under Siege: Engineering Efficiency and Martial Vigor in the War of the Spanish Succession* (Leiden, 2007).

6 What follows here is a condensed version of a more extended discussion in Andrade, *Gunpowder Age*, esp. chs. 4 and 6.

much larger and thicker than their European counterparts. As one European expert in fortification wrote, after surveying China's walls in the mid-twentieth century, before so many of them were razed for development: "In China [...] the principal towns are surrounded to the present day by walls so substantial, lofty and formidable that the medieval fortifications of Europe are puny in comparison."[7]

Whereas medieval European walls, like the Roman walls that preceded them, rarely grew more than three meters thick, standard Chinese walls were an order of magnitude thicker, a practice that went back to Neolithic times.[8] For instance, the capital of a settlement from the Longshan period (3000 to 2000 BC) known as Chengziya (c. 2500 BC), was enclosed by walls eight to ten meters wide.[9] In the Shang period (1600–1046 BC), the city of Zhengzhou had walls ten meters high and more than twenty meters wide at the base, tapering to five meters at the top.[10] To be sure, ancient European earthwork walls could be quite thick, but they were rarely very high, and whereas Europeans began building thinner stone walls by Roman times, Chinese, over the millennia, kept building huge, thick walls. By Ming times (1368–1644), nearly every significant city and town was endowed with them.

Chinese walls were also built differently from those of the West, made not of rubble encased in stone, as were most European walls, but of tamped earth faced with stone or brick. Earthwork absorbs the energy of a projectile, so earthenwork walls might become riddled with holes during an attack, but those holes rarely penetrated deeply, and earthenwork walls tended not to shatter. As a result, it was very difficult to breach Chinese walls, and historians have noted how rare wall breaches were in China's history until the mid-1600s.[11] The breaches that did occur were usually caused not by bombardment but by mining.[12] Unsurprisingly, until the mid-1600s, catapults and guns were generally used not to destroy city walls but to destroy wooden structures and, of course, to kill people.[13]

In contrast, Europe's relatively thin and brittle walls proved vulnerable to gunpowder artillery, which is why, as Europeans developed large guns starting

---

7 Sidney Toy, *A History of Fortification: From 3000 BC to AD 1700* (London, 1955), p. 181.

8 DeVries and Smith, *Medieval Military Technology*, p. 189.

9 Victor F. S. Sit, *Chinese City and Urbanism: Evolution and Development* (Singapore, 2010), p. 39.

10 Ralph D. Sawyer, Mei-chün Sawyer, *Ancient Chinese Warfare* (New York, 2011), p. 125.

11 Peter Lorge, *The Asian Military Revolution: From Gunpowder to the Bomb* (Cambridge, UK, 2008), p. 74.

12 Andrade, "Late Medieval Divergences," 247–276 (see esp. pp. 269–270).

13 Andrade, *Gunpowder Age*, esp. ch. 5.

in the late 1370s, medieval walls began falling with alarming frequency.[14] Indeed, historians have suggested that this phenomenon was linked to the development of the modern European state itself, that gunpowder artillery made feudal castles – and thus feudalism itself – obsolete.[15] They have wondered why a similar process did not occur in China.[16] Historians of China have pointed out that the developments associated with the modern, centralized state

14 Robert D. Smith, "Artillery and the Hundred Years War," pp. 151–160; Rogers, "The Military Revolutions of the Hundred Years War;" Rogers, "'Military Revolutions' and 'Revolutions in Military Affairs;'" Rogers, "Artillery and Artillery Fortress;" Hall, *Weapons and Warfare.*

15 The gunpowder revolution idea can be found in many places, including Adam Smith, Gibbon, and Karl Marx. See Adam Smith, *An Inquiry into the Nature and Causes of the Wealth of Nations* (Oxford, 1869), 2:292; Edward Gibbon, *The History of the Decline and Fall of the Roman Empire* (Dublin, 1788), 12:57; Karl Marx, "Division of Labour and Mechanical Workshop. Tool and Machinery," in Karl Marx, *Economic Manuscripts of 1861–63,* available at Marx & Engels Internet Archive, <https://marxists.catbull.com/archive/marx/works/1861/economic/ch35.htm>, (accessed 23 October 2018). For an intriguing look at early precursors of the idea of "military revolution"-type ideas, see Clifford Rogers, "The Idea of Military Revolutions in Eighteenth and Nineteenth Century Texts," *Revista de História das Ideias* 30 (2009), 395–415. A small sampling of uses of the gunpowder revolution model in more recent historical work would include William McNeill, "The Gunpowder Revolution," *MHQ: The Quarterly Journal of Military History* 3, no. 1 (1990), 8–17; William McNeill, *The Pursuit of Power: Technology, Armed Force, and Society since AD 1000* (Chicago, 1982); John Landers, *The Field and the Forge: Population, Production, and Power in the Pre-Industrial West* (Oxford, 2005); Richard Bulliet, Pamela Crossley et al., *The Earth and its Peoples: A Global History,* vol. II: *Since 1500* (New York, 2008), p. 628. Yuval N. Harari, *Special Operations in the Age of Chivalry, 1100–1550* (Woodbridge, 2007), esp. pp. 1–52. Charles Truxillo, *Crusaders in the Far East: The Moro Wars in the Philippines* (Fremont, 2012), pp. 35–77. John K. Thornton, *Warfare in Atlantic Africa, 1500–1800* (London, 1999), pp. 61–64. James E. McClellan III, Howard Dorn, *Science and Technology in World History: An Introduction* (Baltimore, 2006), pp. 192–195. Geoffrey Parker reserves the term "gunpowder revolution" to refer to strictly military changes, using the term "military revolution" for the social changes (which he believes occurred rather later than medievalists do). See Geoffrey Parker, ed., *The Cambridge Illustrated History of Warfare* (Cambridge, UK, 2008), esp. pp. 2–11 and 106–119. Max Boot allows the term to spread beyond a narrow military capaciousness, in Max Boot, *War Made New: Technology, Warfare, and the Course of History, 1500 to Today* (New York, 2006), pp. 19–105. Other scholars do not use the term "gunpowder revolution" but nonetheless subscribe to the idea itself. Luc De Vos, for example, notes that artillery played an important role in helping different princes "succeed in diminishing feudalism." As he writes, "L'artillerie était une arme couteuse et pour tout dire, bien au dessus des moyens financiers des petits seigneurs féodaux." Luc De Vos, "La bataille de Gavere le 23 juillet 1453: la victoire de l'organisation," in *XXII. Kongress der internationalen Kommission für Militärgeschichte* (Vienna, 1997), pp. 145–157, here: p. 145.

16 Thomas Arnold, *The Renaissance at War* (London, 2001), pp. 18–19; McNeill, "The Gunpowder Revolution," 8–17, esp. p. 8.

had occurred much earlier in China: there was no need for guns to destroy feudalism in China because there was no feudalism in China.[17]

But scholars have paid much less – indeed practically no – attention to the repercussions of wall-building practices. The gun arrived in Europe in the 1320s, and at first guns were used there much as they had been used in China: to kill people and burn structures (early guns appear to have been used not just as projectile-launchers but also as fire-spewers). But starting around 1380, practices began to diverge. Europeans and their near neighbors began making very large guns and using them to breach walls. No such development occurred in China. It's not that the Chinese were unable to build large guns. Chinese metallurgical practices were on a par with (and in some ways superior to) those of Europe, and there is evidence of large Chinese guns.[18] But those guns remained isolated examples. China's walls were simply much more difficult to breach, so wall-smashing artillery was a path not followed in China.

Consider, for example, that around 1490 the best siege guns of Europe were, according to their wielders, capable of creating breaches in walls of eight feet in thickness (2.4 meters).[19] This was certainly a fearsome capacity in Europe, where most walls were less than eight feet thick. But a typical Chinese city boasted walls that were far thicker. The walls of Suzhou, for example, were eleven meters thick at the base and five meters thick at the top.[20] It seems safe to say that the best siege artillery of Europe circa 1500 would have had little success against such fortifications.[21]

---

17 See, most importantly, Lorge, *Asian Military Revolution,* pp. 20–22.

18 See, for example, Cheng Dong 成東, "Ming dai qian qi you ming huo chong chu tan" 明代前期有銘火銃初探, *Wen wu* 文物 5 (1988), 68–79, p. 74; Wang Zhaochun, "Hong wu shi nian da tie pao," in *Zhong hua guo cui ci dian* 中華國粹大辭典, ed. Men Kui 門巋, Zhang Xijin 張燕瑾 (Beijing, 1997), p. 183; Wang Fuzhun 王福諄, "Gu dai da tie pao" 古代大鐵炮, *Zhu zao she bei yan jiu* 鑄造設備研究 3 (2008), 46–56, p. 48.

19 See Philippe Contamine, "L'artillerie royale française à la veille des guerres d'Italie," *Annales de Bretagne* 71, no. 2 (1964), 221–261, here: pp. 222–223.

20 The Chinese measurements for 1352 are as follows: 23 chi high, 35 chi thick at base, and 16 chi wide at the top. I converted to the metric measures using a chi length of 31.6 cm, which is a reasonable guess for the length of a chi during this period. In the late 1360s, records suggest a slightly higher thickness for the top – 18 chi, or about 5.6 meters (3 zhang 5 chi thick at base and 1 zhang 8 chi thick at the top). The figures are from Yinong Xu, *The Chinese City in Space and Time: The Development of Urban Form in Suzhou* (Honololu, 2000), p. 114. Frederick Mote says they were 25 feet thick at the base and 25 feet high, but his estimates are not quite as evidence-based as Xu's. F. W. Mote, "A Millennium of Chinese Urban History: Form, Time, and Space Concepts in Soochow," *Rice University Studies* 59, no. 4 (1973), 35–65, quote at p. 53.

21 To be sure, these French guns of the 1490s were not the massive bombards of the 1440s, but they were just as effective, making up in power and numbers what they lacked in size.

I thus hypothesize (as I have argued at length elsewhere) that the thickness of China's traditional walls created selective pressures against the development of gunpowder artillery of the European type.[22] Siege guns were terrifically expensive to make and to operate, especially early siege guns. In Europe these investments generally paid off because artillery trains were capable of breaching medieval walls fairly rapidly. But China's walls were so thick as to be effectively impervious to gunpowder artillery of this type. Why spend huge sums of money to make guns that were of no use? Large guns were not pursued in China because they were not expedient.

As medieval walls tumbled down, European engineers began designing new walls. Intriguingly, by the late 1400s and early 1500s, those walls were being constructed much like those of China: thick, sloped, and filled with earth.[23] Yet Europeans went even farther in their experiments with fortification. By the early 1500s, as the *trace italienne* (or artillery fortress) emerged, the balance of warfare changed abruptly back to the defense. With their angled bastions and lethal crossfire capacities, the new forts were not only thick and resistant to artillery – like Chinese walls – but they were also far more difficult storm or mine. Attackers had trouble getting close to them.

Historians have argued that the artillery fortress became a key underpinning of European colonialism, allowing small garrisons of Europeans to control far-flung ports overseas.[24] What light does the history of Chinese siegecraft cast on this argument?

2 The Artillery Fortress in China

First, it is important to note that *trace italienne* designs were in fact known in China. In the course of the warlike 1600s, as Chinese officials imported and adapted new Western gun designs, some of them also paid attention to

22 See Andrade, "Late Medieval Divergences," 247–276; Andrade, *The Gunpowder Age*, esp. chs 5 and 6.

23 On early adaptations to artillery, see Kelly DeVries, "Facing the New Military Technology: Non-Trace Italienne Anti-Gunpowder Weaponry Defenses, 1350–1550," in *Heirs of Archimedes: Science and the Art of War through the Age of Enlightenment*, ed. Brett Steele, Tamara Dorland (Cambridge, MA, 2005), pp. 37–71.

24 See especially Geoffrey Parker, "The Artillery Fortress as an Engine of European Overseas Expansion, 1480–1570," in *City Walls: The Urban Enceinte in Global Perspective*, ed. James D. Tracy (Cambridge, UK, 2000), pp. 386–416; Thomas F. Arnold, "War in Sixteenth-Century Europe: Revolution and Renaissance," in *European Warfare, 1453–1815*, ed. Jeremy Black (New York, 1999), pp. 23–44.

Western fortress design.[25] For example, the Ming official Sun Yuanhua (孫元化, 1582–1632) composed a famous military manual, *Xi fa shen ji* (西法神機, c. 1632), which contains a section called "Illustrated Guide to the Artillery Fortress" (銃台圖說).[26] According to Sun, Western-style angled bastions (銳角) were effective because they allowed no place for attackers to shelter: "there is nowhere our [the defenders'] guns cannot reach and the enemy has no way to approach."[27] There is some evidence that Sun actually managed to build Western-style angled bastions during the 1620s, although it is not watertight evidence.[28] Sun was not the only Ming subject championing the *trace italienne.* Two wealthy brothers known as Han Yun and Han Lin (韓雲, dates unknown; 韓霖, c. 1598–c.1649) also wrote treatises praising the angled bastion. As Han Lin wrote, "At present, the towns and prefectures are protected only by square-shaped bastions [敵台皆作方形], and although the vertical sides can protect each other, the fronts are subject to enemy attack. Thus, it is necessary to construct angled bastions, which are highly ingenious [作三角形為妙]."[29] Did the Han brothers build angled bastions? Evidence is suggestive but, again, far from conclusive.

Indeed, until recently, scholars believed that there was no evidence that *trace italienne* designs were actually built in China, but recently a young Chinese scholar has showed that in fact they were.[30] The key to his discovery is a poorly-known figure named Ma Weicheng (馬維城, 1594–1659), who built angled bastions to defend his home county – Xiong County (雄縣, in Hebei Province). Ma may have learned about angled bastions from the Jesuit missionary Adam Schall von Bell, whom Ma apparently knew quite well.[31] Ma seems to

25 On the import of Western gun designs, see Andrade, *Gunpowder Age,* esp. pp. 124–195.

26 Sun Yuanhua 孫元化, *Xi fa shen ji* 西法神機, juan 2. For a general introduction to the *Xi fa shen ji,* see Lin Wenzhao and Deng Yongfang 林文照 and 郭永芳, "Ming mo yi bu zhong yao de huo qi zhua zhu: Xi fa shen ji" 明末一部重要的火器专著：西法神机, *Zi ran ke xue shi yan jiu* 自然科学史研究 6, no. 3 [1987], 251–59.

27 Sun Yuanhua, "Chong tai tu shuo," in *Xi fa shen ji,* cited in Zheng Cheng 鄭誠, "Shou yu zeng zhuang: Ming mo Xi yang zhu cheng shu zhi yin jin" 守圉增壯：明末西洋築城術之引進, *Zi ran ke xue shi yan jiu* 自然科學史研究 30, no. 2 (2011), 129–150, here: p. 133.

28 Kenneth Swope suggests that Sun Yuanhua may not have constructed artillery fortresses and even questions whether Sun Yuanhua knew about artillery fortresses. See Kenneth Swope, *The Military Collapse of China's Ming Dynasty, 1618–1644* (London, 2014), p. 48.

29 Han Lin *Shou yu quan shu* 守圉全書, circa 1638, cited in Zheng Cheng 鄭誠, "Shou yu," p. 139.

30 Zheng Cheng, "Shou yu."

31 This is revealed in a "Biography of Ma Weicheng," *Xiong County Gazeteer* 雄縣志 (Shunzhi Period, 1644–1661), cited in Zheng Cheng, "Shou yu," p. 142. On Schall von Bell, see, most recently, Denis De Lucca, *Jesuits and Fortifications: The Contribution of the Jesuits to Military Architecture in the Baroque Age* (Leiden, 2012), pp. 175–179.

have built angled bastions not just in his home county but also in other parts of China.[32]

Despite Ma's building programs, and despite the considerable attention paid to artillery fortresses in the writings of other Chinese thinkers, it seems that the artillery fortress design did not take root in China. Why? One possible argument is that the intense warfare that accompanied the experiments in fortification came to an end and that in the peace that followed there was no further call for such experimentation.[33] But this is not an entirely satisfactory explanation. Evidence of angled bastions in China seems to be absent after the mid-1640s, yet intensive warfare continued into the early 1680s. The fact is that we still know little about warfare in seventeenth century China, and it is conceivable – indeed I suspect it is likely – that as we learn more we'll find more evidence of geometric defense and angled bastions elsewhere.

But what does seem to be clear is that the armed forces of China had considerable trouble when attacking European-built artillery fortresses. There are three exemplary cases, two involving Dutch-built fortresses and one involving a Russian-built fortress. One of these cases was a short and abortive attempt, and so we'll leave it to the side.[34] But the other two sieges, both of which were drawn out over many months, followed remarkably similar patterns despite the fact that they took place twenty-five years apart and in very different regions: one in subtropical Taiwan, which was held by the Dutch; and the other in subarctic Siberia, in an area claimed by Russia.

In both cases, the forces of China did not follow the sort of procedure a seasoned European commander would probably have followed: dig a network of trenches, set up fortified emplacements to defend the diggers, and creep, step by careful step, toward the defenses, until it was possible to set up batteries capable of breaching the walls (assuming the enemy did not surrender beforehand). Rather, the Chinese attackers in each case appear to have first tried barraging the defenders in order to launch a storm. In both cases that tactic failed and was followed by a considerable amount of more or less fruitless experimentation, until a cordon and wait-and-see tactic was adopted, with mixed results.

Let us start with the earlier of these two cases: the Chinese siege of Zeelandia Castle, which took place from April 1661 to February 1662. (I have written at length about this siege elsewhere, and what follows is a simplification of

32 Zheng Cheng, "Shou yu," pp. 142–144.
33 Zheng Cheng, "Shou yu," pp. 147–148.
34 Interested readers can find a treatment in Tonio Andrade, *Lost Colony* (Princeton, 2011), pp. 316–321.

a complex and well-documented series of events.[35]) Zeelandia was the main Dutch stronghold on Taiwan. Established in 1624, it eventually grew to be a large fortress in the modern Italian style, with angled bastions protecting its walls. Although by European standards it was a substantial fortress, when the Chinese warlord Zheng Chenggong advanced on it in April 1661, he described it as tiny, laughable, minor. As he wrote to the Dutch governor, "All you have left is that little fort, which is like a dead and dried out tree that cannot stand on its own."[36] Or, on another occasion, "How can you people hope to resist me in that little fortress? [...] When I strike with my cannons, the walls will collapse and crumble, without my having even to risk any of my soldiers."[37] Compared to the walls of Chinese cities, he had a point. Moreover, he knew that the Dutch were severely outnumbered. Whereas he had well more than ten thousand troops ready for action, the Dutch had around eleven hundred.

He also bragged about his artillery: "It is true," he wrote to Dutch leaders, "that you people are famous for playing artfully with cannons, but you've never had this many cannons leveled at you. I have brought hundreds of them, ready to use against you."[38] He may have exaggerated their numbers, but there is no doubt that he did have excellent artillery. By this time, Chinese had ample experience with large cannons, a process of adoption and adaption that had begun by the 1520s, accelerating during the Ming-Qing wars that began in the 1620s.[39] This process of interadoption occurred rapidly in the coastal parts of China, and the political and economic organizations of Zheng Chenggong's family played prominent roles in the process.[40] Zheng Chenggong's artillery seems to have been every bit as good as that of the Dutch.

But could he use it as effectively? In terms of accuracy and rapidity of fire, there is no doubt he could. His artillerists proved highly adept at striking their targets and maintaining a rapid rate of fire.[41] But there was far more to "playing artfully with cannons" than quick and accurate shooting, as he learned in May 1661, when he ordered his commanders to attack the Dutch fort.

---

35 See Andrade, *Lost Colony*.

36 Cited in Andrade, *Lost Colony*, p. 163.

37 Cited in Andrade, *Lost Colony*, p. 163.

38 Cited in Andrade, *Lost Colony*, p. 163.

39 See, for instance, Tonio Andrade "Cannibals with Cannons: The Sino-Portuguese Clashes of 1521–1522 and the Early Chinese Adoption of Western Guns," *Journal of Early Modern History* 19 (2015), 1–25; and Tonio Andrade, "The Arquebus Volley Technique in China, c. 1560: Evidence from the Writings of Qi Jiguang," *Journal of Chinese Military History* 4, no. 4 (2015), 115–141.

40 See Andrade, *Lost Colony*, esp. pp. 21–53.

41 In addition to the evidence from this particular battle, which I adduce below, see also Andrade, *Lost Colony*, pp. 218–223.

His attack did not follow the procedure that a European commander would have followed. There was no attempt, for instance, to dig trenches. That's not to say there wasn't digging. Fort Zeelandia stood on a series of dunes. Below them, about two hundred meters away, there stood a city. Between the fort and the city was a field with a few buildings. The diggers at first confined their work to the city itself. Behind the foremost row of buildings they piled up a mountain of sand, which reached at least the second-story windows and stretched the entire length of the street (around 150 meters). Then, during the nearly moonless night before the attack, the workers used this sand to fill gabions, which they then deployed to create a defensive wall for their cannons. The Dutch later counted 25 cannons, including 12-, 18- and 24-pounders. How far was this wall from the castle? It is hard to say, but it was certainly in front of the first row of buildings in the town, so probably around 150 meters away from the fort.

Zheng's barrage started before daybreak, and it went well at first. His gunners fired swiftly, the one charge barely over before the next began.[42] They were disturbingly accurate, seeming to know which house the Hollanders' governor, Frederick Coyet, was sleeping in. Some of the first shots brought its roof down. Many of the besieged thought that the governor had been killed, but in fact he managed to make it out of the damaged building. In an account he wrote later, he described what he did next, writing in the third person (because he wrote under a pseudonym):

> He […] ran to the bulwarks to take a general survey. His practiced eye at once observed the weak position of the enemy's cannon, which were entirely unprotected, and in great danger if attacked. The governor could also see that the enemy – who appeared jubilant over the success of their firing, and very hopeful that a breach would be made in the walls – had wandered in great numbers outside of their barricades and were thus recklessly exposing themselves. He therefore restrained the anxiety of our men, and commanded that not a single shot should be fired in the meantime. All the pieces were then arranged in such a position that their respective shots would cross one another, and were charged with powder, musket-bullets, and large iron nails. The musketeers took up their places along the outskirts of the balustrades; and when at length a suitable opportunity arose, the word of command was given to fire on the unprotected Chinese from above, below, and all sides simultaneously. This

42 National Archive, The Hague, VOC Collection, VOC 1235: 386v, Resolutions of the Council of Formosa, 1661–05–25, morning.

> order was so well executed that, with the first charge, nearly the whole field was strewn with dead and wounded; the enemy being thus taught the lesson not to expose themselves so readily.[43]

Thus, from Coyet's perspective, the Chinese positions were highly exposed. If Zheng's soldiers had dug a proper trench, they would not have been so vulnerable to the fort's fire. Moreover, the wall of sandbasket barricades itself was full of gaps, which Dutch guns could easily penetrate, not just the cannons in the fort itself but also Dutch muskets, firing from positions outside the fort's walls.

The Dutch barrages proved devastating. Indeed, it seems that Chinese officers begged some of their Dutch prisoners to wave to their countrymen and call out, "Stop shooting! Peace! Peace!" but the prisoners wisely demurred.[44] The Dutch gunners obliterated the cannon placements, and once the Chinese had stopped shooting, Dutch musketeers sallied cautiously forth, picking their way through the rubble. They encountered no resistance, just cannons turned on their sides and bodies strewn about. They were able to spike cannons and return with prizes.

Zheng Chenggong seems to have been shocked by this defeat. He immediately withdrew all but a small number of troops, sending the lion's share to the countryside to set up farms, leaving just enough to keep the Dutch bottled up, expecting that hunger would do what his troops couldn't. Thus, the siege became a blockade, which we will discuss later.

But first let us move forward in time and northward to the freezing subarctic, where we will examine the travails of Qing forces, who attempted to capture the Russian fortress of Albazin. It seems that, like Zeelandia Castle, Albazin was an artillery fortress built in the Italian style, with angled bastions.[45] But whereas Zeelandia was a sea fortress in subtropical Taiwan, Albazin stood on the banks of the Amur River, in southern Siberia. Like Zeelandia, its walls were thick earthenwork structures, although they were built in an unusual way: "with clay-earth and tree-roots that were woven and cinched together,

43 Frederick Coyett, "Het Verwaerloosde Formosa," in *Formosa under the Dutch: Described from Contemporary Records*, ed. William Campbell (London, 1903), p. 429.

44 Jiang Shusheng 江樹生, ed. and trans., *Mei shi ri ji: e lan tu di ce liang shi kan Zheng Cheng Gong* 梅氏日記荷蘭土地測量師看鄭成功 (Taipei, 2003), p. 27.

45 On evidence about Albazin as an artillery fortress, see Nicolaas Witsen, *Noord en Oost Tartarye, ofte bondig ontwerp van eenige dier landen en volken* (Göttingen, 1705), 2:622; see also the map drawn by the Prussian military commander Afanasij Ivanovic Bejton circa 1690, in V. F., "A Bejton und seine Karte von Amur," *Imago Mundi* 1 (1935), 47–48.

[and which were] worked in such a way that [the material] became as hard as stone, and unbreakable."[46]

The Qing attack came in July of 1686, and it had been carefully planned. Albazin lay a thousand miles from Beijing as the crow flies and this was forbidding territory, sparsely populated, extremely cold in the winter. The Kangxi Emperor took a personal direct interest, and his commander, Langtan, was equipped with large artillery, a carefully-considered system of supply depots, a fleet of river vessels filled with provisions, and three thousand well-armed troops. In contrast, the Russians had only eight hundred men and far fewer large guns – just eight large bronze cannons and three iron ones. Three thousand troops may seem like a lot, and the ratio of nearly four to one in favor of the Qing seems high, but in sieges, the advantage always goes to the defense. A European commander would probably have felt that far more men were necessary to capture an artillery fortress: ideally, one would have enough for multiple shifts of workers, including trench guards, camp guards, etc. The Qing apparently had the help of local peoples, but even so, their three thousand men proved too few to adequately invest Albazin, especially given that the Qing commanders did not seem to understand how to capture an artillery fortress.

Indeed, it is striking that the Qing forces suffered the same sort of setbacks as the forces of Zheng Chenggong in Zeelandia. Our sources for Albazin are much sparser than for Zeelandia, but it seems, if we examine both European and Chinese sources, that the Qing attacked the fortress quite confidently a number of times, without bothering to open any significant trenches. Unsurprisingly, they were consistently driven back. On 23 July 1686, for instance, the Qing launched a nocturnal assault from two directions, bombarding from the north while attempting a storm from the south.[47] The attempt was decisively repulsed by Russian artillery. As a Russian source notes, "a large-scale barrage of cannons from the town occurred and in the smoke neither the people nor the town could be seen, and the enemy, unable to do anything, retreated and

46 Witsen, *Noord en Oost*, p. 769. A Qing reconnaissance mission similarly reported that the thick, sturdy walls "were made from interspersing trees, with a core of earth as the filling [...] and the outside filled in with clay." [Anonymous], *Ping ding luo cha fang lüe* 平定羅刹方略, 4 juans (circa 1689), quote at juan 3, fos 1v–2r. Online at <http://archive.org/details/02080673.cn> (accessed 23 October 2018). Another European authority writes that the grass, mortar, and tree roots were "set so well together that it was stronger than a normal wall." Gerhard Friederich von Müller, ed., *Sammlung russischer Geschichte des Herrn Collegienraths Müllers in Moscow* (Offenbach am Main, 1779), 5:423.

47 "Lang Tan Zhuan" 郎談傳 [Lang Tan is also known as 郎坦], in E Ertai 鄂爾泰, compiler, *Ba zu tong zhi* 八旗通志, 8 vols (Jilin Sheng Changchun shi 吉林省長春市, 1985 [1739]), juan 153, p. 3886.

stood in small groups below the town behind their gabions."[48] Thus, it seems that although the Qing put up gabion defenses, they did not dig any significant trenches. As a result, although Qing records are reticent about losses, European sources suggest that the carnage was considerable and that the Chinese storm was "driven back with great losses."[49] And just as in the case of the Sino-Dutch siege of 1661–62, the Russians followed their victory with sorties by musketeers, who apparently killed many further Qing troops.[50] In contrast, losses on the Russian side appear to have been slight. This, too, suggests that the Qing troops were unskilled with trench work. Had they opened up trench parallels, as was the practice in Europe, they could have minimized the damage from Russian sorties.

Thus, it seems that in both the Sino-Dutch siege of 1661–2 and the Russo-Qing siege of 1686–7, the forces of China were at first too confident. In each case they launched an initial barrage and attempted a storm, and in each case they were decisively rebuffed, with major losses. In neither case were these initial attacks prepared with effective trench work. European commanders would probably have been far more cautious. By the late 1600s, the practice was to dig three parallels, each one closer to the fortress. Such trenches could shelter besiegers, allowing them to set up batteries and move ever closer to the enemy.[51]

---

48 "Otpiski iz Udinskogo ostrozhka Andreia Iakovleva i Emel'iana Panikadil'nikova k Selengskomu prikaznomu cheloveka, s soobshcheniiem izvestiia poluchennikh im ot raznykh lits ob osade Kitaitsami Albazina i ob ubienii v vremia osada voevody Aleksei Tolbuzina" ["Dispatch from Fort Uda, Andrei Yakovlev and Emelian Panikadil'nikov, to the Selenga chancellory official, with a communication of the news from various people of the siege of Albazin by the Chinese and the slaying, in the time of the siege, of Governor Aleksei Tolbuzin"], in Russia Arkheograficheskaa͡ kommissii͡a, *Akty istoricheskie. Dopolneniia.*, vol. 10: *1682–1700* (Saint Petersburg, 1867), pp. 263–264, quote on p. 264. Thanks to Matthew Payne for the translation.

49 Müller gives the date of this attempt at storming as 1 September, but he is likely wrong about that, because he is also wrong about another important event that occurred during these Russian sallies: the death of the Russian commander, Alexi Tomulsin. Whereas other Europeans had suggested that Tomulsin was killed in July 1686, Müller argued that his death occurred in September: "It was a considerable loss when in this siege the commander Tolbusin was killed. It was not, as Witsen says, the fifth day, when the enemy began shooting, that he was killed, but, as people who were there have testified, it was the beginning of September, when he was struck by an enemy cannonshot." Müller, *Sammlung,* 5:429. Yet according to Qing sources, which tend to be scrupulous in dating, Tomulsin was killed in early fighting, during July of the Julian calendar (which is the calendar Müller and Witsen use to report their dates). "Liang Tan Zhuan," in E Ertai, *Ba zu,* juan 153, p. 3886. Witsen's discussion of Tolmusin's death is at Witsen, *Noord en Oost,* pp. 861–862.

50 Witsen, *Noord en Oost,* pp. 861–862.

51 On the development of the practice of trench parallels, see Ostwald, *Vauban under Siege,* esp. pp. 57–61.

Similarly, if we follow subsequent events in both sieges, we see that the two operations continued to follow similar patterns. In both cases, the forces of China tried various methods to reduce the fortress, groping towards a solution but unaware of – or unwilling to acknowledge – the efficacy of trenches and careful siegeworks. In each case, the Europeans responded effectively, thwarting the forces of China.

In Zeelandia, for instance, Zheng Chenggong was at first content to keep the Dutch blockaded, expecting hunger to do the work while his own soldiers raised crops. The Dutch, however, were able to fish and hunt and forage under the protection of their forts' guns, and, more importantly, they also managed to send for provisions, defying the prevailing seasonal winds to dispatch a ship southward to their headquarters in Batavia (current-day Jakarta, Indonesia). Zheng Chenggong was shocked to see the Dutch reinforcement fleet arrive, since he thought it was impossible to sail against the monsoon, and it appears that he did not quite believe that the ships had been outfitted against him, suspecting instead that the arrival of the fleet was a coincidence.

In any case, once he saw that the Dutch could receive reinforcements and might therefore hold Zeelandia indefinitely, Zheng Chenggong acknowledged that he must take it by force. He set up a new fortified artillery battery. The Dutch neutralized it with a new outwork of their own. He then constructed a new battery at another location, but, again, the Dutch built a new set of walls that neutralized it. In this dance of siegework and counter-siegework, the Dutch continually had the upper hand, and Dutch sources betray the glee that Dutch officials felt at being able so easily to thwart his attempts.

That glee turned to terror, however, when Zheng Chenggong began constructing a new, far more extensive set of fortified batteries. This time he had the help of a high-ranking European, a German sergeant who had defected from the Dutch side.[52] As the sergeant knew from his service in Europe and overseas, there was no shortcut to capturing an artillery fortress. You must creep closer trench by trench, position by position. So he supervised the building of far more systematic siegeworks than Zheng had constructed to that point. Although these siegeworks appear not to have included significant trenchwork, they proved highly effective. They targeted a redoubt that stood on a hill above the main fortress. Zheng Chenggong had attempted to capture this redoubt before, but always unsuccessfully. This time, however, Zheng's gabion walls were far more artfully constructed. Using the dunes for cover, they stretched around the little redoubt, impossible to attack from the main fortress. When eventually Zheng's artillerists opened fire, they quickly breached

52 See Andrade, *Lost Colony*, pp. 278–286.

the redoubt's walls. The Dutch realized resistance was futile and surrendered. Zheng Chenggong received the keys to Zeelandia, which still held all of its considerable treasure and armaments.

Similarly, in Albazin, the Qing made multiple attempts but were consistently frustrated. For example, after failing to storm the fort, they tried an intensive nocturnal bombardment, but, according to Qing sources, "the walls stood strong and could not be reduced."[53] Later, they tried capturing the city's southern defenses.[54] After that failed, they tried building a succession of siegeworks, but each time the Russians neutralized them. The details suggest that just as in the case of Zeelandia, the Qing commanders had trouble understanding where to place their siegeworks and so were consistently frustrated. As a Russian source notes, "the town [Albazin] had been constantly shot by cannons, but the enemy could not gain an advantage."[55]

Eventually the Qing switched to a strategy of blockade. Russian reports note that "the Chinese fortified themselves and put up bulwarks, setting up gabions that were thirty-six feet (six fathoms) high, and on each bulwark were three siege cannons, in addition to another fifteen guns, which stood on the batteries. Around the city they had also dug trenches, as well as various places to live, behind, under, and within their works or fortifications."[56] An early eighteenth-century image of the siege from a seventeenth-century original shows that the Qing counterdefenses grew longer and more massive than the walls of Albazin themselves.[57]

The Russians suffered under this blockade. Whereas the Dutch could forage and hunt and receive food from sea, the Russians found themselves trapped within the walls, and in subarctic Albazin the winter came early and hard. By the beginning of November, four months after the Qing had arrived, the eight hundred Russian men had dwindled to a hundred fifty, an 80% mortality rate. They suffered not from lack of calories – they had plenty of grain – but from diseases of sanitation and malnutrition, particularly scurvy, caused by a lack of vitamin C. Because the Qing had attacked before the construction of Albazin had been fully completed, many of the besieged lived in burrows in the ground. To be sure, the Qing were expiring, too. It seems likely that on the order of 1500 Qing troops died, a mortality rate of fifty percent. By the beginning of December 1686, there were only twelve Russian men healthy enough to man the guns. Their commander, however, "managed miracles; he was able with these few

53 E Ertai, "Lang Tan," p. 3886.
54 E Ertai, "Lang Tan," p. 3886.
55 Müller, *Sammlung*, 5:428.
56 Witsen, *Noord en Oost*, pp. 861–862.
57 Witsen, *Noord en Oost*, p. 662.

people to keep the cannons firing, making it seem as though there were still many people within the fortress."[58]

Despite the success of the blockade, the Russian garrison did not surrender. The Albazin conflict was decided not by arms or hunger but by diplomacy. After representatives of the Russian Czar arrived in Beijing, the Kangxi Emperor ordered the siege to be lifted. In the negotiations that followed, Albazin was eventually delivered to the Qing, while the Russians kept Nerchinsk. Thus, although the Qing did acquire Albazin, they did not do so through conquest.

Technically the Siege of Albazin can thus not be considered a victory for the forces of China, although it is quite likely that Albazin would have fallen eventually. Indeed, although more research is necessary, it seems that Qing forces were on the verge of making another attack when the order for truce arrived. It is likely that this one would have succeeded.

In any case, it seems that the European artillery fortress may well deserve its reputation as an "engine of European expansion."[59] Despite the fact that the artillery fortress itself was known in China, it was not widespread enough for Chinese warmakers to develop the practices necessary to capture it with some degree of calculability. In Europe, those practices were the result of decades of evolution – perhaps even centuries – and they were thus present long before the famous Vauban synthesis that Jamel Ostwald analyses so brilliantly in *Vauban under Siege*.[60] As we have seen, the forces of China could overcome the artillery fortress only with tremendous trouble. The Russian fort was never actually captured; the Dutch one was captured with the help of a European defector.

Still, we must be humble about drawing conclusions. These two cases provide relatively scant information, especially compared to the voluminous evidence we have for sieges in western Europe, where we have hundreds of well-documented cases. Yet there were, so far as I am aware, no other major occasions on which forces of China attacked western artillery fortresses (aside from the third quite minor occasion I alluded to above).

To learn more about Chinese siegecraft and fortification, we must turn to other conflicts in and around China, particularly the warfare of the mid-1600s, when the massive muzzle-loading guns that the Chinese knew as "red barbarian cannons" (紅夷炮) spread rapidly through China, first to Ming forces and then to their enemies, the Qing.[61] Starting in the 1620s, these cannons allowed

58 Witsen, *Noord en Oost*, p. 769.

59 Parker, "The Artillery Fortress."

60 Ostwald, *Vauban under Siege*.

61 For more on the "red barbarian cannon" in China, see Huang Yi-lung 黃一農, "Ming Qing zhi ji hong yi da pao zai dong nan yan hai de liu bu ji qi ying xiang" 明清之際紅夷大砲在東南沿海的流布及其影響, *Zhong yang yan jiu yuan ji li shi yu yan yan jiu suo ji kan*

the systematic breaching of China's thick walls. China's historical sources are so rich that, as scholars explore them, we will likely have cause to reconsider our understanding of Chinese siegecraft during this pivotal period. Similarly, we also still know little about other seminal periods in Chinese military history, as for example the mid-1500s, when designs for powerful muzzle-loading artillery were adopted from the Portuguese (in contrast to the earlier Chinese adoption of the breach-loading *folangji*), and the 1670s, when the formative but understudied War of the Three Feudatories shook the newly-established Qing Dynasty to its foundation.[62]

As we learn more about Chinese siegecraft, we'll doubtless find more challenges to our understanding of Chinese military history and also to our understanding of European military history. For us to truly understand the World of the Siege, whether it comes to representation, discourse analysis, praxis, or technology transfer, we will need to move firmly toward a truly global military history, in which our understanding of Western military history becomes more nuanced and more grounded in comparative perspectives.

Bibliography

Archival Sources

National Archives of the Netherlands, VOC Collection (VOC)

Resolutions of the Council of Formosa, 1661–05–25, VOC 1235: 386v.

Printed Sources

[Anonymous], *Ping ding luo cha fang lüe* 平定羅刹方略, 4 juans (circa 1689). Online at <http://archive.org/details/02080673.cn>, (accessed 23 October 2018).

Arkheograficheskaîa kommissiîa, *Akty istoricheskie. Dopolneniia.*, vol. 10: *1682–1700* (Saint Petersburg, 1867).

中央研究院歷史語言研究所集刊 81, no. 4 (2010), 769–832; Huang Yi-lung 黃一農, "Ming Qing du te fu he jin shu pao de xing shuai" 明清獨特複合金屬砲的興衰, *Qinghua xuebao* 清華學報 41, no. 1 (2011), 73–136; Huang Yi-long 黃一農, "Ou zhou chen chuan yu Ming mo chuan hua de xi yang da pao" 歐洲沉船與明末傳華的西洋大炮, *Zhong yang yan jiu yuan li shi hua yan yan jiu suo ji kan* 中央研究院歷史語言研究所集刊 75, no. 3 (2004), 573–634; Nicola Di Cosmo,"Did Guns Matter? Firearms and the Qing Formation," in *The Qing Formation in World-Historical Time*, ed. Lynn Struve (Cambridge, MA, 2004), pp. 121–166.

62 On the breech-loaders, see Andrade, *Gunpowder Age*, 135–43. On the later Portuguese muzzle-loaders, see Zheng Cheng 鄭誠, "Fa gong kao: 16 shi ji chuan Hua de Ou shi qian zhuang huo pao ji qi yan bian" 發熕考 —16 世紀傳華的歐式前裝火炮及其演變, *Zi ran ke xue shi yan jiu* 自然科學史研究 32, no. 4 (2013), 504–522.

Coyett, Frederick, *Het Verwaerloosde Formosa,* in *Formosa under the Dutch: Described from Contemporary Records*, ed. William Campbell (London, 1903), pp. 59–60.

Gibbon, Edward, *The History of the Decline and Fall of the Roman Empire* (Dublin, 1788).

Müller, Gerhard Friedrich von, ed., *Sammlung russischer Geschichte des Herrn Collegienraths Müllers in Moscow* (Offenbach am Main, 1779).

"Lang Tan Zhuan" 郎談傳, in E Ertai 鄂爾泰, compiler, *Ba zu tong zhi* 八旗通志, 8 vols (Jilin Sheng Changchun shi 吉林省長春市, 1985 [1739]), juan 153.

Marx, Karl, "Division of Labour and Mechanical Workshop. Tool and Machinery," in Marx, Karl, *Economic Manuscripts of 1861–63,* available at Marx & Engels Internet Archive, <https://marxists.catbull.com/archive/marx/works/1861/economic/ch35.htm> (accessed 23 October 2018).

Smith, Adam, *An Inquiry into the Nature and Causes of the Wealth of Nations* (Oxford, 1869).

Sun Yuanhua 孫元化, *Xi fa shen ji* 西法神機 (Late Ming). Online version at <http://echo.mpiwg-berlin.mpg.de/ECHOdocuView?mode=imagepath&url=/permanent/library/P7S49UQD/pageimg> (accessed 23 October 2018).

Witsen, Nicolaas, *Noord en Oost Tartarye, ofte bondig ontwerp van eenige dier landen en volken* (Göttingen, 1705).

Xiong County Gazeteer (Xiong xian xiang tu zhi) 雄縣鄉土志 (China: ~1905). Online version available at <http://ctext.org/library.pl?if=gb&file=107829&page=64> (accessed 23 October 2018).

Sun Zi 孫子, *Bing Fa* 兵法.

Secondary Sources

Andrade, Tonio, "The Arquebus Volley Technique in China, c. 1560: Evidence from the Writings of Qi Jiguang," *Journal of Chinese Military History* 4, no. 4 (2015), 115–141.

Andrade, Tonio, "Cannibals with Cannons: The Sino-Portuguese Clashes of 1521–1522 and the Early Chinese Adoption of Western Guns," *Journal of Early Modern History* 19 (2015), 1–25.

Andrade, Tonio, *The Gunpowder Age: China, Military Innovation, and the Rise of the West in World History* (Princeton, 2016).

Andrade, Tonio, "Late Medieval Divergences: Comparative Perspectives on Early Gunpowder Warfare in Europe and China," *Journal of Medieval Military History* 13 (2015), 247–276.

Andrade, Tonio, *Lost Colony* (Princeton, 2011).

Arnold, Thomas, *The Renaissance at War* (London, 2001).

Arnold, Thomas, "War in Sixteenth-Century Europe: Revolution and Renaissance," in *European Warfare, 1453–1815*, ed. Jeremy Black (New York, 1999), pp. 23–44.

Boot, Max, *War Made New: Technology, Warfare, and the Course of History, 1500 to Today* (New York, 2006).

Bulliet, Richard; Crossley, Pamela et al., *The Earth and its Peoples: A Global History*, vol. II: *Since 1500* (New York, 2008).

Cheng Dong, 成東, "Ming dai qian qi you ming huo chong chu tan" 明代前期有銘火銃初探, *Wen wu* 文物 5 (1988), 68–79.

Contamine, Philippe, "L'artillerie royale française à la veille des guerres d'Italie," *Annales de Bretagne* 71, no. 2 (1964), 221–261.

Cosmo, Nicola Di, "Did Guns Matter? Firearms and the Qing Formation," in *The Qing Formation in World-Historical Time*, ed. Jeremy Black (Cambridge, MA, 2004), pp. 121–166.

DeVries, Kelly, "The Technology of Gunpowder Weaponry in Western Europe during the Hundred Years' War," in *XXII. Kongress der internationalen Kommission für Militärgeschichte* (Vienna, 1997), pp. 285–299.

DeVries, Kelly; Smith, Robert D., *Medieval Military Technology* (2nd ed., Toronto, 2012).

DeVries, Kelly; Smith, Robert D., *The Artillery of the Dukes of Burgundy, 1363–1477* (Woodbridge, 2005).

DeVries, Kelly, "Facing the New Military Technology: Non-Trace Italienne Anti-Gunpowder Weaponry Defenses, 1350–1550," in *Heirs of Archimedes: Science and the Art of War through the Age of Enlightenment*, ed. Brett Steele, Tamara Dorland (Cambridge, MA, 2005), pp. 37–71.

Hall, Bert S., *Weapons and Warfare in Renaissance Europe: Gunpowder, Technology, and Tactics* (Baltimore, 1997).

Harari, Yuval N., *Special Operations in the Age of Chivalry, 1100–1550* (Woodbridge, 2007).

Huang Yi-long, 黃一農, "Ou zhou chen chuan yu Ming mo chuan hua de xi yang da pao" 歐洲沉船與明末傳華的西洋大炮, *Zhong yang yan jiu yuan li shi hua yan yan jiu suo ji kan* 中央研究院歷史語言研究所集刊 75, no. 3 (2004), 573–634.

Huang Yi-long, 黃一農, "Ming Qing du te fu he jin shu pao de xing shuai" 明清獨特複合金屬砲的興衰, *Qing hua xue bao* 清華學報 41, no. 1 (2011), 73–136.

Huang Yi-long, 黃一農， "Ming Qing zhi ji hong yi da pao zai dong nan yan hai de liu bu ji qi ying xiang" 明清之際紅夷大砲在東南沿海的流布及其影響, *Zhong yang yan jiu yuan ji li shi yu yan yan jiu suo ji kan* 中央研究院歷史語言研究所集刊 81, no. 4 (2010), 769–832.

Jiang Shusheng, 江樹生, ed. and trans., *Mei shi ri ji: He lan tu di ce liang shi kan Zheng Cheng Gong* 梅氏日記荷蘭土地測量師看鄭成功 (Taipei, 2003).

Landers, John, *The Field and the Forge: Population, Production, and Power in the Pre-Industrial West* (Oxford, 2005).

Lin Wenzhao and Deng Yongfang 林文照 and 郭永芳, "Ming mo yi bu zhong yao de huo qi zhua zhu: Xi fa shen ji" 明末一部重要的火器专著：西法神机, *Zi ran ke xue shi yan jiu* 自然科学史研究 6, no. 3 [1987], 251–259.

Lorge, Peter, *The Asian Military Revolution: From Gunpowder to the Bomb* (Cambridge, UK, 2008).

Lucca, Denis De, *Jesuits and Fortifications: The Contribution of the Jesuits to Military Architecture in the Baroque Age* (Leiden, 2012).
McClellan III, James E.; Dorn, Howard, *Science and Technology in World History: An Introduction* (Baltimore, 2006).
McNeill, William, "The Gunpowder Revolution," *MHQ: The Quarterly Journal of Military History* 3, no. 1 (1990), 8–17.
McNeill, William, *The Pursuit of Power: Technology, Armed Force, and Society since AD 1000* (Chicago, 1982).
Mote, F. W., "A Millennium of Chinese Urban History: Form, Time, and Space Concepts in Soochow," *Rice University Studies* 59, no. 4 (1973), 35–65.
Parker, Geoffrey, ed., *The Cambridge Illustrated History of Warfare* (Cambridge, UK, 2008).
Parker, Geoffrey, "The Artillery Fortress as an Engine of European Overseas Expansion, 1480–1570," in *City Walls: The Urban Enceinte in Global Perspective*, ed. James D. Tracy (Cambridge, UK, 2000), pp. 386–416.
Rogers, Clifford, "The Artillery and Artillery Fortress Revolutions Revisited," in *Artillerie et fortification: 1200–1600*, ed. Nicolas Prouteau, Emmanuel de Crouy-Chanel, Nicolas Facherre (Rennes, 2011), pp. 75–80.
Rogers, Clifford, "The Idea of Military Revolutions in Eighteenth and Nineteenth Century Texts," *Revista de História das Ideias* 30 (2009), 395–415.
Rogers, Clifford, "'Military Revolutions' and 'Revolutions in Military Affairs:' A Historian's Perspective," in *Toward a Revolution in Military Affairs? Defense and Security at the Dawn of the Twenty-First Century*, ed. T. Gongora, H. von Riekhoff (Westport, 2000), pp. 21–35.
Rogers, Clifford "The Military Revolutions of the Hundred Years War," *The Journal of Military History* 57, no. 2 (1993), 241–278.
Sawyer, Ralph D.; Sawyer, Mei-chün, *Ancient Chinese Warfare* (New York, 2011).
Sit, Victor F. S., *Chinese City and Urbanism: Evolution and Development* (Singapore, 2010).
Smith, Robert D., "Artillery and the Hundred Years War: Myth and Interpretation," in *Arms, Armies and Fortification in the Hundred Years War*, ed. Anne Curry, Michael Hughes (Woodbridge, 1994), pp. 151–160.
Smith, Robert D., "All Manner of Peeces: Artillery in the Late Medieval Period," *Royal Armouries Yearbook* 7 (2002), 130–138.
Smith, Robert D., *Rewriting the History of Gunpowder* (Sundby Lolland, 2010).
Swope, Kenneth, *The Military Collapse of China's Ming Dynasty, 1618–1644* (London, 2014).
Thornton, John K., *Warfare in Atlantic Africa, 1500–1800* (London, 1999).
Toy, Sidney, *A History of Fortification: From 3000 BC to AD 1700* (London, 1955).

Truxillo, Charles, *Crusaders in the Far East: The Moro Wars in the Philippines* (Fremont, 2012).

V. F., "A Bejton und seine Karte von Amur," *Imago Mundi* 1 (1935), 47–48.

Vos, Luc De, "La bataille de Gavere le 23 juillet 1453: la victoire de l'organisation," in *XXII. Kongress der internationalen Kommission für Militärgeschichte* (Vienna, 1997), pp. 145–157.

Wang Fuzhun, 王福諄, "Gu dai da tie pao" 古代大鐵炮, *Zhu zao she bei yan jiu* 鑄造設備研究 3 (2008), 46–56.

Wang Zhaochun, 王兆春, *Shi jie huo qi shi* 世界火器史 (Beijing 2007).

Wang Zhaochun, "Hong wu shi nian da tie pao," in *Zhong hua guo cui ci dian* 中華國粹大辭典, ed. Men Kui 門巋 and Zhang Xijin 張燕瑾 (Beijing, 1997), p. 183.

Xu, Yinong, *The Chinese City in Space and Time: The Development of Urban Form in Suzhou* (Honolulu, 2000).

Zheng Cheng, 鄭誠, "Fa gong kao: 16 shi ji chuan Hua de Ou shi qian zhuang huo pao ji qi yan bian" 發熕考—16 世紀傳華的歐式前裝火炮及其演變, *Zi ran ke xue shi yan jiu* 自然科學史研究 32, no. 4 (2013), 504–522.

Zheng Cheng, 鄭誠, "Shou yu zeng zhuang: Ming mo Xi yang zhu cheng shu zhi yin jin" 守圉增壯：明末西洋築城術之引進, *Zi ran ke xue shi yan jiu* 自然科學史研究 30, no. 2 (2011), 129–150.

CHAPTER 10

"The Enterprises and Surprises That They Would Like to Perform": Fear, Urban Identities, and Siege Culture during the French Wars of Religion

Brian Sandberg

1 Introduction*

Urban communities became sites of religious violence as confessional politics and religious rivalries destabilized France following the tragic death of King Henri II in a jousting accident in Paris in 1559. Calvinism had been growing dramatically in the kingdom during the 1550s, as perhaps ten percent of French subjects had embraced the Reformed religion—including numerous artisans, merchants, bankers, and nobles who dominated urban politics.[1] Royal officials, Catholic clergy, and municipal leaders all referred to Calvinists as members of the "so-called Reformed religion" (*Religion prétendue reformée*), accusing them of heresy. Calvinism remained officially illegal throughout the kingdom, but local bans and heresy trials did not halt the spread of the Reformed religion. Catholic municipal leaders were troubled by evidence that French Calvinists, or Huguenots, were holding clandestine worship services in individuals' homes in many towns and cities. Calvinist nobles who owned urban residences or visited cities posed a serious threat to public order, since many of them had military experience and traveled with armed entourages. French Catholics knew that members of the "so-called Reformed religion" lived in their midst in urban centers, and they feared that the Huguenots might attempt to take over city governments in order to accomplish their much-desired religious reform program. Such fears were especially acute in the capital city of Paris, which had a predominantly Catholic population, but with known Calvinists living in the city and its surrounding suburbs.[2]

* I would like to thank the Institut d'Études Avancées de Paris, Fulbright Scholar Program, and Northern Illinois University for their support of my research.

1 Philippe Charèyre, "Démographie et minorités protestantes," *Bulletin de la Société de l'Histoire du Protestantisme Français* 148, no. 4 (2002), 867–889.

2 Philip Benedict, *Christ's Churches Purely Reformed: A Social History of Calvinism* (New Haven, 2002), pp. 127–151; Philip Benedict, "Confessionalization in France? Critical Reflections and New Evidence," in *Society and Culture in the Huguenot World, 1559–1685,* ed. Raymond

 | DOI:10.1163/9789004395695_011

Even before fighting broke out, fears of conspiracies destabilized urban communities across France. In March 1560, a group of Calvinist nobles met secretly near Amboise, where the royal court was then based, hoping to deliver a confession to the young king François II and perhaps liberate him from Catholic influence, but their plans were uncovered and their leaders were arrested. An affiliated group of Calvinist nobles and their followers allegedly attempted to overwhelm one of the city gates at Amboise during a surprise attack, but were repulsed by royal guards. In the immediate aftermath of this failed attack, a number of conspirators were executed. The Conspiracy of Amboise confirmed the fears of French Catholics, who believed that the Huguenots aimed to kidnap the king and coerce him to convert as a prelude to subjugating Paris and the entire kingdom. The executions heightened Calvinists' fears of religious repression and persecution, as episodes of religious violence broke out in a number of communities in different regions.[3]

The growing concerns by both Catholics and Calvinists over religious disunity, persecution, and sedition eroded the urban identities of cities, which had long presented themselves as cohesive *bonnes villes* (good cities) with firmly ensconced civic liberties. Most major urban centers in France possessed stone city walls, which defined urban boundaries and civic identities, in addition to some bastioned fortifications. Despite such symbols of civic pride, many communities were seriously divided by religious divisions and confessional politics. These religio-political divisions undermined civic loyalties, unsettled municipal governments, and potentially compromised urban defenses.[4]

The sense of insecurity only worsened with the death of François II in December 1560, the succession of his younger brother Charles IX, and the formation of a regency government. As royal authority deteriorated, a series of iconoclastic attacks, riots, and murders erupted in urban areas in 1561–1562. When François de Lorraine, duc de Guise, attempted to suppress a clandestine Reformed worship service at Vassy in March 1562, violence broke out and his troops killed dozens of Calvinists. News of the massacre at Vassy spread rap-

A. Mentzer, Andrew Spicer (Cambridge, UK, 2002), pp. 44–61; Barbara B. Diefendorf, *Beneath the Cross: Catholics and Huguenots in Sixteenth-Century Paris* (Oxford, 1991), pp. 49–63.

3 Arlette Jouanna, Jacqueline Boucher, Dominique Biloghi, Guy Le Thiec, eds., *Histoire et dictionnaire des Guerres de Religion* (Paris, 1998), pp. 60–67; Diefendorf, *Beneath the Cross*, pp. 55–56.

4 Michael Wolfe, "Walled Towns during the French Wars of Religion (1560–1630)," in *City Walls: The Urban Enceinte in Global Perspective*, ed. James D. Tracy (Cambridge, UK, 2000), pp. 317–348; Bernard Chevalier, *Les Bonnes Villes de France du XIVe au XVIe siècles* (Paris, 1982).

idly, prompting outraged Huguenots across the kingdom to mobilize and instigating the first in a series of religious wars.[5]

Throughout the French Wars of Religion (1562–1629), urban centers were repeatedly targeted by field armies and localized military forces. Huguenot and Catholic parties attempted to seize walled cities and towns in order to protect co-religionaries, permit worship services, and assert moral control over urban populations. The French Wars of Religion are especially remembered for urban massacres, such as the notorious Saint Bartholomew's Day Massacre of August 1572, which have been examined closely by Barbara Diefendorf, Denis Crouzet, Arlette Jouanna, and other historians.[6] Natalie Zemon Davis has famously interpreted urban massacres and other contemporary episodes of crowd violence as "religious riots."[7] However, the military operations organized with the objective of seizing urban centers have not received the same attention as the religious riots and massacres. Contemporaries referred to these military operations as *prises* (town seizures) – referring to the surprise attacks, uprisings, blockades, formal sieges, bombardments, and assaults against urban centers.

This chapter will analyze the patterns of town seizures and the language of *prises* during the French Wars of Religion, problematizing the concept of siege warfare by considering the diverse the military practices that Huguenots and Catholics employed as they attempted to control urban centers. The first half of the chapter examines the practices of urban warfare and the emergence of a language of *prises* in the early religious wars. The section begins with a close analysis of the First War of Religion (1562–1563), which established enduring patterns of town seizures, but also arguably created a new language for describing siege warfare within the kingdom. French military correspondence, siege narratives, and political pamphlets produced in the aftermath of this first religious war provide the main sources for examining the chaotic urban struggles and the descriptions of town seizures. I develop an analysis of these sources during the religious wars of 1566–1571, as the languages of *prises* coalesced. The

5 Mack P. Holt, *The French Wars of Religion, 1562–1629* (2nd ed., Cambridge, UK, 2005), pp. 44–49; Diefendorf, *Beneath the Cross*, pp. 56–63.

6 Allan A. Tulchin, *That Men Would Praise the Lord: The Triumph of Protestantism in Nîmes, 1530–1570* (Oxford, 2010); Barbara B. Diefendorf, *The Saint Bartholomew's Day Massacre: A Brief History with Documents* (Boston, 2009); Arlette Jouanna, *The St. Bartholomew's Day Massacre: The Mysteries of a Crime of State*, trans. Joseph Bergin (Manchester, 2007); Denis Crouzet, *La nuit de la Saint-Barthélemy: Un rêve perdu de la Renaissance* (Paris, 1994); Emmanuel Le Roy Ladurie, *Carnival in Romans*, trans. Mary Feeney (New York, 1980).

7 Natalie Zemon Davis, "Rites of Violence," in *Society and Culture in Early Modern France* (Stanford, 1975), pp. 152–187.

second half of the chapter focuses on several case studies of town seizures during specific phases of the religious wars: Montaigu and Cahors (1579–1580), Marseille (1588–1598), and Cheylard (1621). This approach allows an examination of continuities and subtle shifts in the patterns of siege warfare and in the strategies of narrating town seizures. I argue that the conditions of confessional conflict and civil warfare in France gradually transformed the practices of siege warfare and the language of *prises*.

2 Town Seizures and the Emergence of a Language of *Prises*

2.1 *Siege Warfare during the First Religious War (1562–1563)*

As religious warfare broke out in March 1562, Huguenot forces seized dozens of towns and major cities across France through political maneuvers and military attacks, cleansing churches and establishing Reformed municipal councils. Many of these communities had mixed-confessional populations with significant Calvinist minorities, allowing for political and military assistance for Huguenot attempts to seize control of cities. A Huguenot army gained control of the key city of Orléans, and then advanced up and down the Loire River to capture Beaugency, Blois, Tours, Angers, Sancerre, and La Charité-sur-Loire, as well as nearby Poitiers and Bourges. Huguenot troops under the baron des Adrets stormed Lyon in a night attack in April, pillaging Catholic churches and smashing the sacred images – an event described by Catholics as the sack of Lyon. Huguenots in Normandie seized Rouen, Caen, Dieppe, and Le Havre. Meanwhile, Calvinist forces in southern France secured control of Agen, Béziers, Montauban, Montpellier, Nîmes, Millau, and numerous small towns across Guyenne and Languedoc – sometimes through internal uprisings from within communities and at other times through external military attacks. Regional nobles played an important role in many of these town seizures by exerting influence from their châteaux and conducting localized military operations. When Huguenots seized towns, they installed Calvinist municipal leaders who often transformed religious spaces, closed monasteries, expelled Catholic clergy (and sometimes Catholic residents, too), and built new fortifications.[8] In the southeast, Huguenots took Valence, Mornas, Orange, Sisteron, and Gap. Meanwhile, the main Huguenot army led by Louis I de Bourbon, prince de Condé, and Gaspard de Châtillon, admiral de Coligny, threatened to blockade Paris in October 1562, creating panic in the capital.[9]

8 Janine Garrisson, *Protestants du Midi, 1559–1598* (new edition, Toulouse, 1991), pp. 167–170.
9 Jouanna et al., *Histoire et dictionnaire des Guerres de Religion*, pp. 110–120.

The town seizures by Huguenot forces galvanized a militant Catholic reaction throughout the kingdom in late 1562 and early 1563. Local Catholics and civic guards defeated urban uprisings by Huguenots in Bordeaux, Toulouse, and Aix-en-Provence. Catholic nobles responded to the Calvinist wave of seizures by blockading and besieging numerous towns and cities controlled by the Huguenots. Catholic forces along the Rhône retook Orange in June, brutally sacking the city. Catholic forces in the Loire valley reconquered Tours and Blois in July, massacring Calvinists and dumping their bodies in the river. Catholics then retook Poitiers and Bourges in August. Provençal Catholics besieged Sisteron, which fell in September. A Catholic army under Antoine de Bourbon, roi de Navarre, besieged Rouen in September, finally storming its defenses in late October and pillaging the city for three days.

The royal family organized a major field army, composed of Catholic troops, to mount a counteroffensive against Calvinist forces. The main Catholic and Huguenot armies fought an indecisive battle at Dreux in December 1562, in which both sides suffered heavy losses. The Calvinist prince de Condé was captured and the Huguenot army withdrew toward its base of operations at Orléans following the battle. The Catholic army was also seriously disorganized by the death of the maréchal Jacques d'Albon, seigneur de Saint-André, and the capture of the connétable Anne de Montmorency.[10] The main royal army, now under the command of the duc de Guise, advanced on the Huguenot stronghold of Orléans, investing the city in early February 1563. The Huguenot defenders of Orléans had reinforced the city's fortifications, however, and mounted a strong defense. As Catholic besieging forces prepared to launch an assault, a Calvinist noble assassinated the duc de Guise, shocking many contemporaries.

The death of Guise and the difficulties of conducting multiple sieges encouraged the royal family to conclude peace negotiations with the Huguenots, resulting in the compromise Peace of Amboise in March 1563. The mass public funeral of the slain duc de Guise in Paris nonetheless reinforced the capital's Catholic identity and demonstrated the strength of Catholic militancy.[11] The First War of Religion ended with the odd spectacle of a bi-confessional royal army besieging Le Havre, which was defended by a Huguenot and English garrison, following England's military intervention in support of their fellow Protestants.[12] Nonetheless, this first religious war established patterns of military

10 James B. Wood, *The King's Army: Warfare, Soldiers, and Society during the Wars of Religion in France, 1562–1576* (Cambridge, UK, 1996), pp. 184–204.

11 Diefendorf, *Beneath the Cross*, pp. 70–72.

12 See: Brian Sandberg, "Charles IX comme Roi très chrétien: L'image du roi et le problème de l'hérésie," unpublished paper presented at *Miroirs de Charles IX: Images, Imaginaires, Symboliques*, Institut National d'Histoire de l'Art, Paris, November 2011.

mobilization, town seizures, defensive fortification, siege warfare, and negotiations that would endure throughout the long French Wars of Religion.

2.2 *Siege Narratives as Sources (1560s–1570s)*

Printers responded to the religious warfare of 1562–1563 by publishing polemical pamphlets, many of which contained accounts of military campaigns and sieges.[13] As the religious "troubles" extended into a series of Wars of Religion stretching from 1562 to 1629, hundreds of siege narratives were published as printed pamphlets, providing key sources on the practices and perceptions of siege warfare in early modern France. These narratives reported a range of military operations against urban centers, including surprise attacks, uprisings, blockades, sieges, and assaults. I will treat all of these narratives as "siege narratives" because of their focus on the objectives of seizing towns and their use of the language of *prises*.

Authors and editors of printed siege accounts often signaled a narrative approach by entitling their works as a *discours, histoire, rélation,* or *récit* of a particular surprise attack, blockade, or siege. Other texts' titles focused on descriptions or phases of siege warfare: *entreprise, sommation, état du siège, assaut, prise,* or *réduction.* Lengthy formal sieges often prompted the publication of multiple editions of siege narratives, with descriptions of new siegeworks, the latest sorties, and preparations for assaults simply added onto the text of a previously printed pamphlet. Research by Michael Wolfe, Jeffrey Sawyer, Luc Racaut, and Amy Houston on siege narratives and print propaganda during the religious wars has revealed some of the representational techniques and polemical strategies used by authors in depicting siege warfare.[14] These studies tend to analyze printed siege narratives in the context of polemical literature

13 [Anonymous], *Histoire comprenant en brief ce qui est advenu depuis le partement des Sieurs de Guise, connestable, & autres, de la Cour, estant à Sainct Germain, jusques à ce temps* (Orléans, 1562); Antoine du Plain, *Cantique nouveau: contenant le discours de la guerre de Lyon, de l'assistance que Dieu a faite à son Eglise audit lieu, durant le temps de son affliction en l'an 1562 ; plus un cantique spirituel* (Lyon, 1563).

14 On siege narratives, see: Amy Houston, "Defending the City, Defending the Faith: The Sieges of the French Civil Wars, 1552–1628" (Ph.D. dissertation, Harvard University, 2010); Michael Wolfe, "Writing the City Under Attack During the French Wars of Religion," in *Situazioni d'Assedio / Cities Under Siege / États de Siege,* ed. Lucia Carle, Antoinette Fauve-Chamoux (Montalcino, 2002), pp. 197–203; Wolfe, "Walled Towns during the French Wars of Religion (1560–1630)," pp. 317–348. On printed pamphlets and propaganda, see: Hélène Duccini, *Faire voir, faire croire: l'opinion publique sous Louis XIII* (Seyssel, 2003); Luc Racaut, *Hatred in Print: Catholic Propaganda and Protestant Identity During the French Wars of Religion* (Aldershot, 2002); Jeffrey K. Sawyer, *Printed Poison: Pamphlet Propaganda, Faction Politics, and the Public Sphere in Early Seventeenth-Century France* (Berkeley, 1990).

or "pamphlet wars", suggesting that siege narratives did not really form a distinct literary genre, since they were often inserted into religious pamphlets, political pamphlets, civic histories, and other printed accounts.

However, these printed siege narratives have not been adequately studied in conjunction with the many extant manuscript accounts of sieges. Thousands of additional siege accounts are embedded in manuscript correspondence, instructions, *mémoires*, capitulations, municipal histories, city council deliberations, military reports, and logistical documents held in national, regional, and municipal archives in France. Some of these manuscript sources provide diverse perspectives on well-known town seizures and sieges that were covered by the authors of printed pamphlets. Other manuscript accounts relate numerous surprise attacks, blockades, and formal sieges of towns and cities during the religious wars. Examining various printed and manuscript siege accounts together allows us to consider siege narratives as a distinct genre of writing, with common descriptive elements, narrative tropes, stylistic conventions, and cohesive language.

I argue that interpreting the complexities of siege warfare during the French Wars of Religion requires a transversal approach to analyze these varied manuscript and printed sources in combination with one another. Recent research by Denis Crouzet, Nicholas Le Roux, Mark Greengrass, and others on religious politics during the reign of Henri III (1574–1589) suggests the possibilities of such an approach, as does Pierre-Jean Souriac's localized research on religious warfare in the Toulouse region.[15] James B. Wood has constructed a detailed study of the rhythms of siege warfare at the 1573 siege of La Rochelle based on diverse sources. A Catholic army under Henri de Valois, duc d'Anjou, carried out an intense 117-day siege of this key Calvinist stronghold, launching at least eight major assaults and losing almost 4,700 combat casualties before abandoning the siege.[16] We could usefully apply this method to other major sieges of the French Wars of Religion, but this effort is beyond the scope of the current chapter. Instead of reduplicating Wood's efforts here, I want to examine evidence of the practices of siege warfare through the language of *prises* in printed and manuscript narratives, focusing on depictions of attempted town seizures that are not as well-known as the siege of La Rochelle.[17]

15 Pierre-Jean Souriac, *Une guerre civile: Affrontements religieux et militaires dans le midi toulousain (1562–1596)* (Seyssel, 2008).

16 Wood, *The King's Army*, pp. 246–274.

17 This approach complements recent work on small war in early modern Europe. See: Simon Pepper, "Aspects of Operational Art: Communications, Cannon, and Small War," in *European Warfare, 1350–1750*, ed. Frank Tallett, D. J. B. Trim (Cambridge, UK, 2010), pp. 181–202.

2.3 *The Language of* Prises *in the Second and Third Religious Wars (1566–1571)*

French authors and combatants utilized the concepts of *prises* (seizures) to refer to attempts to capture towns and cities utilizing a varity of tactics, including surprise attacks, blockades, formal sieges, bombardments, and assaults. The related term *entreprises* (which can be translated as surprise or storm, but also as undertaking or initiative) could refer to military operations in general, but was normally used more narrowly to describe surprise attacks, stratagems, direct assaults, and negotiated capitulations. As the religious wars expanded, a growing number of political and religious pamphlets circulated news of town seizures. Exploring the conceptual language of *prises* reveals the practices of siege warfare during the religious wars.

The early phases of the religious wars created a pattern of "flipping" religio-political control of towns, producing a culture of fear of *prises*. Surprise attacks and assaults, which were sometimes described as *prises par force*, accompanied outbreaks of religious conflict throughout the entire period of the religious wars. Larger armies organized formal sieges of major cities, towns, fortresses, and châteaux during most campaigning seasons. The violence and disorder of town seizures and sieges produced lasting fears that linked conspiracy, treason, sedition, and heresy in early modern French political culture. The language of *prises* could be used to celebrate one's own conquest of an enemy town, to narrate a town seizure in a neutral fashion, or to accuse opponents of illegitimately seizing a city and associate them with the twin crimes of heresy and rebellion (*lèse-majesté*).

Appreciating the violence of these siege operations can extend Penny Roberts suggestion that contests between Catholics and Calvinists over control of religious sites represented a central aspect of the religious wars.[18] Successful seizures established control of urban spaces and populations, permitting access to religious sites. Calvinists who took control of predominantly Catholic or mixed-religious towns normally aimed to "cleanse" existing churches through iconoclasm and "reform" them into God's churches.[19] Calvinists often expelled Catholic clergy and closed monasteries, oratories, and shrines, which they regarded as offensive. Meanwhile, Catholics sought to defend or restore religious sites to the Roman, Apostolic, and Catholic religion – a complex pro-

---

18 Penny Roberts, "The Most Crucial Battle of the Wars of Religion? The Conflict over Sites for Reformed Worship in Sixteenth-Century France," *Archiv für Reformationsgeschichte* 89 (1998), 247–267.

19 For a contemporary account, see: Claude de Sainctes, *Discours sur le saccagement des églises catholiques, par les hérétiques anciens, & nouveaux calvinistes, en l'an mil cinq cens soixantedeux* [...] (Paris, 1563).

cess that often involved cleansing, reconstruction, and reconsecration. Châteaux could also be considered religious sites, since they normally contained chapels that could be used as places of worship. Catholic and Calvinist nobles thus often acted as key protectors of their co-religionaries by allowing them to use their personal chapels for worship services and devotions, especially in the religiously mixed regions of southern France.

The *prises* of the religious wars thus did not simply represent political "regime changes." A change in the military control of a town could result in sweeping religious and social transformations that could disrupt everyday urban life and reorient civic identities – especially in *mi-partie* communities that had mixed-confessional populations. Siege warfare allowed occupying armies to impose permanent garrisons, conduct municipal elections, establish political control, expel or convert members of the opposing confession, and enforce a new religious settlement. *Prises* often resulted in executed or banished municipal officers, divided elite clienteles, exiled bishops and preachers, harassed parish clergy, resettled judges and law courts, and numerous religious refugees.

Prises and defenses of towns seem to have been key components in the strategies that religio-political *partis* used to advance their confessional agendas and political goals. The terms of truces and religious peaces granted certain *places de sûreté* (or security towns) to each *parti* or clientele. Huguenots also negotiated for specific *lieux de culte* (worship sites), where Calvinist practice would be permitted. The Huguenot *places de sûreté* have been studied somewhat, but little work has been done on the broader role of security towns.[20] Fear of surprise attacks sometimes led to the construction of citadels outside of city walls and the installation of military garrisons at many towns. The building of citadels posed a serious problem for debate in confessionalized political culture, however. New fortification construction was often cited in pamphlets and correspondence as a violation of religious peaces. City councilors and provincial officials often called for the destruction of fortifications that had been recently built by their confessional opponents.[21]

Noble correspondence and military treatises discussed in detail the problem of conducting *entreprises* and defending against them. François de Bonne, duc de Lesdiguières, insists in his military treatise on the need for speed when launching *entreprises,* although his use of the term implies all sorts of surprise

20 Pierre-Jean Souriac, "Les places de sûreté protestantes: Reconnaissance et déclin de la puissance politique et militaire du parti protestant (1570–1629)" (Mémoire de Maîtrise, Université de Toulouse II-Le Mirail, 1997).

21 For an example, see: Consuls d'Aix-en-Provence to Jean-Louis de La Valette duc d'Épernon, Aix-en-Provence, 4 February 1594, Bibliothèque nationale de France (BNF), Dupuy 62, f° 72–73.

attacks.[22] Treatises on fortification and siegecraft produced during the French Wars of Religion discussed diverse aspects of siege warfare – including surprise attacks, blockades, *pétards*, disease, logistics, approach trenches, battery placement, mining, assaults, defensive fire, and countermining. For example, Antoine de Ville's 1628 treatise on fortifications divides his discussion of siege attacks into "attacks by surprise" and "attacks by force." In his classification, "attacks by surprise" include treasonous acts, seditious movements, surprises, escalades, and directs assaults on gates using *pétards* and other equipment. In contrast, Ville's long section on "attacks by force" deal with formal siege operations. When discussing surprise attacks, Ville provides advice on how to select appropriate targets for surprise attacks and how to win over garrisons through treachery.[23]

The language of *prises* could be used polemically to brand opponents as traitors and criminals. An anonymous Catholic pamphlet published in 1568 attacks Gaspard de Châtillon, admiral de Coligny, one of the principal Huguenot military leaders, precisely by criticizing his *entreprises*. The author condemns Coligny's *entreprises* as treacherous, expressing doubt on the efficacy of Colingy's leadership: "I do not see that the *entreprises* of the admiral will succeed (which God will never permit)."[24] The pamphlet labels the admiral "the principal chief of the rebellion" and a "sworn enemy of the king and his people."[25] The admiral's armed *parti* is described as full of "desperate men and malcontents."[26] The incensed pamphleteer complains that the Huguenots re-

22 The original reads: "Il est à notter qu'il ne faut jamais dilayer à faire voz affaires en ce qui est de la guerre, et faut user de dilligence et secret à faire les entreprises ou surprendre son ennemy et ne rien entreprendre de difficille soubz espoir de chose plus grande sans estre pourveu de tous cas necessaires." François de Bonne, duc de Lesdiguières, "Discours de lart militaire faict par monseigneur le connestable de Lesdiguieres," BNF, Mss. fr. 23042, f° 40. [Published in *Actes et correspondance du connétable de Lesdiguières*, ed. Louis Archambaud Douglas, J. Roman (Grenoble, 1878).]

23 The author focuses on examples from the ancient world and from the Dutch Revolt, rather than from the French Wars of Religion. Antoine de Ville, *Les Fortifications du chevalier Antoine de Ville Tholosain, avec l'ataque & defence des places* (Lyon, 1628).

24 The original reads: "je ne voy point que quand bien les entreprinses dudit Amiral reüssiroient (ce que Dieu ne permettra jamais)." [Anonymous], *Responce a un certain escrit, publié par l'Admiral et ses adherans, pretendans couurir & excuser la rupture qu'ils ont faite de l'Edict de Pacification, & leurs nouveaux remuemens et entreprises contre l'Estat du Roi, le bien & repos de ses subjectz* (Paris, 1568).

25 "Depuis le commencement des troubles nous advouons & recongnoissons ledit Amiral le principal chef de toute la rebellion, autheur & conducteur des factieux, ennemy conjuré du roy & de son peuple, & perturbateur du repos public." [Anonymous], *Responce*.

26 The original reads: "il esblouit les yeux des plus liguez & associez à son party, remis pour la troisiesme fois a la campagne, ralliant tous les desesperez & malcontens, remuant la terre & les enfers pour confondre & renverser cest estat." [Anonymous], *Responce*.

mained unpunished, even though they had raised troops against the king and "have attempted three or four times to kill the king, his brothers, and the queen, their mother," referring to failed surprise attacks. The Huguenot militants are referred to as "assassins" who "have sold the king's towns and gates to foreign and domestic enemies of the crown, inviting them into communities, to let these foreign harpies pillage and ravage the entire kingdom."[27] The author thus effectively links Coligny's alleged *entreprises* with his use of mercenary soldiers – who were routinely accused by contemporaries of pillaging, rape, and murder. The pamphlet concludes by complaining that the Huguenots have been pardoned for "murder" and for all the "oppressions and cruelties" that they have committed, alluding to the violent seizures of Amboise, Orléans, and Nîmes.[28]

Narratives of *prises* could have lasting significance well beyond their immediate polemical contexts. Jérémie Foa's recent work on the *surprise* of Besançon in 1575 brilliantly reconstructs how local Catholics circulated manuscript accounts and employed processions to keep the memory of the "treason" and "pillage" of the town by Huguenots alive.[29] The language of *prises* that described military operations intended to seize towns thus comprised an array of concepts and practices that included blockades, raids, internal uprisings, stratagems, surprise attacks, direct assaults, and formal sieges.

27 The original reads: "& apres qu'ils ont entrepris par trois & quatre fois d'exterminer le roi, Messeigneurs, & la Roine leur mere, & que le May & tels autres assassins ont a l'article de la mort, nommé & reue le ceux desquels ils estoient attiltrez & subornez, apres que par une long licence ils se sont permis de vendre les villes & portes du roy, aux naturels & anciens ennemis de la couronne, les introduisant aux plus interieures parties d'icelles, de faire piller & ravager tout le royaume par des harpyes estrangeres, & apres que le meurdre des Princes, lieutenans generaux, officiers, & magistrats de justice, & pasteurs de l'Eglise, & tous autres excez leur ont esté remis & pardonnez, & que iusques a ce jourd'huy le peuple a faict joug a leurs oppressions & cruautez, c'est l'ors que par maniere de risee & mocquerie, ils se plaignent d'auoir esté rudement traictez du roy." [Anonymous], *Responce.*

28 [Anonymous], *Responce.*

29 Jérémie Foa, "La Polémique autour de la surprise de Besançon (1575)," unpublished paper given at *Usages et stratégies polémiques en Europe au temps de l'humanisme du* XIVe *au milieu du* XVIIe *siècle,* Centre Roland Mousnier, Université de Paris IV, 13–15 February 2014.

3 Case Studies of Town Seizures during the Religious Wars

3.1 *The* Prises *of Montaigu and Cahors during the* Guerre des Amoureux (*1579–1580*)

The multilayered language of *prises* was employed by Huguenots as well. The correspondence of Henri de Bourbon, roi de Navarre, provides intriguing descriptions of *prises* during the Seventh War of Religion (1579–1580), also known as the *Guerre des Amoureux*. Henri de Navarre had maintained a neutral stance at the outset of renewed fighting in 1579, but he refused to transfer Huguenot *places de sûreté* to royal control, as required by the Peace of Bergerac.[30] Navarre finally took up arms in April 1580, justifying his mobilization through letters and a printed pamphlet.[31]

In one of his letters, the roi de Navarre emphasizes that Catholic attacks on Calvinist-controlled towns compelled him to mobilize. Navarre writes that "the misbehaviors, tricks, enterprises, surprises, robberies, massacres, injustices, and all sort of violations that the enemies of this State and of the public peace and tranquility have engaged in since the edict for the peace and Conference of Nérac can serve as my defense."[32] Here, the roi de Navarre distinguishes between *prises*, which he seems to regard as legitimate responses to his enemies' illegitimate acts – which include *entreprises* and *surprises*. Navarre went on to specify that "the enemies of this State [...] have seized [the towns] we have abandoned, armed the places that we have disarmed, closed those that we have opened, surprised the others that were no longer guarded, chased out those that [the towns] had received, killed or murdered those who were defenseless."[33] Finally, he lists the Huguenot communities that have recently

30 Ronald S. Love, *Blood and Religion: The Conscience of Henri IV* (Montreal, 2001), pp. 112–113.

31 Henri de Bourbon roi de Navarre to "messieurs de la noblesse," 15 April 1580, in *Recueil des lettres missives de Henri IV*, vol. 1, ed. Berger de Xivrey (Paris, 1843–1876), pp. 288–295.

32 The original reads: "Les desportements, artifices, entreprises, surprises, voleries, massacres, injustices, toutes especes de contraventions dont les ennemis de cest Estat et du repos et tranquillité publique ont usé depuis l'edict de la paix et conference de Nerac, me peuvent servir de deffense." Henri de Bourbon roi de Navarre to "messieurs de la noblesse," 15 April 1580, in *Recueil des lettres missives de Henri IV*, 1:288–295.

33 The original reads: "les ennemis de cest Estat, impatiens de notre ayse et repos, se sont saisis de ce que nous avons delaissé, armé les places que nous avons desarmées, fermé celles que nous avons ouvertes, surpris les aultres qui n'estoient plus guardées, chassé dehors ceulx qui les ont receues, tué ou meurtry ceulx qui n'estoient plus en deffence." Henri de Bourbon roi de Navarre to "messieurs de la noblesse," 15 April 1580, in *Recueil des lettres missives de Henri IV*, 1:288–295.

suffered violence at the hand of Catholic forces.[34] Henri de Navarre embraced his role as protector of his Calvinist co-religionaries and called for justice for all of these "criminal" attacks.[35]

As Navarre assembled a field army, a Huguenot noble and his clients in Poitou captured several Catholic soldiers from the garrison at the château de Montaigu, and then coerced them to assist the Huguenots in gaining entrance to the château around 15 March 1580. Soon thereafter, the Huguenots were able to extend their control over the adjacent town of Montaigu through another attack.[36] These surprise attacks reveal the potential for stratagems to succeed in taking towns during the religious wars.

Henri de Bourbon, roi de Navarre, refers to the taking of Montaigu in a letter written to François de Bourbon, duc de Montpensier, less than a month after the surprise attack on the château. "Regarding the *prise* of Montaigu, there is nothing strange," Navarre claimed. He justifies the seizure of Montaigu by complaining that "what has been taken and usurped from me has not yet been restored."[37] Navarre then lists his châteaux that have been seized by Catholic troops: "my château de Montignac-le Comte, where all of my titles and papers concerning my comté de Périgord and vicomté de Limoges are located, [and] my château de Montron, which have been detained for several years, [as well as] my château de Puy-Normand, which was taken by surprise five or six

34 The original reads: "Villeneufve d'Agenés, qui fut repris incontinent, en peut rendre bonne preuve; et la ville d'Agen, la ville de Leuserte, la ville de Langon, et les inhumanitez plus que barbares qui ont esté commisses trois jours aprés la conference, sans aulcun respect de la presence de la Royne, mere du Roy, qui tant avoit pris de travail pour icelle; la prise de La Reolle et celle de Montagnac, celle de Figeac, aussy faicte contre la foy publicque et respectivement donnée par les habitans d'une et d'aultre religion ; l'entreprise sur Pamiers par mille ou douze cens hommes de guerre, à descouvert ; la prinse de Soureze et le massacre commis de sang froid ; le forcement des femmes et filles, et de nouveau la revolte du gouverneur de la Reolle pratiquée par les dicts ennemis." Henri de Bourbon roi de Navarre to "messieurs de la noblesse," 15 April 1580, in *Recueil des lettres missives de Henri IV*, 1:288–295.

35 Henri de Bourbon roi de Navarre to "messieurs de la noblesse," 15 April 1580, in *Recueil des lettres missives de Henri IV*, 1:288–295.

36 Théodore Agrippa d'Aubigné recounted the seizure of Montaigu in his *Histoires* and in his *Mémoires*. Théodore Agrippa d'Aubigné, *Histoire Universelle, Tôme Sixième, 1579–1585*, vol. IX, ed. Alphonse de Ruble (Paris, 1892), pp. 1–8; Théodore Agrippa d'Aubigné, *Mémoires de Théodore Agrippa d'Aubigné*, ed. Ludovic Lalanne (Paris, 1854), pp. 488–490.

37 The original reads: "De la prise de Montagu, il ne la fault trouuer estrange. Ce qui a esté pris & usurpé sur moy ne ma encor esté rendu." Henri de Bourbon roi de Navarre to Louis II de Montpensier, duc de Montpensier, Nérac, 12 May 1580, BNF, Mss. fr. 3344, f° 27–28. A different copy of this letter is published in Xivrey, *Recueil des lettres missives de Henri IV*, 1: 300–302.

months ago."[38] The language of *prises* could thus justify one's own seizures of town or châteaux, especially when justified as retaliation for previous town seizures by one's enemies.

Navarre then alleges that the Catholic attacks on his châteaux are part of a broader pattern of unjust seizures of Calvinist towns. He complains of "the enterprises and surprises that [the Catholics] want to perform on every [place] that belongs to me and on all that those of the Religion hold."[39] The roi de Navarre simultaneously asserts his seigneurial privilege, his noble status, his prerogative as army commander, and his position as Huguenot leader. However, at the same time, Navarre reassures the duc de Montpensier that he is open to discussing a ceasefire and that he is willing to include the status of Montaigu in any negotiations that might ensue.[40]

Soon after writing this letter, Henri de Navarre led his field army to besiege Cahors, a strategically important town in Guyenne, on 28 May 1580. Navarre's troops stormed Cahors three days later, on 1 June, leading Henri de Navarre to describe the *prise* as "miraculous, because after having become master of one part [of the town], we had to acquire the rest of it foot by foot, barricade by barricade."[41] The taking of Cahors and other towns forced new peace negotiations, which produced the Peace of Fleix in November 1580. Philippe Duples-

38 The original reads: "comme mon chasteau de Montignac le Conte, auquel sont tous mes tiltres & papiers, qui concernent mon conté de Perigort & viconté de Limoges, mon chasteau de Montron, qui sont detenuz y a plusieurs annees, mon chasteau de Puynormand, surpris depuis cinq ou six moys." Henri de Bourbon roi de Navarre to Louis II de Montpensier, duc de Montpensier, Nérac, 12 May 1580, BNF, Mss. fr. 3344, f° 27–28. A different copy of this letter is published in Xivrey, *Recueil des lettres missives de Henri IV*, 1:300–302.

39 The original reads: "Les entreprises & surprises qu'on a voulu generalem^t f.e sur tout ce qui may.tient [m'appartient?] et sur ce que ceulx de la religion tiennent, dont jay fait une sommaire delcaraõn a mess^rs de la noblesse, laquelle je vous enuoye, attendant une plus ample & particuliere que jespere verrez bien tost." Henri de Bourbon roi de Navarre to Louis II de Montpensier, duc de Montpensier, Nérac, 12 May 1580, BNF, Mss. fr. 3344, f° 27–28. A different copy of this letter is published in Xivrey, *Recueil des lettres missives de Henri IV*, 1:300–302.

40 The original reads: "Ce pendant je ne doubte point que ceulx dud^t Montagu ne facent la guerre, et qu'on ne la leur face aussi. Mais sil se faict quelques suspension darmes, comme il a pleu a monsieur den faire f.e ouuerture, je seray trescontent que lad^t place y soys comprise." Henri de Bourbon roi de Navarre to Louis II de Montpensier, duc de Montpensier, Nérac, 12 May 1580, BNF, Mss. fr. 3344, f° 27–28. A different copy of this letter is published in Xivrey, *Recueil*, 1:300–302.

41 The original reads: "Je croy que vous aurés esté bien esbahi de la prise de ceste ville; elle est aussy miraculeuse, car aprez avoir esté maistre d'une partie, il a fallu acquerir le reste pied à pied, de barricade en barricade. Et puis que Dieu m'a faict la grace de l'avoir, je desire la conserver, et y establir quelque beau reglement." Henri de Bourbon roi de Navarre to monsieur Scorbiac, Cahors, 1 June 1580, in *Recueil des lettres missives de Henri IV*, 1:304.

sis-Mornay claimed that Navarre's troops immediately relinquished Montaigu, Cahors, Saint-Émilion, and Mende, as required by the peace, "even though the forces of His Majesty remained in Dauphiné against those of the Religion [meaning the Huguenots]."[42] For Huguenots such as Henri de Navarre, then, *prises* were justified to liberate Calvinist populations suffering from persecution. Calvinists clearly were able to utilize the concept of town seizures in their own polemical language and political writings.

3.2 *Marseille and the Wars of the Catholic League* (*1588–1595*)

Marseille offers fascinating evidence of the changing dynamics of *prises* during the League wars of the late sixteenth century. The port city had a diverse religious and ethnic population that had already experienced waves of urban violence during the earlier phases of the religious wars. The rise of a series of localized Catholic Leagues in the late 1570s intensified Catholic religious politics, creating new divisions within France's broad Catholic majority population between Leaguers and *politiques* (Catholic moderates). The wars of the Catholic League in the 1580s and 1590s made attempts at controlling cities even more confusing for communities like Marseille that were trying to weather the storms of religious violence.[43]

Following the assassinations of Henri I de Lorraine, duc de Guise, and his brother at the Estates of Blois in 1588, numerous cities and towns declared their support for the Catholic League. Additional communities joined the next year, after the assassination of Henri III made Henri de Navarre, who was already the pre-eminent Calvinist leader, the heir presumptive to the throne of France. Many Leaguer municipal leaders openly recognized the cardinal Charles de Bourbon as king Charles X and swore public oaths of allegiance to the Catholic League. The religio-political dynamics shifted and radicalized, allowing new opportunities for *prises* and *entreprises* at Anjou, Tours, and Orléans in 1588, and then numerous town seizures across southern France in 1589. Histories of the French Wars of Religion normally focus on the siege of Paris in 1590–1591 as defining siege warfare in this phase of the religious wars, but the smaller surprise attacks and town seizures were also highly significant to the ongoing military operations.

---

42 Philippe Duplessis-Mornay, quoted in Hugues Daussy, *Les huguenots et le roi: le combat politique de Philippe Duplessis-Mornay, 1572–1600* (Geneva, 2002), p. 216.

43 On the League wars in Marseille and southeastern France, see: Fabrice Micallef, *Un désordre européen: La compétition internationale autour des « affaires de Provence »* (*1580–1598*) (Paris, 2014); Stéphane Gal, *Grenoble au temps de la Ligue: Étude politique, sociale et religieuse d'une cité en crise* (*vers 1562 – vers 1598*) (Grenoble, 2000); Le Roy Ladurie, *Carnival in Romans*.

The Provençal countryside surrounding Marseille became severely divided between Catholic Leaguers, Calvinists, *politique* supporters of Henri IV, and other *partis*. In 1585, Charles de Casaulx, who was captain of the quarter of La Blanquerie in Marseille, had attempted to seize control of the city for the Catholic League. This internal uprising failed and Casaulx was forced to flee to Aix-en-Provence, a nearby Leaguer center. When a Savoyard army intervened to support the Catholic League in Provence in 1590, Marseillais Leaguers seized control of the city and recalled Charles de Casaulx to lead the municipal government of Marseille. A *parti* of city notables gradually coalesced around the religious politics of supporting Henri IV, leading their opponents to call them *bigarrats* (a pejorative term that was applied to local *politiques*). When the *bigarrats* managed to gain control of Marseille's municipal government again in October 1590, Casaulx and his supporters organized a new *coup d'état* in February 1591.

Marseille civic leaders seem to have attempted to establish a distinct political position of neutrality in the chaotic League wars. An attempted surprise attack on Marseille by Charles-Emmanuel I de Savoie, duc de Savoie, in the summer of 1591 failed, leading to a brief siege of Marseille by Savoyard forces in late 1591. Casaulx was elected *premier consul* in October 1592, seemingly solidifying his control over the Marseillais municipal government. Gaspard de Pontevès, comte de Carcès, and Leaguer troops again attempted to take Marseille by surprise in August 1592, producing a day of clashes known as the *journée des brûlats*. Later, Jean-Louis de La Valette, duc d'Épernon, and his forces attacked Marseille in April 1593. As a result of all this violence, a number of Marseillais merchants relocated to the ports of Toulon and La Ciotat, while other residents fled to Aix-en-Provence and other Provençal towns in the interior. Dominican monks apparently attempted to assassinate Casaulx at some point in this period.

Henri IV's abjuration in July 1593 increased the political pressure on towns that were openly supporting the Catholic League or were attempting to maintain neutrality. Spanish troops arrived to support Marseille in 1595, as the city conducted negotiations with Philip II and embraced a more extreme Catholic religious politics.[44] Charles de Lorraine, duc de Guise, son of the victim of the 1588 assassination at Blois discussed above, then became provincial governor of Provence in October 1595 and decided to launch a siege operation to take Marseille. Guise was able to win the loyalties of Pierre Bayon de Libertat, a noble who commanded the porte Réale, the most important gate in Marseille. Libertat assassinated Charles de Casaulx as Guise's troops approached the city.

44 Micallef, *Un désordre européen*, pp. 235–246.

The next day, Libertat allowed the duc de Guise and his troops to enter Marseille.[45] A surprise attack with internal support thus permanently ended Marseille's Leaguer phase and ushered in a new *politique* regime friendly to Henri IV.

In the late 1590s, the Catholic League organization began to fragment and Leaguer cities' ability to mount effect siege defenses gradually degraded. Henri IV's powerful forces, composed of Huguenot and *politique* soldiers, conducted numerous sieges to subdue Leaguer cities across France, even as reinforced royal garrisons elsewhere increasingly made new surprise attacks by Leaguers difficult. Historians often set these sieges into the context of royal authority and "absolutist" state-building. For example, Annette Finley-Croswhite stresses that "The Wars of Religion created a situation in which the king had to placate the towns in order to pacify France. Yet in rebuilding a royal alliance with the towns, Henry IV also took every opportunity to strengthen his royal authority."[46] Indeed, the erosion of Leaguer military strength seems to have been largely due to the inability of Leaguer forces to seize new towns or to organize effective siege defenses against Henri IV's armies in the late 1590s. Henri IV's siege of Amiens in 1597 finally forced Leaguer and Spanish forces to accept Henri IV's royal status and negotiate the Peace of Vervins and the Edict of Nantes in 1598. Marseille's experience of *prises* during the League wars was somewhat anomalous, but still fit into the broader context of military interventions and campaigns of submission across the kingdom.

3.3 *Cheylard during the Latter Stages of the Religious Wars (1621)*

Although the Edict of Nantes diminished full-scale religious warfare for a time, localized religious violence and attempted town seizures continued well into the seventeenth century. Religious tensions remained serious around the scattered Huguenot *places de sûreté* and in the mixed-confessional provinces of Languedoc and Guyenne, as well as a few predominantly Huguenot towns in the Loire valley. In these regions, nobles continued to launch small-scale military operations to attempt surprise attacks long after the promulgation of the Edict of Nantes, which was never fully implemented. Localized conflicts occurred as mixed-religious communities and regions grappled with the complicated provisions of the Edict, which called for even predominantly Calvinist communities to re-establish the Catholic mass in parish churches and to

---

45 Béatrice Hénin, "Le dictateur Casaulx assassiné," in *Histoire de Marseille en treize événements*, ed. Philippe Joutard (Saint-Etienne, 1988), pp. 86–101.

46 Annette Finley-Croswhite, *Henry IV and the Towns: The Pursuit of Legitimacy in French Urban Society, 1589–1610* (Cambridge, UK, 1999).

permit access to Catholic religious orders. Then, following the assassination of Henri IV in May 1610, new episodes of larger-scale civil conflict and religious warfare erupted across southern France. Despite a gradual erosion of Huguenot military power in the early seventeenth century due to conversions to Catholicism, Calvinist forces were still able to seize control of a number of towns in southern France during the 1614–1615 civil conflict, the 1617 civil war, and the 1618–1620 Wars of Mother and Son.[47]

Political pamphleteers began to describe town seizures in new ways during the religious wars of the 1620s. After Louis XIII occupied Béarn in 1620 and restored Catholic practice there, alarmed Calvinists across France formed new political unions and mobilized significant military forces. Many Calvinist-majority towns began constructing new bastioned fortifications and expelled their Catholics residents. Meanwhile, Huguenot militants attempted to seize control of contested *mi-partie* communities. The pamphlet *La trahison descouverte … de la ville de Cheylar* (1621) represents an interesting example of a narrative of an attempted *prise* from the latter phases of the religious wars in France.[48]

La trahison descouverte is a printed pamphlet in the form of a letter by Anne de Lévis, duc de Ventadour, who served as *lieutenant-général* in the province of Languedoc. The pamphlet was published in 1621, soon after the outbreak of a new religious war, which would engulf most of the 'Huguenot Crescent,' across southwestern France, where Calvinists and Catholics lived in close proximity to each other – sometimes in mixed-confessional communities. Cheylard (or Cheylar) was a small town with a predominantly Calvinist population located in the Vivarais region of Languedoc, where the Catholic Lévis family held significant influence. According to the pamphlet, around midnight, the Reformed inhabitants had "introduced the enemies of the king, disturbers of the public peace into our said town of Cheylar[d]." The Calvinist militants then besieged the adjoining château: "trying during two days and two nights to take by force our château of the said Cheylard with petards, ladders, mantlets, and other siege weapons." However, the governor of the château, the sieur du Bourg, successfully defended it against the Calvinist attack, forcing them to retreat.[49]

The duc de Ventadour claimed to be directly responsible for the garrison at the château du Cheylard and the sieur du Bourg was one of the duc de Ventadour's own clients, suggesting motives for the publication of this account in

47 Brian Sandberg, *Warrior Pursuits: Noble Culture and Civil Conflict in Early Modern France* (Baltimore, 2010).

48 Anne de Lévis, duc de Ventadour, *La trahison descouverte, et foy faussée de ceux de la religion pretendue réformée de la ville de Cheylar au pays de Viverois* […] (Lyon, 1621).

49 Lévis, *La trahison descouverte*.

order to celebrate the defense of the château. The duc de Ventadour accuses the Calvinist inhabitants of Cheylard of rebellion for having joined the Calvinist militants in their assault on the château and their iconoclastic attacks on the Catholic church within the town. Ventadour indicates that he has ordered that "the walls, gates, and towers of the said town will be entirely razed and demolished" in punishment for the rebellion. This was a rather typical response, since military forces that succeeded in seizing towns during the religious wars frequently ordered at least partial demolitions of bastioned fortifications.

While the duc de Ventadour was dealing with Cheylard, Louis XIII led a royal army to suppress Huguenot rebellion in Guyenne and Languedoc in 1621. Pamphleteers describing the military operations aiming to retake rebellious cities and towns championed these efforts as royal *entreprises* that were favored by God, especially when led by Louis XIII in person.[50] Siege narratives from this campaign emphasized the seizures of Clérac, Nérac, and Négrepelisse by Huguenots as justifications for the subsequent sieges by royal troops. At the same time, sources discussed a projected *entreprise* on La Rochelle in 1621 that never fully materialized that year.

Huguenots launched additional surprise attacks during the religious wars of the 1620s. Benjamin de Rohan, seigneur de Soubise, carried out several attacks described as *entreprises* in 1622. A Catholic garrison in Montpellier discovered the plans of Henri II de Rohan, duc de Rohan, for a surprise attack on the mixed-confessional city in 1628, thwarting Rohan's attempt to seize the city in an *entreprise*. The seizure of Chomérac was one of the final successful *prises* by Huguenots during the religious wars.[51] Local Catholic nobles and royal armies gradually seized Calvinist-dominated cities in southern France and eroded Huguenot military strength during the religious wars of the 1620s. The epic siege of La Rochelle in 1627–1628 reduced the largest Calvinist stronghold. By the end of 1629, Huguenot cities and military forces were no longer able to resist Catholic forces, bringing the religious wars to an end.

---

50 "Les heureux succez dõt les entreprises de sa Majesté sont favorisez de toutes parts, au desadvantage de ses ennemis, meritent d'estre publiez, pour entendre la loüange à Dieu qui est l'autheur de ses sainctes intentions, & le vengeur de ses rebelles, et pour en donner l'honneur à ceux qui par le merite de leurs services, se l'acquierent justement." [Anonymous], *La prise du fort de Corconne, en Languedoc. Par Monsieur le marquis de Fossez, gouverneur de la ville & citadelle de Montpellier. Avec le recit veritable de ce qui s'est passé au restablissement des anciennes armes de ladite ville* (Paris, 1628), BNF, 8° Lb36 2589.

51 *La Prise de la ville de Chaumerac en Vivarests. Par Monseigneur le duc de Montmorency. Avec l'execution de six vingts des rebelles qui ont esté penduz à la veuë du Pousin. Et le pillage & bruslement du chasteau de Mauras, & autres maisons qui pouvoient favoriser le passage des rebelles de Privas audit Pousin* (Paris, 1628), BNF, 8° Lb36 2634.

4 Conclusion

Early modern siege warfare is often depicted as positional warfare, revolving around slow, methodical siege operations by a besieging army and a defending garrison. The many town seizures of the French Wars of Religion offer a different vision of siege warfare that was highly dynamic and unpredictable. Seizures of towns and cities – whether through surprise attacks or prolonged sieges – had the potential to radically transform communities and alter the religious destiny of the entire society. In the context of religious violence and civil war, the conduct of sieges involved much more than military engineering. The language of *prises* and *entreprises* continued to shift throughout the French Wars of Religion, as siege warfare practices and religio-political parties evolved. During the latter stages of the wars, royal propagandists increasingly praised Catholic *prises* as Godly acts, while condemning Huguenot seizures as seditious and rebellious. The town seizures of the religious wars ultimately destroyed Huguenot military resistance and ensured that France would restore its identity as a relatively unified Catholic kingdom.

Bibliography

Archival Sources

Bibliothèque nationale de France (BNF)

BNF, Dupuy 62.

BNF, Mss. fr. 3344.

BNF, Mss. fr. 23042.

Printed Sources

Actes et correspondance du connétable de Lesdiguières, ed. Louis Archambaud Douglas, J. Roman (Grenoble, 1878).

[Anonymous], *Responce a un certain escrit, publié par l'Admiral et ses adherans, pretendans couurir & excuser la rupture qu'ils ont faite de l'Edict de Pacification, & leurs nouveaux remuemens et entreprises contre l'Estat du Roi, le bien & repos de ses subjectz* (Paris, 1568).

[Anonymous], *La Prise de la ville de Chaumerac en Vivarests. Par Monseigneur le duc de Montmorency. Avec l'execution de six vingts des rebelles qui ont esté penduz à la veuë du Pousin. Et le pillage & bruslement du chasteau de Mauras, & autres maisons qui pouvoient favoriser le passage des rebelles de Privas audit Pousin* (Paris, 1628), BNF, 8° Lb^{36} 2634.

[Anonymous], *La prise du fort de Corconne, en Languedoc. Par Monsieur le marquis de Fossez, gouverneur de la ville & citadelle de Montpellier. Avec le recit veritable de ce qui s'est passé au restablissement des anciennes armes de ladite ville* (Paris, 1628), BNF, 8° Lb36 2589.

[Anonymous], *Histoire comprenant en brief ce qui est advenu depuis le partement des Sieurs de Guise, connestable, & autres, de la Cour, estant à Sainct Germain, jusques à ce temps* (Orléans, 1562).

Aubigné, Théodore Agrippa d', *Histoire Universelle, Tôme Sixième, 1579–1585*, vol. IX, ed. Alphonse de Ruble (Paris, 1892).

Lévis, Anne de, duc de Ventadour, *La trahison descouverte, et foy faussée de ceux de la religion pretendue réformée de la ville de Cheylar au pays de Viverois...* (Lyon, 1621).

Plain, Antoine du, *Cantique nouveau: contenant le discours de la guerre de Lyon, de l'assistance que Dieu a faite à son Eglise audit lieu, durant le temps de son affliction en l'an 1562 ; plus un cantique spirituel* (Lyon, 1563).

Recueil des lettres missives de Henri IV, vol. 1, ed. Berger de Xivrey (Paris, 1843–1876).

Sainctes, Claude de, *Discours sur le saccagement des églises catholiques, par les hérétiques anciens, & nouveaux calvinistes, en l'an mil cinq cens soixantedeux* [...] (Paris, 1563).

Ville, Antoine de, *Les Fortifications du chevalier Antoine de Ville Tholosain, avec l'ataque & defence des places* (Lyon, 1628).

Secondary Sources

Benedict, Philip, *Christ's Churches Purely Reformed: A Social History of Calvinism* (New Haven, 2002).

Benedict, Philip, "Confessionalization in France? Critical Reflections and New Evidence," in *Society and Culture in the Huguenot World, 1559–1685*, ed. Raymond A. Mentzer, Andrew Spicer (Cambridge, UK, 2002), pp. 44–61.

Charèyre, Philippe, "Démographie et minorités protestantes," *Bulletin de la Société de l'Histoire du Protestantisme Français* 148, no. 4 (2002), 867–889.

Chevalier, Bernard, *Les Bonnes Villes de France du XIVe au XVIe siècles* (Paris 1982).

Crouzet, Denis, *La nuit de la Saint-Barthélemy: Un rêve perdu de la Renaissance* (Paris, 1994).

Daussy, Hugues, *Les huguenots et le roi: le combat politique de Philippe Duplessis-Mornay, 1572–1600* (Geneva, 2002).

Davis, Natalie Zemon, "Rites of Violence," in *Society and Culture in Early Modern France: Eight Essays by Natalie Zemon Davis* (Stanford, 1975), pp. 152–187.

Diefendorf, Barbara B., *Beneath the Cross: Catholics and Huguenots in Sixteenth-Century Paris* (Oxford, 1991).

Diefendorf, Barbara B., *The Saint Bartholomew's Day Massacre: A Brief History with Documents* (Boston, 2009).

Duccini, Hélène, *Faire voir, faire croire: l'opinion publique sous Louis XIII* (Seyssel, 2003).

Finley-Croswhite, Annette, *Henry IV and the Towns: The Pursuit of Legitimacy in French Urban Society, 1589–1610* (Cambridge, UK, 1999).

Foa, Jérémie, "La Polémique autour de la surprise de Besançon (1575)," unpublished paper given at *Usages et stratégies polémiques en Europe au temps de l'humanisme du XIVe au milieu du XVIIe siècle,* Centre Roland Mousnier, Université de Paris IV, 13–15 February 2014.

Gal, Stéphane, *Grenoble au temps de la Ligue: Étude politique, sociale et religieuse d'une cité en crise (vers 1562–vers 1598)* (Grenoble, 2000).

Garrisson, Janine, *Protestants du Midi, 1559–1598* (new edition, Toulouse, 1991).

Hénin, Béatrice, "Le dictateur Casaulx assassiné," in *Histoire de Marseille en treize événements,* ed. Philippe Joutard (Saint-Etienne, 1988), pp. 86–103.

Holt, Mack P., *The French Wars of Religion, 1562–1629* (2nd ed., Cambridge, UK, 2005).

Houston, Amy, "Defending the City, Defending the Faith: The Sieges of the French Civil Wars, 1552–1628" (Ph.D. dissertation, Harvard University, 2010).

Jouanna, Arlette, *The St Bartholomew's Day Massacre: The Mysteries of a Crime of State,* trans. Joseph Bergin (Manchester, 2007).

Jouanna, Arlette; Boucher, Jacqueline; Biloghi, Dominique; Le Thiec, Guy, eds., *Histoire et dictionnaire des Guerres de Religion* (Paris, 1998).

Ladurie, Emmanuel Le Roy, *Carnival in Romans,* trans. Mary Feeney (New York, 1980).

Love, Ronald S., *Blood and Religion: The Conscience of Henri IV* (Montreal, 2001).

Micallef, Fabrice, *Un désordre européen: La compétition internationale autour des « affaires de Provence » (1580–1598)* (Paris, 2014).

Pepper, Simon, "Aspects of Operational Art: Communications, Cannon, and Small War," in *European Warfare, 1350–1750,* ed. Frank Tallett, D. J. B. Trim (Cambridge, UK, 2010), pp. 181–202.

Racaut, Luc, *Hatred in Print: Catholic Propaganda and Protestant Identity During the French Wars of Religion* (Aldershot, 2002).

Roberts, Penny, "The Most Crucial Battle of the Wars of Religion? The Conflict over Sites for Reformed Worship in Sixteenth-Century France," *Archiv für Reformationsgeschichte* 89 (1998), 247–267.

Sandberg, Brian, "Charles IX comme Roi très chrétien: L'image du roi et le problème de l'hérésie," unpublished paper presented at *Miroirs de Charles IX: Images, Imaginaires, Symboliques,* Institut National d'Histoire de l'Art, Paris, November 2011.

Sandberg, Brian, *Warrior Pursuits: Noble Culture and Civil Conflict in Early Modern France* (Baltimore, 2010).

Sawyer, Jeffrey K., *Printed Poison: Pamphlet Propaganda, Faction Politics, and the Public Sphere in Early Seventeenth-Century France* (Berkeley, 1990).

Tulchin, Allan A., *That Men Would Praise the Lord: The Triumph of Protestantism in Nîmes, 1530–1570* (Oxford, 2010).

Souriac, Pierre-Jean, *Une guerre civile: Affrontements religieux et militaires dans le midi toulousain (1562–1596)* (Seyssel, 2008).

Souriac, Pierre-Jean, "Les places de sûreté protestantes: Reconnaissance et déclin de la puissance politique et militaire du parti protestant (1570–1629)" (Mémoire de Maîtrise, Université de Toulouse II-Le Mirail, 1997).

Wolfe, Michael, "Walled Towns during the French Wars of Religion (1560–1630)," in *City Walls: The Urban Enceinte in Global Perspective,* ed. James D. Tracy (Cambridge, UK, 2000), pp. 317–348.

Wolfe, Michael, "Writing the City Under Attack During the French Wars of Religion," in *Situazioni d'Assedio / Cities Under Siege / États de Siege,* ed. Lucia Carle, Antoinette Fauve-Chamoux (Montalcino, 2002), pp. 197–203.

Wood, James B., *The King's Army: Warfare, Soldiers, and Society during the Wars of Religion in France, 1562–1576* (Cambridge, UK, 1996).

CHAPTER 11

Summary: Under Siege? Defining Siege Warfare in World History

Peter H. Wilson

This short chapter considers the extent to which the term 'siege' constitutes a valuable category of historical analysis, rather than simply a colloquial metaphor. Siege warfare is usually defined by reference to specific techniques used to defend and capture defined strategic points, usually fortresses and important settlements.[1] Little consideration is given to whether such definitions work for periods and places other than those to which they are immediately applied. In short, few have asked whether sieges, like battles, are culturally or temporally specific. This chapter intends to stimulate debate by drawing on examples from across the world, primarily during early modernity and the nineteenth century with a brief outlook to the present to ask whether this form of warfare still persists.

It argues for a distinction between use of the term 'siege' as a tool for historical analysis across time, space and cultures, and the term's historical usage in specific contexts. As Anke Fischer-Kattner points out in the Introduction, the rhetoric of a siege has its own history, related to, but also distinct from the military events to which it has been applied. We can only understand the significance of that rhetoric, as well as the conduct and impact of such operations (attack and defence of fixed positions), if we define what we mean by the term 'siege.' Historical analysis requires a definition that is sufficiently robust and consistent to permit comparisons across time and place, but not so rigid as to impose an ahistorical timeless category on what is always a messy and ambiguous historical record. Any definition is likely to remain imperfect, but the task of constructing one is nonetheless rewarding, since it sharpens our understanding of what distinguishes a siege from other kinds of fighting, including

1 Examples of technical definitions include Bernhard von Poten, ed., *Handwörterbuch der gesamten Militärwissenschaften*, 7 vols (Bielefeld and Leipzig, 1887), 1:159–160; Christopher Duffy, *Fire and Stone: The Science of Fortress Warfare 1660–1860* (Newton Abbot, 1975), pp. 94–101. For colloquial and popular usage see <http://www.oxforddictionaries.com/definition/english/siege> (accessed 23 October 2018); http://www.merriam-webster.com/dictionary/siege (accessed 23 October 2018); <http://www.macmillandictionary.com/dictionary/british/siege> (accessed 23 October 2018); <https://en.wikipedia.org/wiki/Siege> (accessed 23 October 2018).

 | DOI:10.1163/9789004395695_012

positional warfare more generally. It also helps identify what might be time, place and culture-bound about a siege and whether this form of warfare is disappearing even though the term remains current in popular usage.

I would like to start with a story which is deliberately picked from outside the timeframe of this volume. Sometimes, looking at the unfamiliar can throw new perspectives on the known. My story comes from the Mexican Revolution and concerns an event popularly known as the *Decena Tragica*, or Ten Tragic Days. The name is important, since it is neither called a siege nor a battle, yet has elements commonly associated with both, as well as other features.[2]

After the initial overthrow of Porfirio Diaz in 1911, the new liberal president, Francisco Madero, struggled to assert his authority over a variety of conservative and more radical opponents. On the morning of 9 February 1913, dissidents in the Federal Army sprung Bernardo Reyes and Felix Diaz, Porfirio's nephew, from the Santiago jail, where they had been incarcerated after an earlier rebellion. Reyes and Diaz then headed downtown, expecting little resistance, but General Lauro Villar, commanding the guard at the National Palace, refused to join the insurrection. Reyes was killed and Villar badly wounded in the ensuing firefight, before Diaz retired to the southwest of the city where he occupied the Ciudadela, the country's main arsenal. The rebels, who were now dubbed Felicistas, proceeded to open indiscriminate artillery fire on the city.

Alerted by loyal officers, Madero left his official residence on the outskirts and hurried to the National Palace, meeting Victoriano Huerta on the way. Huerta, the highest ranking general, was officially convalescing from an eye infection, but offered to take command in view of Villar's wounds. Having rallied government forces, Huerta surrounded the Ciudadela by blocking off the surrounding streets with barricades and occupying houses. Government forces then launched two days of attacks on the Felicistas using the *Rurales*, or Mexican Mounties who were mown down by machine gun fire, raising suspicions that Huerta was deliberately trying to remove troops known to be loyal to Madero. Both sides continued to shell each other ineffectively for eight more days. There are clear indications that neither wanted to hurt the other, and most of the shell fire hit civilians, killing over 1,000. Meanwhile, the US ambassador, Henry Lane Wilson, brokered talks between Huerta and the Felicistas, officially to end the confrontation, but really to facilitate a coup. With foreign backing seemingly secure, Huerta arrested Madero and vice president José

2 René De La Pedraja, *Wars of Latin America, 1899–1941* (Jefferson, 2006), pp. 178–182; Peter Calvert, *The Mexican Revolution 1910–1914: The Diplomacy of Anglo-American Conflict* (Cambridge, UK, 1968), pp. 131–156; Alan Knight, *The Mexican Revolution*, 2 vols (Cambridge, UK, 1986), 1:480–490.

Maria Pino Suárez at noon on 18 February. Madero was bullied into resigning the next day, allowing Huerta to become provisional president. Madero and Pino Suárez were then murdered three days later whilst ostensibly being escorted to safety by government troops. This action fatally compromised Huerta and ensured a resumption of widespread revolution which troubled Mexico into the early-1920s.

Though brief, this summary reveals the complexity of the events which included a successful jail break; a failed assault on a defended position (the National Palace); diplomatic double-cross; a coup; and political assassination. The complex operations around the Ciudadela certainly involved elements of what we might expect in a siege. It was not a fortress, but was certainly a defensible position. It was encircled by government forces who blockaded it, albeit deliberately ineffectively. The attackers also undertook other actions normally associated with a siege, including erecting their own protective fortifications, and bombarding and assaulting those of the defenders. Yet, all this took place within the wider enclosed space of Mexico City, while the ten days also saw street fighting involving irregulars and civilian volunteers, as well as regular troops and paramilitary police.

Other examples could no doubt be found, but the point is that historical events are frequently complex and resist simple categorisation. Contemporary naming practices are certainly helpful, and in most cases before the twentieth century, what was called a *siege* by participants and contemporary observers generally conforms to how later commentators have used that term. However, some events were called both 'battles' and 'sieges,' such as the Prussian-Danish struggle over the Dybbol redoubt in 1864.[3] More importantly, contemporary naming practices were rarely explicit about the reasons why some engagements were classed as sieges, while others were not.

Definitions serve many useful functions. As historians, we need precision for analysis and to communicate findings. Definitions help alert us to what is general, distinctive or atypical. They can certainly be problematic, risking either imposing artificial homogeneity or being so loose as to be virtually useless. Arguably, the *process* of trying to define something is often more useful than the actual outcome, as it helps refine questions, even if it does not always provide definitive dictionary-style definitions. Neither colloquial usage nor the technical definition of a siege withstands closer inspection as a universal concept applicable across historical time and place, but both do point to elements

3 Charles Lowe, "The Redoubts of Düppel," in *Battles of the Nineteenth Century*, ed. Archibald Forbes et al., 7 vols (London, 1896–1901), 1:224–233; Tom Buk-Swienty, *1864: The Forgotten War that Shaped Modern Europe* (London, 2015).

which can be helpfully integrated within a perhaps less precise, yet more flexible analytical category.

The following argues that a definition of siege warfare is best conceived as a set of interlocking, mutually dependent elements, rather than a simple list of supposedly distinctive characteristics. Some of these elements are present in other forms of combat (notably battle and other forms of positional warfare), but assume a different significance when mutually combined within the context of siege warfare. Each element is itself a variable, the exact size, shape and significance varies through interaction with the others. Readers will soon see that a precise, universally-applicable definition is an illusion, as it is impossible to demarcate absolutely between different types of military operation. However, the main thrust of the argument here is that previous attempts at definition have been overly preoccupied with the tactics of attack and defence, and that it would be helpful to supplement these elements by considering wider operational aspects and the effects of sieges upon participants.

1 Scale

The first element is that of physical *scale* and has two aspects. One is size in its simplest sense. The colloquial use of 'siege' pays little attention to the question of scale, with the term being applied for hostage situations, police sieges of houses or banks and the like. From a military context, these situations are generally encountered as small parts of something much bigger, for instance in street fighting during combat in urbanised areas like the example of Mexico City in 1913. For our purposes then, we must confront the problem of the scale of the action to see where there is a minimum size required for it to warrant the term 'siege.' This is extremely hard to pin down, and is similar to the problem of distinguishing a battle from a 'skirmish' or some other smaller scale combat. Political-science-style quantification, such as that used by the Correlates of War Project, is unhelpful here and there are no compelling reasons to fix rigid thresholds in terms of the size of the place being besieged or the number of human participants.[4] Instead, I would argue that this should be treated similarly to definitions of 'battle' in that size is relative to the capacity of the societies engaged to defend or besiege a particular place. This capacity in turn is related to their form of military organisation, weaponry and resource mobilisation capabilities.[5]

4 <http://www.correlatesofwar.org/> (accessed 23 October 2018).

5 For an attempt at such a definition for European warfare around 1700 see Jamel Ostwald, *Vauban under Siege: Engineering Efficiency and Martial Vigor in the War of Spanish Succession* (Leiden, 2007), pp. 348–358.

A siege is always a meaningful incident with it being sufficiently large and important to be classed as such; though like battles, there are 'major' and 'minor' sieges. The capture of fortresses and towns were considered on par with major battlefield victories and deemed worthy of being celebrated in laudatory imperial accounts, as Pratyay Nath and Kahraman Şakul demonstrate in their discussions of Mughal and Ottoman sieges. Sigrun Haude's discussion of besieged Bavarian Catholic clerics during the Thirty Years War makes it clear that these priests and nuns experienced sieges as distinct events. Yet, their beginnings and ends were becoming fuzzy in the personal accounts left by the clerics as they merged into periods of lessening or escalating tension. Jamel Ostwald illuminates how accounts of sieges could use certain rituals or symbolic acts to mark key stages in the events, both as a record of progress (or failure), as well as points where key decisions had to be made. The initial arrival of the besiegers in sufficient strength, their opening of trenches and the siting of batteries, and the first time a 'practicable' breach had been blown in the walls were all important stages in a siege when it was considered honourable, under specific circumstances, to accept the besiegers' summons to surrender. Likewise, the abandonment of a siege camp, or the arrival of a relief force clearly marked the end to a siege, as might the defenders' acceptance of terms or their massacre following a successful assault. As Anke Fischer-Kattner's comparison of the iconography of siege operations and surrender makes clear, these markers defining a siege as a specific event were fairly common throughout the early modern world and indeed persisted into the nineteenth century. By contrast, positional warfare involving struggles over fortified positions generally lacks such clear temporal markers, such as the trench warfare of the First World War which is usually subsumed within campaigns, such as the Western Front.

We can be only a little more precise with the issue of duration which is the second aspect of scale. A siege is almost always longer than a battle, since (as we shall see) *delay* is another key element of our definition. Again, the actual time will vary depending on the period and part of the world we are studying, but most battles were over within a day prior to the nineteenth century, whereas sieges usually extend beyond the limits of normal human endurance, in the sense of what participants can do without receiving food and water. This logistical aspect is a major part in how sieges are conducted, both by defenders and attackers. As with the other elements, it is impossible to fix it with absolute precision, since endurance will depend on the availability of food and water, both in terms of the defenders' stockpiled supplies, and the besiegers' access to resupply while operations are on-going. Human endurance also varies biologically and can be influenced by factors like climate, environment and the seasons. However, it is not culturally specific, meaning that the question of endurance is general across time and space as part of the element of scale that defines siege warfare. Early modern Europe saw several important prolonged sieges, notably those of Ostend (1601–4) and Breda (1624–5) during the Dutch Revolt (1568–1648). In the first example, the Dutch were able to resupply the garrison by sea, prolonging the siege which cost the Spanish around 40,000 dead. In the second case, the Spanish were able to starve the Dutch into submission after 275 days, but only through constructing a vast circle of entrenchments extending to 96 redoubts, 45 batteries and 37 forts, to isolate Breda completely.[6]

2 Objective

The second defining element is that the *objective* of a siege is always directly related to the defence or capture of a recognised place. Battles are also often about places, such as key bridges or mountain passes, but they do not have to include these to be classed as *battles*. Sieges, by contrast, require a place with some kind of man-made protective defences, but these do not need to be permanent. Many different kinds of fortified places have been besieged, including towns and cities without formal fortifications, as well as fortified settlements and purely military fortresses.[7] In all cases, however, the besieged place needs to be of sufficient size to meet our requirements of 'scale.' The significance of place is revealed in how participants and subsequent observers have named the event. In the case of our opening example, the fighting was not called the siege of the Ciudadela, National Palace or even Mexico City. Likewise, various battles have several names particularly when there were several proximate villages or towns to choose from, as was the case with Blenheim/Höchstädt in August 1704, or Lützen/Großgörschen in May 1813. Such ambiguity does not exist in the case of a siege, with any variations resulting from linguistic differences, as in the case of Lérida (Spanish)/Lleida (Catalan) besieged in 1644.

The objective of a siege is mutually determined by attackers and defenders. In the case of the former, the place besieged remains significant, regardless of their actual objectives. They may be attacking because they desire the place

6 [Anonymous], *Belägerung der Statt Ostende* ([no place], 1604); Erik Swart, "Defeat, Honour and the News: The Case of the Fall of Breda (1625) and the Dutch Republic," *European History Quarterly* 46 (2016), 6–26. A modern map of the Spanish entrenchments is available at: J. P. M. Roze, C. W. A. M. Eimermann, *De Belegering van Breda door Spinola, 1624–1625*, <https://erfgoed.breda.nl/upload/iblock/493/spinola600dpi.pdf> (accessed 23 October 2018).

7 James D. Tracy, ed., *City Walls: The Urban Enceinte in Global Perspective* (Cambridge, UK, 2009); Martha D. Pollack, *Cities at War in Early Modern Europe* (Cambridge, UK, 2010).

itself, to keep or to destroy it. Additionally or alternatively, they may want what is inside the place, be it the inhabitants or material resources, or for what is around it like territory. The place might have to be taken or destroyed, because it is in the attackers' way. Or they may be attacking simply to break the enemy's will to resist. Regardless of their precise reason, the attackers can only achieve their objective through capturing the besieged place. Again, this relates to scale. The attackers are engaged in what is, for them, a significant operation, rather than simply a raid to frighten or harass the defenders. If the attackers did not want to take the place, there would be no siege.

For the defenders, their resistance is sufficiently important to prompt them to offer substantial and prolonged resistance – something which again relates to the upcoming point about delay. Resistance does not have to be a heroic last stand, but needs to entail an effort to hold the place for some time. This raises some cultural specifics, because not all peoples have shared the western concept of a fixed position. For example, the New Zealand Maoris during the 1840s–60s built defensive stockades, called 'Pas,' to delay the British and cause casualties, but were never intended to be held indefinitely. Like the opening example of Mexico City, the engagements of the six-month long Tauranga campaign in 1864 demonstrate the limits of any definition, since they combined elements of what are typically called battles as well as sieges. British forces attempted to trap the Maoris and bring them to battle. The Maoris either retreated, or fought delaying actions from behind their stockades. Actions such as that at Gate Pa, in April 1864, are variously described as 'fights,' 'assaults' or 'battles,' yet the defending Maoris are also labelled as a 'garrison,' echoing the confusion in how the engagement at Dubbol in the same year has been labelled.[8]

3 Boundaries

At this point, most discussions focus on tactics to distinguish sieges from battles, especially the methods used by the besiegers to get inside the place they are attacking or at least to compel the defenders to give up.[9] The colloquial use of siege generally implies that the besieged are surrounded, such as hostage-takers encircled by police. Encirclement certainly features in the classic sieges

8 James Belich, *The New Zealand Wars* (London, 1988), pp. 178–188; James Cowan, *The New Zealand Wars: A History of the Maori Campaigns and the Pioneering Period*, vol. 1: *1845–1864* (Wellington, 1922), pp. 421–440; A. Hilliard Atteridge, "The taking of the Gate Pah," in *Battles of the Nineteenth Century*, ed. Archibald Forbes et al., 7 vols (London, 1896–1901), 1:276–283.

9 For example Duffy, *Fire and Stone*, pp. 94–101.

identified by military history, with their lines of contravallation surrounding the defenders and circumvallation protecting the siege works from potential relief efforts. However, there are numerous famous sieges which did not involve a full investment. This was often the case when the place under attack could be resupplied by sea, as in the aforementioned-case of Ostend in 1604, or the failed attempt by the imperial army under Wallenstein to capture Stralsund in 1628 during the Thirty Years War. The French monarchy faced similar problems during its attack on the port city of La Rochelle, held by Huguenot rebels in 1629. In this case, they only managed to cut off supplies by constructing a huge barrier across the harbour.[10]

The same could occur during sieges of inland positions, particularly if the besiegers lacked the manpower to encircle the defenders completely, or the object of attack was so large that neither the attackers nor defenders could mount active operations along the entire circumference. The French attack in April–July 1849 on the Roman Republic defended by Garibaldi involved the defence and capture of the Janiculum hill on the west side only. Once this was taken, the city surrendered, since the French could have shelled it.[11] Likewise, during the eleven-month siege of Sebastopol 1854–5 the Allies merely blockaded the harbour and invested the south side of the town. The Russians were able to resupply the garrison, plus evacuate most of their forces when the southern defences were breached. Less well-known, but perhaps more informative for present purposes is the siege of Humaita 1866–8 during the War of the Triple Alliance, otherwise called the Paraguayan War 1864–70. Known at the time as the 'Sebastopol of South America,' Humaita was in fact a relatively small, weak fort, supplemented by several lines of outer defences (trenches) enhancing the natural topography of jungle and lagoons. Having repelled the initial Paraguayan invasions of their territories, Brazil, Argentina and Uruguay resolved to capture Humaita which blocked their advance towards the Paraguayan capital. Operations began in April 1866 and involved a complex mix of six major battles, traditional siege operations including sapping, bombardments, assaults and starvation blockades, plus numerable raids and skirmishes. It was not until 2 November 1867 that the Allies managed to cut communications along the Rio Paraguay to Asuncion. It took another five months before they finally surrounded Humaita, and then not perfectly as most of the garrison broke out in July 1868 and only surrendered after being trapped in the

10 Herbert Langer, *Stralsund 1600–1630* (Weimar, 1970); D. P. O'Connell, *Richelieu* (London, 1968), pp. 160–161, 168–182.

11 George Macaulay Trevelyan, *Garibaldi's Defence of the Roman Republic* (London, 1907).

Chaco marshes.[12] Contemporaries and subsequent commentators labelled all these cases as 'sieges' even though the besieged were not fully encircled for all or most of the time. We could add further examples if we extended the view into the twentieth century where the lines between siege and battle become still more blurred with several notable actions being described as both like Verdun (1916) and Dien Bien Phu (1953–4).[13]

The reason why these events were still understood as sieges by the participants is because the partial investment was combined with other defining elements (of which, more shortly). Moreover, while the investment may not have been complete, it remained a possibility, because of the defenders' immobility through their commitment to a static position. For example, during the siege of Rome, French cavalry patrols ranged around the city, while the allied Neapolitan army advanced from the south-east intending to extend the investment, only to be routed by Garibaldi who sortied out of the city to meet them at Velletri. Likewise, the Allied army launched an amphibious raid to try and cut supplies into Sebastopol, while the main Turkish force occupied Eupatoria to the north partly to assist these efforts. And, as we have seen, the Paraguayan defenders of Humaita were finally encircled once the Brazilian fleet broke past their batteries and steamed up the river. Throughout, operations remained focused on the capture of a single place, clearly demarcating the roles of attacker and defender. The battles and other actions forming part of these extended operations all involved the defenders' sorties or efforts to break the siege by attacking the besiegers' camp and positions, or besiegers' attempts to tighten the noose by taking outlying positions.

The preceding helps identify inside/outside *boundaries* as the third element in defining a siege. In all the cases we have just reviewed, operations were characterised by a distinction between a besieged 'inside' and an external 'outside' of the attackers' works and positions. These boundaries are different to the lines or fronts characterising a battle. One army may be encircled either prior or during a battle, but its deployment is still broadly governed by a more fluid sense of 'front' and 'rear,' rather than a more fixed distinction of 'inside/outside' demarcated by protective defences be they the defenders' fortifications,

---

12 George Thompson, *The War in Paraguay* (London, 1869), pp. 121–277; Charles J. Kolinski, *Independence or Death: The Story of the Paraguayan War* (Gainesville, 1965), pp. 114–157.

13 Alistair Horn, *The Price of Glory: Verdun 1916* (London, 1993); William F. Buckingham, *Verdun 1916: The Deadliest Battle of the First World War* (London, 2016); Bernard B. Fall, *Hell in a Very Small Place: The Siege of Dien Bien Phu* (New York, 2002); Jules Roy, *The Battle of Dien Bien Phu* (New York, 2001); Howard R. Simpson, *Dien Bien Phu: The Epic Battle America Forgot* (Lincoln, 2008); Martin Windrow, *The Last Valley: Dien Bien Phu and the French Defeat in Vietnam* (London, 2005).

or those of the surrounding besiegers. Such boundaries create specific problems for the defenders, especially given that a significant proportion of those 'inside' are often civil inhabitants rather than soldiers.

The combination of duration and place (necessitating defence of a fixed location) mean that besieged are forced to consider questions which most field armies can dispense with: how far do the civilians inside need to be protected? And how far can they be trusted? The latter is especially a dilemma for defenders who are also occupiers or a colonial power. As the contributions by Anke Fischer-Kattner and Sigrun Haude make clear, civilians were frequently trapped in besieged towns during early modernity. While they could assist in defence, for example by repairing damaged walls, their consumption of food and other resources threatened the viability of long-term resistance. However, their safety was generally important to the defenders' honour and a major reason for holding the town against the enemy, and it was well-known that they risked the besiegers' wrath if the defence collapsed. However, as Philip Stern, Brian Sandberg and Eric Dursteler show, early modern wars frequently lacked clear divisions between the contending parties, and civilian inhabitants were not always solidly behind a military defence. Some might sympathise with the besiegers and could act as a fifth column undermining the defence. There might also be religious or social divisions within the inhabitants, such as in the Thirty Years War when the patricians of German cities like Stralsund and Magdeburg besieged by the imperial army favoured negotiation, whereas the poorer inhabitants frequently urged resistance in the knowledge that they lacked the resources to buy off soldiers' demands for plunder or accommodation should the besiegers be allowed to enter the city.[14]

4 Attack and Defence

Boundaries also help determine another defining element of a siege: the fairly clear and stable distinction between *attack and defence*. While this distinction rests on the existence of clearer boundaries, it also assumes autonomy as a separate characteristic defining the differing roles of the belligerents. The approach and objectives shaping attack and defence vary more in battles than sieges. Battles can have an attacker and a defender, but equally both sides might attack, either deliberately, or through an unplanned encounter. Similarly, both

14 Peter H. Wilson, *Europe's Tragedy: The Thirty Years War* (London, 2009), pp. 423–424, 428–431, 467–471.

sides can employ attack and defence at different times and on different parts of the same battlefield. Operationally, sieges are more static, as one side is clearly defending, while the other is attacking. Both can employ offensive and defensive tactics such as sorties, bombardment, mining and various countermeasures, but these are always related to advancing or retarding the primary operational objective of capturing the place.

Things could certainly become complex if a relief army approached, as exemplified by the famous 'triple siege' of Turin 1639–40 during a civil war in the duchy of Savoy which in turn was part of the Franco-Spanish War of 1635–59 and related to the Thirty Years War (1618–48). Turin was defended by Madame Reale with 2,000 French guards in the citadel who were blockaded by 12,000 pro-imperial Savoyard forces under her rival Prince Tommaso after July 1639. Tommaso was then surrounded by a 19,000-strong French relief force under Harcourt, who arrived in May 1640. A similarly-sized Spanish army under Leganes proceeded to encircle Harcourt. Thus, there were now three forces engaged in three inter-connected sieges: Madame Reale besieged by Tommaso who was in turn besieged by Harcourt who was besieged by Leganes. Tommaso conducted his siege of Madame Reale by constructing entrenchments facing Turin, whilst building a second line facing outwards to protect himself from Harcourt. The latter also dug two lines to attack Tommaso and defend himself from Leganes. Trapped in the middle, Tommaso eventually agreed terms in September 1640 and the siege collapsed.[15]

5 Delay

The fifth defining element is *delay*. This is obviously related to the question of scale noted above, since delay lengthens the duration of a siege, but it also distinguishes how sieges are conducted in ways differentiating them for other kinds of engagements. The crucial issue in siege warfare is time. The defenders are trying to delay the attacker, perhaps because this is their objective, such as the Maoris who were fighting a war of attrition, or because they wish to distract the enemy to prevent interference with their own operations elsewhere. Alternatively, delay might be directly due to the desire to hold the place. Prolonged defence might force the attackers to abandon the siege altogether, or it might buy time for a relief such as occurred at Vienna which resisted the Ottomans for two months in 1683 until rescued by an international army

15 Alessandro di Saluzzo, *Histoire militaire du Piémont*, 5 vols (Turin, 1859), 4:121–141.

assembled from Poland and the Holy Roman Empire.[16]. Delay can be present in battle too, because these can also be fought defensively to achieve similar strategic objectives. Similarly, delay can be sought at operational level by other methods such as evasive manoeuvers, scorched earth or guerrilla tactics. However, delay assumes a much more prominent role in defining how a siege is fought, because of the focus of a siege to capture a specific place. For instance, the defenders cannot shift their defence to another location, if the fighting is going badly, since they are committed to defending their chosen protected site, at least for some significant time.

Delay is also significant for how contemporaries perceive a siege and thus how they view it as such, rather than as some other kind of operation. It might be argued that this aspect is less watertight since the early twentieth century, especially because battles have become longer, while media innovation has accelerated the transmission of information, meaning the public is aware before the event is over.[17] Nonetheless, the longer duration of sieges, at least before the twentieth century, focused public attention on operations whilst these were still in progress as was the case at Colchester in 1648 described by Anke Fischer-Kattner. By contrast, battles made their impact directly through their outcome as news spread of victory or defeat. Observers had more time to become conscious of the siege as it developed and their perceptions could impart significance to events well beyond the original military objective. For example in 1857–8, the mutinous Sepoys deliberately exploited British anxiety about the safety of Europeans and Indo-Europeans, as well as the question of British prestige. The potential fall of towns threatened the aura of imperial omnipotence and could rally additional support for the mutineers. Likewise, the British command felt obliged to continue the siege of Delhi 1857 (incidentally, another partial investment), because of the symbolic significance it had assumed in the mutiny.[18]

Delay impacts on the conduct of a siege, which is another argument for seeing it, rather than tactics, as a defining characteristic. The defenders employ tactics to frustrate the attackers' objective of a quick capture. These methods mostly relate to fortifications, which are either pre-prepared or improvised to slow down the enemy and keep them outside the boundary, but they can extend to 'offensive' actions like sallies to wreck enemy trenches and otherwise add to the delay. Conversely, the attackers' methods are intended to neutralise

16 John Stoye, *The Siege of Vienna* (London, 2011); Thomas M. Barker, *The Eagle and the Crescent: Vienna's Second Siege in its Historical Setting* (New York, 1967).

17 Athina Karatzogianni, ed., *Violence and War in Culture and the Media: Five Disciplinary Lenses* (London, 2012).

18 Saul David, *The Indian Mutiny 1857* (London, 2003).

the defences and speed up the process of getting inside. Most of what they do is unwelcome and forced upon them by the defender. They are obliged to protect themselves from defensive fire or counter-attacks, whilst advancing their efforts to cross the boundary to the inside.

During the classic era of European siege warfare, between the later fifteenth and early twentieth centuries, these methods have included constructing batteries and sapping towards the enemy defences, which were reduced by bombardment or mining.[19] Bombardment could also extend to 'shock and awe,' trying to hasten the end by demoralising the defenders. As Tonio Andrade indicates, these methods featured in China which otherwise employed explosive technology differently than Europeans, the Ottomans, Mughals and Safavids.[20] Assaults were used once a 'practicable' breach had been made in the defences, assuming the besieged did not surrender first. Though costly, assaults might be deemed necessary by the attackers primarily to minimise further delay after what had often been a lengthy process. Sometimes, assaults were launched without any prior preparation to surprise the defenders and remove the need for a siege altogether. The same end might be achieved by tricks, such as sneaking men in disguised as civilians or contacting sympathisers within a town who might open the gates, as in the case of Clissa in 1596 described by Eric Dursteler. Likewise, defenders might be induced to capitulate by threats, such as to destroy their property outside their town. Huguenot and Catholic nobles used all three methods to take the often poorly fortified and isolated towns and castles in the French countryside during the Wars of Religion discussed in Brian Sandberg's chapter. All these methods were motivated by the desire to reduce delay to a minimum due to strategic, logistical or political reasons. Blockades were usually a method of last resort, when either previous attempts to break inside had failed, or the besiegers simply lacked the manpower or equipment to do anything else. Again, blockades related to time, since they worked by increasing the pressure on the defenders' mental and material capacity to hold out.

Here we can see the significance of delay in defining a siege. Places which are taken by assault, or through surprise, ruse or negotiations alone are not usually called 'sieges' by the participants, but are known by other names (if at all) like 'coup de main' or simply 'capture' as noted by Brian Sandberg. Yet, the attackers have achieved the same objective as if there had been a siege. This

---

19 These methods are itemised in Simon Pepper, Nicholas Adams, *Firearms and Fortifications: Military Architecture and Siege Warfare in Sixteenth-Century Siena* (Chicago, 1986); Ostwald, *Vauban under Siege*, pp. 350–355.

20 Douglas E. Streusand, *Islamic Gunpowder Empires: Ottomans, Safavids and Mughals* (Philadelphia, 2011).

holds true for both modernity and early modernity, as well as for operations outside Europe. An example is the British capture of Magdala, the seemingly impregnable mountain fortress of Emperor Theodore (Tewodros) II of Abyssinia.[21] The place fell after a short assault in April 1868, because the Abyssinian counter attack had been defeated the day before in the battle of Arogi and, unknown to the British, most of the defenders had subsequently fled. To contemporaries, the assault was simply part of the 'battle of Magdala,' rather than a siege. Likewise, participants have distinguished between a siege and a battle fought to break it as, for example, Pavia (1525), Freiburg (1644), or Turin (1706).[22]

Assault and surprise thus only become 'siege' tactics within the context of delay when they are employed once a siege has begun in order to short circuit an otherwise lengthy process. Conversely, 'siege' techniques can feature in the conduct of battles. Field fortifications were used long before the development of trench warfare from the mid-nineteenth century, as were the techniques used to overcome such defences, such as sapping, mining, bombardment and assault.[23] The presence of all these features on the Western Front of the First World War has led to widespread comparison of the fighting there with early modern and nineteenth-century siege warfare.[24] The main combats, like those on the Somme in 1916, remain, however 'battles' because the other elements we have identified with sieges were absent. Most notably, trench warfare was characterised by lines or fronts, rather than the inside/outside boundaries of sieges.[25] It is best characterised as positional warfare rather than as siege warfare more specifically defined. Similarly, the Allied naval blockade of Germany during the same conflict only loosely conforms to a 'siege' on the broadest strategic sense of a struggle over endurance. This example returns us to the opening issue of *scale*, suggesting an upper as well as a lower threshold in distinguishing a siege from some other kind of operations.

---

21 Brian Bond, ed., *Victorian Military Campaigns* (London, 1970); Philip Marsden, *The Barefoot Emperor: An Ethiopian Tragedy* (London, 2008); Darrell Bates, *The Abyssinian Difficulty: The Emperor Theodorus and the Magdala Campaign 1867–68* (Oxford, 1979); Frederick Myatt, *March to Magdala: Abyssinian War, 1868* (London, 1970).

22 Jean Giono, *Le désastre de Pavie* (Paris, 2012); Hans Helmut Schaufler, *Die Schlacht bei Freiburg im Breisgau 1644* (Freiburg, 1979); Clemente Assum, *L'assedio di Torino (maggio–settembre 1706) e la battaglia di Torino (7. Settembre 1706)* (Turin, 1926); Dario Gariglio, *1706: L'assedio di Torino* (Turin, 2005).

23 Anthony Saunders, *Trench Warfare 1850–1950* (Barnsley, 2010).

24 For an example of the conflation of trench and siege warfare: Bruce Watson, *Siege: A Comparative Study* (Westport, 1993), pp. 5–6.

25 General Staff Intelligence, *British Trench Warfare 1917–18: A Reference Manual* (London, 2014).

The foregoing has already highlighted the question whether sieges are still part of contemporary warfare.[26] This question cannot be addressed definitively here, but exploring it briefly is nonetheless useful in bringing our attention back to change over time, and might help us should we decide to identify characteristics of early modern siege warfare. Changes in weaponry have certainly rendered the defence of conventional fixed positions very difficult: artillery and aerial bombardment have increased considerably in accuracy and potential, especially in the last twenty years. It looks unlikely that we will see a siege where a conventional regular army defends a sizable position, such as a fortress or major urban settlement, against a similarly conventional opponent.

Despite repeated media references to 'sieges,' it is questionable whether the recent and current fighting in cities such as Fallujah or Aleppo can count as such. Operations are prolonged, but lack the clear boundaries which seem to demarcate participants and the inside/outside element of siege warfare. One reason is that modern cities are simply far larger than any in past human history, rendering conventional methods of attack and defence difficult to apply.[27] Another is that the combatants are often irregulars or insurgents who deliberately blur the distinctions between themselves and civilians. Regular forces also frequently do not observe such distinctions. Consequently, there is often no obvious military infrastructure to be fought over. Tactical methods associated with siege warfare may well be present, including bombardment and blockades to starve both defenders and civilians, but as we have seen such practices feature in other forms of warfare.[28] Meanwhile, attacks on small outposts, such as those occupied by the British army during the Helmand campaign in Afghanistan, appear to fit more in the category of raids and harassing tactics, than full sieges.

The boundaries distinguishing inside and outside have been eroded by the changing scale of warfare and the technology used to wage it. Combat now takes place simultaneously in multiple overlaid, interlocking yet distinct spaces. Individual fighters or groups of combatants might be physically trapped in specific confined spaces akin to the defended positions found in early modern and nineteenth-century sieges, yet they can still engage with and be engaged by much more distant opponents. For instance, they might be able to strike at enemies far away through cyber warfare or post propaganda online.

---

26 For example, the term 'siege' is missing from the *Department of Defense Dictionary of Military and Associated Terms Joint Publication 1–02 as through 15 February 2016* <http://www.dtic.mil/doctrine/new_pubs/jp1_02.pdf> (accessed 23 October 2018).

27 The influence of mega cities on contemporary warfare is noted by Wayne E. Lee, *Waging War: Conflict, Culture and Innovation in World History* (Oxford, 2016), pp. 499–500.

28 Bing West, *No True Glory: A Frontline Account of the Battle for Fallujah* (New York, 2005).

Conversely, they can be hit by air strikes or missiles launched far from the location under attack, or indeed targeted by drones operated remotely from another continent.

This short discussion suggests there are five elements to defining sieges as a distinct form in warfare. *Scale* indicates that a siege is an identifiable military activity with sufficient importance to the belligerents to be recognised as a distinct event within a broader continuum of operations. It is large enough to be an event in itself, rather than simply part of something else, like a protracted struggle over buildings such as Hougomont farm at the battle of Waterloo, yet sufficiently small as not to constitute an entire strategy, like the Allied blockade of Germany in the First World War. It also lasts sufficiently long for human endurance to be a factor in its conduct and outcome. Both sides share a common *objective* in that they are fighting over the possession of a clearly defined and mutually recognised protected site. The objective helps demarcate clear *boundaries* between a defended 'inside' and an 'outside' realm of the attackers, and any potential relief for the besieged. The two sides have clearly demarcated roles of *attack and defence* which influence their choice of tactics. *Delay* is the strategic objective of the defence and influences how both sides conduct the siege and how the combat is perceived, as well as affecting the wider significance attached to its outcome. Perhaps most fundamentally for the definition, these five elements are mutually dependent and only assume full significance when combined with each other. Individual aspects (types of objective, tactics employed etc.) can be found in other forms of combat and are, in themselves, not unique to sieges.

The preceding suggests that siege warfare is a distinct form of war, which might be combined with others, such as battle, raiding or guerrilla warfare. The five defining elements pinpoint features common across cultures, places and periods, even if there have been numerous variations in how they have been practiced, and even more in their representation and rhetoric. Many of these variations are found most obviously in how the methods of attack and defence have changed over time and space, notably as the belligerents' weapons and resources have developed, and as the size and character of inhabited settlements have evolved. As the discussion also suggests, the pace and impact of these changes grew with industrialisation and urbanisation, as well as with developments in weaponry and technology, and have probably reached the point where they are ending sieges as a distinct form of warfare, even if the rhetoric of a siege remains current in the media and popular consciousness.

Bibliography

Printed Sources

[Anonymous], *Belägerung der Statt Ostende* ([no place], 1604).

Department of Defense Dictionary of Military and Associated Terms Joint Publication 1–02 as through 15 February 2016, <http://www.dtic.mil/doctrine/new_pubs/jp1_02.pdf> (accessed 23 October 2018).

Dictionary Entries "siege", <http://www.oxforddictionaries.com/definition/english/siege> (accessed 23 October 2018); <http://www.merriam-webster.com/dictionary/siege> (accessed 23 October 2018); <http://www.macmillandictionary.com/dictionary/british/siege> (accessed 23 October 2018); <https://en.wikipedia.org/wiki/Siege>_(accessed 23 October 2018).

General Staff Intelligence, *British Trench Warfare 1917–18: A Reference Manual* (London, 2014).

Roze, J. P. M.; Eimermann, C. W. A. M., *De Belegering van Breda door Spinola, 1624–1625*, <https://erfgoed.breda.nl/upload/iblock/493/spinola600dpi.pdf> (accessed 23 October 2018).

Secondary Literature

Assum, Clemente, *L'assedio di Torino (maggio–settembre 1706) e la battaglia di Torino (7. Settembre 1706)* (Turin, 1926).

Atteridge, A. Hilliard, "The taking of the Gate Pah," in *Battles of the Nineteenth Century*, ed. Archibald Forbes et al., 7 vols (London, 1896–1901), 1:276–283.

Bates, Darrell, *The Abyssinian Difficulty: The Emperor Theodorus and the Magdala Campaign 1867–68* (Oxford, 1979).

Barker, Thomas M., *The Eagle and the Crescent: Vienna's Second Siege in its Historical Setting* (New York, 1967).

Belich, James, *The New Zealand Wars* (London, 1988).

Bond, Brian, ed., *Victorian Military Campaigns* (London, 1970).

Buckingham, William F., *Verdun 1916: The Deadliest Battle of the First World War* (London, 2016).

Buk-Swienty, Tom, *1864: The Forgotten War that Shaped Modern Europe* (London, 2015).

Calvert, Peter, *The Mexican Revolution 1910–1914: The Diplomacy of Anglo-American Conflict* (Cambridge, UK, 1968).

Cowan, James, *The New Zealand Wars: A History of the Maori Campaigns and the Pioneering Period*, vol. 1: *1845–1864* (Wellington, 1922).

The Correlates of War Project, <http://www.correlatesofwar.org/> (accessed 23 October 2018).

David, Saul, *The Indian Mutiny 1857* (London, 2003).

Duffy, Christopher, *Fire and Stone: The Science of Fortress Warfare 1660–1860* (Newton Abbot, 1975).
Fall, Bernard B., *Hell in a Very Small Place: The Siege of Dien Bien Phu* (New York, 2002).
Gariglio, Dario, *1706: l'assedio di Torino* (Turin, 2005).
Giono, Jean, *Le désastre de Pavie* (Paris, 2012).
Horn, Alistair, *The Price of Glory: Verdun 1916* (London, 1993).
Karatzogianni, Athina, ed., *Violence and War in Culture and the Media: Five Disciplinary Lenses* (London, 2012).
Knight, Alan, *The Mexican Revolution*, 2 vols (Cambridge, UK, 1986).
Kolinski, Charles J., *Independence or Death: The Story of the Paraguayan War* (Gainesville, 1965).
Langer, Herbert, *Stralsund 1600–1630* (Weimar, 1970).
Lee, Wayne E., *Waging War: Conflict, Culture and Innovation in World History* (Oxford, 2016).
Lowe, Charles, "The Redoubts of Düppel," in *Battles of the Nineteenth Century*, ed. Archibald Forbes et al., 7 vols (London, 1896–1901), 1:224–233.
Marsden, Philip, *The Barefoot Emperor: An Ethiopian Tragedy* (London, 2008).
Myatt, Frederick, *March to Magdala: Abyssinian War, 1868* (London, 1970).
O'Connell, D. P., *Richelieu* (London, 1968).
Ostwald, Jamel, *Vauban under Siege: Engineering Efficiency and Martial Vigor in the War of Spanish Succession* (Leiden, 2007).
Pedraja, René De La, *Wars of Latin America, 1899–1941* (Jefferson, 2006).
Pepper, Simon; Adams, Nicholas, *Firearms and Fortifications: Military Architecture and Siege Warfare in Sixteenth-Century Siena* (Chicago, 1986).
Pollack, Martha D., *Cities at War in Early Modern Europe* (Cambridge, UK, 2010).
Poten, Bernhard von, ed., *Handwörterbuch der gesamten Militärwissenschaften*, 7 vols (Bielefeld and Leipzig, 1887).
Roy, Jules, *The Battle of Dien Bien Phu* (New York, 2001).
Saluzzo, Alessandro di, *Histoire militaire du Piémont*, 5 vols (Turin, 1859).
Saunders, Anthony, *Trench Warfare 1850–1950* (Barnsley, 2010).
Schaufler, Hans Helmut, *Die Schlacht bei Freiburg im Breisgau 1644* (Freiburg, 1979).
Simpson, Howard R., *Dien Bien Phu: The Epic Battle America Forgot* (Lincoln, 2008).
Stoye, John, *The Siege of Vienna* (London, 2011)
Streusand, Douglas E., *Islamic Gunpowder Empires: Ottomans, Safavids and Mughals* (Philadelphia, 2011).
Swart, Erik, "Defeat, Honour and the News: The Case of the Fall of Breda (1625) and the Dutch Republic," *European History Quarterly* 46 (2016), 6–26.
Thompson, George, *The War in Paraguay* (London, 1869).
Tracy, James D., ed., *City Walls: The Urban Enceinte in Global Perspective* (Cambridge, UK, 2009).

Trevelyan, George Macaulay, *Garibaldi's Defence of the Roman Republic* (London, 1907).
Watson, Bruce, *Siege: A Comparative Study* (Westport, 1993).
West, Bing, *No True Glory: A Frontline Account of the Battle for Fallujah* (New York, 2005).
Wilson, Peter H., *Europe's Tragedy: The Thirty Years War* (London, 2009).
Windrow, Martin, *The Last Valley: Dien Bien Phu and the French Defeat in Vietnam* (London, 2005).

Index

www.ingramcontent.com/pod-product-compliance
Ingram Content Group UK Ltd.
Pitfield, Milton Keynes, MK11 3LW, UK
UKHW022224051026
15624UKWH00002B/343

9 789004 395688